An Integrated Approach to
New Food
Product Development

An Integrated Approach to
New Food
Product Development

Edited by
Howard R. Moskowitz
I. Sam Saguy
Tim Straus

CRC Press
Taylor & Francis Group
Boca Raton London New York

CRC Press is an imprint of the
Taylor & Francis Group, an **Informa** business

CRC Press
Taylor & Francis Group
6000 Broken Sound Parkway NW, Suite 300
Boca Raton, FL 33487-2742

© 2009 by Taylor and Francis Group, LLC
CRC Press is an imprint of Taylor & Francis Group, an Informa business

---

Library of Congress Cataloging-in-Publication Data

An integrated approach to new food product development / editors Howard R. Moskowitz, I. Sam Saguy, and Tim Straus.
    p. cm.
    "A CRC title."
    Includes bibliographical references and index.
    ISBN 978-1-4200-6553-4 (hard back : alk. paper)
    1. Food--Research. 2. New products. 3. Food industry and trade. 4. Food--Marketing.
I. Moskowitz, Howard R. II. Saguy, Israel, 1946- III. Straus, Tim. IV. Title.

TX367.I56 2009
338.4'7664--dc22   2009013711

Visit the Taylor & Francis Web site at
http://www.taylorandfrancis.com

and the CRC Press Web site at
http://www.crcpress.com

1006215695

Contents

PART I Setting the Agenda for Successful New Product Development

p.6,
↑ p·7 & 20

PART II Defining and Meeting Target Consumer Needs and Expectation

p. 49 , p.66

PART III   The Right Food

PART IV   Proper Packaging and Preparation

PART V   *Positioned Correctly at the Shelf and in the Media*

PART VI   *Meet Corporate Logistics and Financial Imperatives*

Foreword

At Nestlé, we believe that "sharing is winning." This simple but powerful concept covers a plethora of topics, domains, and disciplines such as science, technology, consumers' insights, in-depth business world understanding and, of course, the evolving dynamic global marketplace. The richness of knowledge included in this book maps out the innovative terrain and sketches tools for an integrated approach to new product development. In order to create value with consumers and customers, the challenge of new products demands sensitivity to, and integration of, many disciplines. Applying knowledge founded on solid scientific grounds and principles is the only feasible approach to renovate business practices and create innovative new ones, in order to succeed in the worlds of today and tomorrow. Learning is a lifelong journey. Only time will reveal how successfully this book will help that learning, making it real, and applying it in practice.

During the recorded history of mankind, food has evolved from helping people survive to becoming a vehicle to deliver much more than simple survival nutrients. The knowledge underlying the industrialization of modern foods is one of the most successful, yet least heralded achievements of science.

At Nestlé, we incorporate many disciplines and processes to enable us to predict new trends, envision long-term opportunities, perfect our science, and conceive new products that will drive our top and bottom line performance. As a worldwide leader in the food industry and in our journey to be the No. 1 nutrition, health, and wellness company, our vision is very broad, covering consumer trends as they evolve around the globe and impact various geographic sectors. Our strategy is science-based and consumer-centric, which simultaneously satisfy the aspects of food products, vision, and leadership. We are also looking at meeting future challenges, such as personalized nutrition, obesity and weight management, and healthy aging. We continuously and extensively investigate opportunities for food and health, such as food for the brain, life quality, wellness, performance, and personalization addressing the future information on genetic makeup or predisposition to food-related diseases. We look at foods that can ameliorate such conditions as obesity and diabetes, foods that enhance mood, and foods that are healthful and convenient, all with the goal of combining advances in science, betterment of human conditions, and, of course, growth in all aspects of our businesses. Most importantly, it is our utmost responsibility to bring all this to our consumers around the world in an understandable and convenient way, hence our increasing focus on combining product solutions with convenient solutions.

It is the awesome responsibility of the food industry to address the macrotrends that impact the very lives of our global population. As the population of the world is aging with an increasing number of adults over age 65 and a decreasing number of children under age 5, the macrostrategies must be altered to adapt to the appropriate nutrition. The food industry must ensure that nutrition, health, and wellness become, and ultimately remain, central driving forces that firmly and significantly influence the strategic paths of all major food companies.

Sustainable winning in the marketplace is not straightforward, unless it is founded on well-defined strategy, values, leadership, and know-how. It requires an umbrella of overarching practices and topics. To enhance the success rate in new product development, it calls for innovation in all aspects of the food chain. Innovation must address the needs of expanding profits with specific new product lines, through the far greater vision as practiced by Nestlé, the world's largest food company. Facets in this vision span the range from the DNA revolution and how it could be incorporated into new product development, to alternative processing technologies, all grounded in sound scientific principles. Yet, vision alone cannot suffice. Vision in business requires management at all levels, applied well, productively, and inspirationally to all levels of the company as well as to life science, marketing, consumer science, sales, and many others.

It is my hope, on a personal basis, and speaking for my company, Nestlé, on a business and science foundation, that this book, written by experts who share their knowledge and experience, will have an impact on the industry's needs. I hope that this book will make that necessary, significant contribution to the industry, optimizing the overall new product development process in terms of speed and time to market. This will allow better utilization of our resources, and higher quality and healthier foods for the body and soul, just as the wisdom contained therein leads to much higher success rate(s) in the marketplace.

Doctor Werner J. Bauer
Executive Vice President
Nestlé S.A.
Chief Technology Officer
Head of Innovation, Technology and Research & Development
Vevey, Switzerland

Preface

How did new product development evolve as the frenetic and complex "innovate or die" process that today drives most food and beverage companies? Financially, everyone appreciates the necessity of new products to a company's lifeblood and very existence. Reported figures indicate that over 50% of current profits are associated with products that have been on the shelf for less than five years. When assessing a company's value, Wall Street closely examines a company's new product successes (or failures). And so, the "beat goes on."

So, how (and why) did we arrive at the multidisciplined, interdepartment processes employed throughout the industry today? What invisible hand guided, or perhaps failed, to lead us on a consistent basis?

The history of food is filled with new product development achievements. Initially, all foods were consumed at their source—an animal was slaughtered and consumed in one sitting, water was drunk from a stream, and milk from the teat or carried in some "organic" vessel. Fruit, vegetables, nuts, and some grains were picked and eaten. Some were carried to shelter to be consumed later.

As time went on, man discovered better ways to preserve and transport food. Bread was invented over 10,000 years ago as a method to transport grain. Beer was invented for the same purpose over 5,000 years ago. Cheese was discovered when a herder was trying (once again) to transport milk in a cow's stomach and found it turned into cheese. Hence, a new and better way to transport milk was invented.

Basic food inventions were then innovated to meet a specific need. Exodus tells of the innovation of bread into unleavened bread so the Jewish people could flee Egypt before the Pharaoh changed his mind.

The English transformed their traditional English ale into IPA (Indian pale ale) by including more hops and a higher alcohol content to preserve the beverage so that it could make the journey with their troops to India during the Raj. Canning allowed fruits and vegetables to be preserved from season to season. Innovations in ultrahigh temperature (UHT) sterilization and aseptic packaging improved upon earlier forms of heat treatment, providing milk with an extremely long shelf life. And freeze-drying was created as a way for man to eat on extended space flights.

Food technologists and food scientists, whether professionals in a society or just "plain folks doing their job," have invented various food preservation systems designed to make food safe and reduce spoilage. Taken as a whole, their contributions have probably done more to enhance the health and longevity of man than any medicine ever created.

But most food innovation today is not geared to free a fleeing person or ensure the morale of an army far from home, nor is most modern food innovation designed to eliminate human hunger. Today, most modern nations are blessed with an abundance of food. In turn, food innovation has taken on a completely different role. Food innovation today is about commerce—building shares, increasing income, gaining marketplace dominance, and growing corporate profits. And by nature of commerce, the process of developing new food products has come to rely on a number of different disciplines:

- Consumer insights and consumer trend tracking
- Health, nutrition, and well-being
- Culinary arts
- Food formulating and sensory analyses
- Processing and technology
- Package engineering and graphic design
- Shelf life evaluation and enhancement
- Manufacturing systems development
- Innovation partnerships and open innovation
- Sales and trade relations
- Marketing, advertising, and promotion

Invention alone is not enough. Most companies are locked into commercial combat with other companies that share their categories, distribution outlets, and certainly their customers' wallets and stomachs. Intracategory competitors compete for sales, shares, and shelf space within their defined categories. Today's supermarket retailers also battle with foodservice outlets for customer dollars. And inter-category competition goes through cycles where suddenly a previously accepted commercial "acquaintance" blindsides to evolve into a direct competitor. Consider the case of the caffeinated candy bar competing with a beverage for refreshment and replenishment.

Although the consumer's stomach has a finite capacity, the marketplace has a near infinite potential for the marketer who can "break the code," to identify, perfect, and introduce an innovative new product that will attract and maintain consumer usage, meet and surpass his expectations, build volume, and enhance corporate profits.

It is no longer just a story of technical invention that is necessary to succeed in the market, although having a technical advantage is important in maximizing sales, shares, and profitability. Today, it is all about understanding a consumer's need and expectation and creating the right products to fulfill them. It is about developing a superior tasting food that meets the consumer's preparation capabilities, so as to encourage repeat usage. It is about ensuring that the product is packaged and promoted properly in order to stimulate trial. It needs also to address health and well being, sustainability, and consider various aspects of the environment. The product must also meet shelf life and pricing requirements set by distribution outlets.

So, while early food development was relatively simple—just eliminating hunger and harm—today's food companies must deal with the vagaries of the consumer and trade's real and perceived wants and needs. We must understand nutrition and diet trends, sort out the short-lived opportunities, and efficiently identify those opportunities that have the potential for sustained business. We must understand our direct competitors, their strengths, and available assets. We must also be wary of the invader who seeks to steal the usage occasion on which our product is based and, right under our nose, create a totally new set of competitors.

That brings us to why we are offering this book.

Through our respective professional experiences, we have found the interrelation of the different disciplines when developing a new food or beverage product to be paramount for success. Too often, really great new products fail in the marketplace, not because they were bad products, but because they were incorrectly packaged, poorly positioned, mispriced, or badly promoted. And we have seen products that were spot on in their marketing, but the food itself simply did not perform.

Most new product development texts are concerned with either just the "marketing" or the "technical" aspects of new product innovation. Few attempt to present the crucial interdependence of all the disciplines in being successful. Many of these texts are academically driven, whereas we have blended academic "ivory tower" with the experience of seasoned "real market" practitioners, which when combined provide concepts and tools for successful new product innovation.

To serve this notion, this book begins with the concept of a success equation that discusses the interrelationship of the key tasks (not disciplines) that are critical to new product success. The chapters that follow are then devoted to discussion on each aspect of completing the key task from the perspective of different disciplines. Our intent is to provide the reader with a fuller appreciation of the importance of each task, and how each discipline contributes to and then optimizes the holistic success of the product.

We hope this book will enable you to think differently about how you approach new product development. A book alone can neither prevent failure nor guarantee success. Unfortunately, failure with new products is far too frequent and is merely buried into the cost of doing business. But success raises your company's prestige, profits, and fosters continued success! We hope the ideas that our authors present help you to grow to the next level of success and guide you into newer, more profitable ways in the world of foods and beverages.

Howard Moskowitz
I. Sam Saguy
Tim Straus

Acknowledgments

First and foremost, we are extremely proud of the accomplishments of our contributors. They went through tremendous effort to capture and record their expertise, in order to share their lifework in the food and beverage industry with readers. Our authors readily accepted the challenge; few realized the hard work that inevitably goes into actually delivering a finished chapter. Despite being incredibly busy with their professions, they diligently put their ideas on paper in a way that is accessible to professionals and novices alike. We hope the experience and expertise contained herein will enhance the whole process of new product development, making it more efficient and more successful in the increasingly competitive marketplace.

We would also like to acknowledge Howard's exceptionally able editorial assistant, Linda Lieberman. It is fair to say that without Linda's daily guidance and nurturing, this book might never have been born. Conceived yes, but it was Linda who mothered it, along with Suzanne Gabrione, director of marketing at Moskowitz Jacobs, Inc. Thank you.

Howard's acknowledgment: How does an editor acknowledge the hundreds of people who, over the years, are the real "bricks and mortar" of a book? Well, it is not easy. Perhaps now is the most appropriate opportunity to acknowledge two individuals who, at the start of my career, were really the prime movers for this book, though they could not know it at the time. These early mentors, Drs. Harry Jacobs and John Kapsalis, guided me when we all worked together at the U.S. Army Natick Laboratories. Through them, I was introduced to the world of food, from which my lifelong interest arose. And, finally, let me thank my wife Arlene, my two sons and daughters-in-law, and my ten grandchildren. It is for you, really, the wind beneath my sails, that I persevere and create. Thank you.

Sam's acknowledgment: There are many people who, over the years, shaped and contributed to my professional life. Perhaps now is the most appropriate opportunity to acknowledge two individuals who made the most impact: Professor Marcus Karel at MIT, my early mentor, for being the bright beacon that guided me through exciting scientific collaboration, and for immense inspiration and encouragement; and Dr. James R. Behnke, for opening the doors to the Pillsbury Company, the American food industry, and new product development, and for his friendship that grew over the years. Special thanks to my wife, Irit, of over 39 years, for being a great inspiration and so understanding in putting up with the extra long hours spent "playing" and laboring in front of the computer. To my children Ami and Dan, Ami's wife, Hagit and my four incredible grandchildren Aviv, Bar, Mor, and Maayan perfecting the joy of grandfathering and enriching our life with happiness and pride. Special thanks also to Howard Moskowitz who introduced me to consumer research, for being a lifelong inspiration and such a great role model, and to Tim Straus for the innovative marketing, fresh ideas, and vision.

Tim's acknowledgment: An opportunity to coauthor and edit a book on new food and beverage product development would be impossible without the experience of working with clients in a broad array of categories. So to all the consumer packaged goods (CPGs), foodservice, restaurant, and ingredient clients I have had the pleasure to assist, a heartfelt thank you. Without your programs, the skills and insights outlined in this text would have never been possible. Last, Frances Jackson at Turover Straus who assisted in the preparation of materials deserves a debt of gratitude. And thanks to my wife, grown children, family, and friends who are amused, sometimes amazed, and always supportive of my "unusual profession." Thank you all.

Editors

Howard Moskowitz is the president of Moskowitz Jacobs Inc., founded in 1981. He is an experimental psychologist in the field of psychophysics (the study of perception and its relation to physical stimuli) and inventor of world-class market research technology.

Dr. Moskowitz graduated from Harvard University in 1969 with a PhD in experimental psychology. Prior to that, he graduated from Queens College (New York), Phi Beta Kappa, with degrees in mathematics and psychology. He has written/edited 16 books, published over 300 refereed and conference articles, lectures widely, serves on the editorial board of major journals, and mentors numerous students. He is the coauthor of the recently published book *Selling Blue Elephants: How to Make Great Products That People Want Even Before They Know They Want Them* (Wharton School Publishing, 2007). In less than ten months, the book had brisk sales in six continents, was on international bestseller lists, had been endorsed by top universities and corporations, and enjoyed wide media coverage. It has been republished and translated in eight countries, with another seven licensed and in production.

His latest efforts focus on four areas:

1. Using experimental design of messaging for juror selection (with law firms), and for package/shelf/Web design (dynamic landing page optimization)
2. Using experimental design to understand and optimize customer experience (with Steve Onufrey of Unisys)
3. Understanding the mind of the "high-end customer," (he and Alex Gofman are writing a new book with Dr. Stefano Marzano and Marco Bevolo of Philips Design [Eindhoven])
4. Creating the Blue Elephants Center in Shanghai and Beijing, China, to promote the use of experimental design of ideas and products, for innovation and world competitiveness

Among his contributions to market research is his 1975 introduction of psychophysical scaling and product optimization for consumer product development. In the 1980s, his contributions were extended to health and beauty aids. In the 1990s, the concept development approach was introduced to pharmaceutical research. His research/technology developments have led to concept and package optimization

(IdeaMap®, MessageMap® for pharma), integrated and accelerated development (DesignLab®), and the globalization and democratization of concept development for small and large companies through an affordable, transaction-oriented approach (IdeaMap.Net®).

Moskowitz has won numerous awards, such as the 2001, 2003, 2004, and 2006 ESOMAR awards for his innovation in Web-enabled, self-authored conjoint measurement and for weak signals research in new trends analysis and concept development. Self-authored concept technology brings concept/package design development and innovation into the realm of research, substantially reducing cost, time, and effort for new product and service development. In 2005, he received the Parlin Award, the highest award from the American Marketing Association. In 2006, he was given the first Research Innovation Award by the Advertising Research Foundation and was also inducted into the Hall of Fame of New York's Market Research Council.

I. Sam Saguy is a professor of food science and technology at Robert H. Smith Faculty of Agriculture, Food and Environment, The Hebrew University of Jerusalem, where he held the position of the faculty director of research affairs. He teaches graduate courses on new product development, kinetics and quality loss during processing and shelf life, and an undergraduate course on unit operations in food engineering and food technology.

He is a graduate of the Technion, The Israeli Institute of Technology, in chemical engineering (BSc), food engineering, and biotechnology (MSc and DSc); he also holds a diploma in business administration from The Israeli Institute of Productivity, Tel Aviv, Israel. He was a visiting lecturer at MIT and of the Graduate School, Department of Food Science, Rutgers–The State University of New Jersey. Professor Saguy is a scientific reviewer of many professional journals publishing on food-related topics, and serves on several editorial scientific boards.

Dr. Saguy has gained outstanding international recognition both in academia and in the food industry, and has established a record of synergistic interface of academic research with engineering, R&D, and involvement in all the facets of new product development from the concept to the marketplace. Being one of the few people who have maintained their academic status while also spending many years working for several major food corporations, he consults for numerous food and biotechnology companies in the United States, Europe, India, and Israel. Dr. Saguy has contributed more than 100 papers in reviewed scientific journals, 6 patents, 2 edited books, 14 book chapters, and numerous speeches at scientific meetings. He is the coeditor (with E. Graf) of the book *New Product Development: From Concept to the Market Place* (New York: Van Nostrand Reinhold, 1991).

A fellow of the Institute of Food Technologists, United States, Sam was also recognized by the Israeli Food Industry Association with a lifetime achievement award for his academic achievements and contributions.

 **Tim Straus** is an innovative marketing specialist experienced in the development of new products and new distribution channels and in the setup of new ventures within food organizations. For 25 years, he has applied his unique combination of marketing talents to help a full range of companies grow along existing lines and successfully enter new categories.

As principal and chief marketing officer for The Turover Straus Group, Tim has assisted several food industry clients in mapping out strategic paths for new product initiative and growth. These efforts have been successfully implemented by Hunt-Wesson, The National Dairy Board, The Quaker Oats Company, National Cattlemen's Beef Association, Interstate Bakeries (Hostess), Aurora Foods (Mrs. Butterworth's, Mrs. Paul's), Sara Lee Foods, Farmland National Beef, KFC, Pizza Hut, Subway, and many others.

Prior to forming The Turover Straus Group in 1993, Tim served as the senior vice president/director of client service for a leading food consulting firm, where Tim's new product experience included work with Tyson Foods, Continental Baking, JR Simplot, Campbell Soup, and other food companies. Tim's early career experience included working with advertising agencies supporting McDonald's and Procter & Gamble.

Contributors

Todd K. Abraham
Research, Development and Quality
Kraft Foods, Inc.
Glenview, Illinois

Jose M. Aguilera
Department of Chemical and
    Bioprocess Engineering
Universidad Católica de Chile
Santiago, Chile

Ana Balasa
Department of Food Biotechnology
    and Food Process Engineering
Berlin University of Technology
Berlin, Germany

Jacqueline H. Beckley
The Understanding & Insight Group,
    LLC
Denville, New Jersey

Marco Bevolo
Independent Author and Conference
    Speaker
Eindhoven, the Netherlands

Doerte Boll
Department of Food Biotechnology
    and Food Process Engineering
Berlin University of Technology
Berlin, Germany

Aaron L. Brody
Packaging/Brody, Inc.
Duluth, Georgia

Paul Chaudury
Formerly Innovation
Sara Lee North America
Downers Grove, Illinois

J. Peter Clark
Consultant to Process Industries
Oak Park, Illinois

Max Cochet
Kalypso LP
Beachwood, Ohio

Stacey A. Cox
Product Research & Guidance
H.J. Heinz Company
Pittsburgh, Pennsylvania

Grant Davidson
Philips Design
Eindhoven, the Netherlands

Robert Delaney
Culinary Research & Development
H.J. Heinz Company
Pittsburgh, Pennsylvania

Lynn Dornblaser
CPG Trend Insight
Mintel International Group Ltd.
Chicago, Illinois

Jeffrey Ewald
Optimization Group, Inc.
Ann Arbor, Michigan

J. Bruce German
Department of Food Science
    and Technology
University of California
Davis, California

and

Nestlé Research Centre
Lausanne, Switzerland

Alex Gofman
Moskowitz Jacobs Inc.
White Plains, New York

Henry Jäger
Department of Food Biotechnology
    and Food Process Engineering
Berlin University of Technology
Berlin, Germany

Dietrich Knorr
Department of Food Biotechnology
    and Food Process Engineering
Berlin University of Technology
Berlin, Germany

Leon Levine
Leon Levine & Associates, Inc.
Albuquerque, New Mexico

Tammo de Ligny
Philips Design
Eindhoven, the Netherlands

Andrea Maier
Food Consumer Interaction Department
Nestle Research Center
Lausanne, Switzerland

Alexander Mathys
Department of Food Biotechnology
    and Food Process Engineering
Berlin University of Technology
Berlin, Germany

Herbert L. Meiselman
US Army Natick Research Development
    and Engineering Center
Natick, Massachusetts

and

Consulting and Training Services
Rockport, Massachusetts

Howard R. Moskowitz
Moskowitz Jacobs Inc.
White Plains, New York

Esma Oba
Department of Food Biotechnology
    and Food Process Engineering
Berlin University of Technology
Berlin, Germany

Micha Peleg
Department of Food Science
Chenoweth Laboratory
University of Massachusetts
Amherst, Massachusetts

Phillip S. Perkins
Bush Brothers & Company
Knoxville, Tennessee

Chip Perry
Kalypso LP
Beachwood, Ohio

Cornelia A. Ramsey
Division of Community Engagement
Center for Clinical and Translational
    Research
Virginia Commonwealth
    University
Richmond, Virginia

Michele Reisner
Moskowitz Jacobs Inc.
White Plains, New York

I. Sam Saguy
Institute of Biochemistry, Food Science
    and Nutrition
Robert H. Smith Faculty of Agriculture,
    Food and Environment
The Hebrew University of Jerusalem
Rehovot, Israel

Carole Schmidt
Doyle Research Associates
Chicago, Illinois

Tim Straus
The Turover Straus Group, Inc.
Springfield, Missouri

Helmut Traitler
Innovation Partnerships
Nestec Ltd.
Glendale, California

Bryan Urbick
Consumer Knowledge Centre Limited
Edgware, United Kingdom

Mario Valdovinos
Tyson Foods
Springdale, Arkansas

Martinus A.J.S. van Boekel
Department of Agrotechnology
    and Food Sciences
Wageningen University
Wageningen, the Netherlands

Marcus Volkert
Department of Food Biotechnology
    and Food Process Engineering
Berlin University of Technology
Berlin, Germany

Heribert J. Watzke
Nestec Ltd.
Vevey, Switzerland

David Christopher Wolf
Strategic Innovation
The Turover Straus Group, Inc.
Springfield, Missouri

Dave Zino
Culinary Center
National Cattlemen's Beef
    Association
Chicago, Illinois

Part I

Setting the Agenda for Successful New Product Development

| Success | = | Defining and meeting target consumer needs and expectations | × | The right food | × | Proper packaging and preparation | × | Positioned correctly at the shelf and in the media | × | Meet corporate logistics and financial imperatives |

1 The New Product Success Equation: Building Success into the New Product Development Process (a.k.a. An Objective-Based Product Development Process)

Tim Straus

CONTENTS

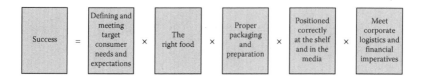

1.1   NEW PRODUCT IMPERATIVE

It is an undisputed fact that new products are the lifeblood of most food companies. Furthermore, although new product development (NPD) is tied to a company's lifeline, most companies have not agreed on a unified approach toward NPD. The different expectations of professionals and departments in NPD lead to this confusing disparity, which often generate completely different, occasionally contradictory expectations. One thing that all agree upon, whether from business case histories or from academic opinion, is that the "best" and "optimal" procedures should be based on a generous amount of data and inputs from both NPD practitioners and market performance. In other words, NPD success should be based on empirical observation and history.

It is generally accepted that there is no Wizard of Oz "yellow brick road" to be followed to assure NPD success. New product failure in the marketplace and the accompanying costs boggle the mind, especially when we take into account the loss both in sunk costs and market opportunity. Couple this failure in NPD with the inescapable fact that the business world now perceives the food and beverage sector to be a sluggish, saturated market, with less glamor than say cosmetics/beauty aids on the one hand, or financial services on the other. Finally, to cap it all, in mid-2008, escalating commodity costs have driven profits down and have forced food companies and restaurant chains to reevaluate their offerings.

The complexity of food and beverage NPD is to a significant part traceable to the spectrum of tasks and professions, which play crucial roles in all the phases of the process. Failure in any part of the spectrum might ripple down into failure of the NPD effort. Many textbooks and articles try to shore up the many different parts of this spectrum by focusing on critical areas such as consumer research, concept development, innovation, multifunctional teams, marketing, management, sales, advertising, etc. The bottom line, however, remains the same. The overall integration of the jigsaw pieces representing the various functions taking part in NPD is far from resolved.

1.2 INDUSTRY'S NEW PRODUCT "PRACTICES"—"STAGE GATE" AND ORGANIZATIONAL EXAMPLES

The industry uses many processes to assist NPD. Stage-Gate™ methods and portfolio analyses, and even the CEO's wife are used by many corporations. Indeed, the processes for NPD may shift and change depending upon which CEO is in power, what the current zeitgeist and fashion for consulting companies may be, and what specific computer technologies for management of information have managed to excite the corporation's "C-suite."

In general, most companies bring either a "process-oriented" or an "organization-oriented" approach to their new product programs. The particular process reflects the organizational structure of the company. For example, in many larger companies, the Stage-Gate system is often employed. This process-oriented system helps ensure that these larger companies concentrate their efforts on prioritized initiatives, and that each department can process the work required to bring specific initiatives to the market.

The Stage-Gate system defines the processes as shown in Figure 1.1. The Stage-Gate system specifies the flow of tasks to follow and the precise order to follow. This structured approach is important for companies, especially those that have to harness creativity in a massively competitive environment. The rigid procedure ensures adherence to quality in the process. Here are the steps (Cooper, 2001):

1. *Discover*: Prework designed to discover opportunities and generate ideas.
2. *Idea screen*: Screen the output of the discovery stage to ensure they meet certain acceptability goals with the consumer.

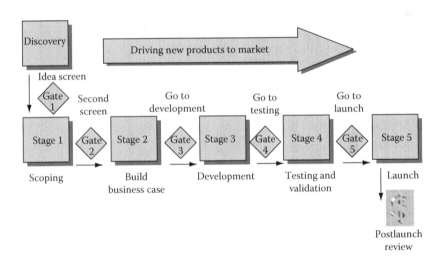

FIGURE 1.1 The Stage-Gate model. (Cooper, R.G., *Winning at New Products,* 3rd Edn., Perseus Publishing, Cambridge, MA, 2001. With permission.)

3. *Scope*: A quick preliminary investigation of the project. This step comprises primarily desk research.
4. *Build business case*: A much more detailed investigation involving primary research—both market and technical—leading to a business case, including product and project definition, project justification, and a project plan.
5. *Develop*: The actual detailed design and development of the new product and the design of the operations and production process.
6. *Test and validate*: Tests or trials in the marketplace, laboratory, and plant to identify and validate the proposed new product and its marketing and production/operations.
7. *Launch*: Commercialize—beginning of full operations or production, marketing, and selling.

Each area is completed by its respective department, and then internally evaluated. If and when certain metrics are met, the product moves to the next stage of development.

In contrast to the Stage-Gate process, many companies use an "Organizational and Discipline Oriented" approach to pass the product from Consumer Insights-to-Marketing-to-R&D/Packaging-to-Engineering-to-Manufacturing-to-Sales. This Discipline orientation typically manifests itself via the following implementation scenarios:

1. The Marketing Research/Insights Team uncovers an "insight" that can be exploited with new products.
2. The Marketing/Brand Management Team ideates and prepares a series of product concepts that "address" the targeted insight. The team may be assisted by members of Market Research, R&D, and outside consultants. The concepts are then given to Market Research to screen for those individual concepts that "exceed the norm."
3. The most promising concepts are then given to R&D to begin development work. At some point, the R&D Team returns to Marketing with prototypes of the concepts.
4. Marketing writes a marketing plan that outlines the launch parameters and financial scenarios for the launch and long-term business.
5. R&D and Operations commercialize the products and scale-up production.
6. The plan and product are tested in some sort of simulated or small test market for validation.
7. The product is then launched into the broader marketplace.

This organizational system uses the expertise of each team member's professional discipline to define their role, tasks, area of responsibility, and sequence in the entire process.

1.3   SEEKING A BETTER SYSTEM

The development of new food products is a wonderfully complex endeavor that utilizes the skills, patience, assets, and commitment of an entire company. Some in the organization view the development of a new product as "art" where the brilliant idea is conjured in the brain of the inventor and brought to fruition by force of personality and perseverance.

Others view the development as a staged process where each step must be executed in a perfect sequence to ensure that all bases are covered and each department of the organization has completed their dutiful assignments. Most will agree, however, that a successful effort in NPD requires the melding and interaction of various professional disciplines to ensure that each new opportunity is well conceived and executed.

The question that many active companies ask is "what is the best process for our company to use that will help to insure our new products succeed in the marketplace." The purpose of this chapter is to lay out the New Product Success Equation for building success into the new product process. We have already looked at two processes, the Stage-Gate system and a typical Organizational/Discipline approach as an illustration that corporations are serious about "getting the process right." Now we move to specifics of the process.

We will begin this discussion by asking you to recall whether you have heard one or several of these now-standard statements that connect with product success. Over the years, each of these statements must have been uttered thousands of times, by corporate professionals worldwide.

1. "We use Stage-Gate, so we know we will be successful."
2. "I know a new product success when I see one. My gut is always right."
3. "The chairman wants this to work, so let's make it work."
4. "We exceeded the concept test results norm, so we have a winner."
5. "We started with 200 concept ideas and narrowed it to this winning idea."
6. "BASES tests indicate this will be an $80 million brand, so we should proceed with the launch and marketing plan as written."
7. "With our brand name on this, it will succeed."
8. "Wal-Mart says they will carry it, so we are on our way to success."

The illustrative statements uttered above preceded both new product successes and complete failures.

Quite often, a postmortem of most new product launches reveal a very specific understanding of the elements that drove success and the shortcomings that led to ultimate failure. And when one considers the high rate of marketplace failures, even by truly professional organizations staffed by clearly competent professionals, we have to surmise that the process being employed does not address the important factors that drive success or result in failure.

1.4 FACTORS RELATING TO NEW PRODUCT FAILURES

From my experience in observing, working with, and competing against other companies in developing and introducing new products, I have found that failures usually indicate a breakdown of one or more important components of the new product. These errors or oversights are often more elusive than they appear. New product failures in the Food and Beverage Industry are usually associated with one or more of the following six factors, which we first list and then explicate.

1.4.1　Factor 1: Off-Base Targeting

Targeting and knowing exactly who the target is constitutes the basis for all product successes. Problem areas typically include:

1. There was no clear understanding of the target and their existing needs, shopping, food preparation, and product usage habits.
2. The product did not optimize its appeal to the specific target user. Instead, it tried to appeal to a much wider target. The product actually "missed the boat," failing to appeal to those for whom it had been primarily intended.
3. The target did not represent a true potential user. A full understanding of their potential product usage was clouded by other issues, including the consumer's failure to use products in the category, lack of appropriate appliances to prepare the product, household sizes, and the specific needs or peculiarities of others present in the household.
4. There was no existing product (or products) targeted for replacement by the new product? For these products, what were the dynamics underlying purchase, usage, portioning, buying rate/purchase cycle, and pricing? And specifically how did the new product offer something better?

McDonald's and Breakfast

When McDonald's first introduced breakfast during the early- to mid-1970s, it launched a handheld menu consisting of Egg McMuffin sandwiches, Danish, coffee, milk, and juice.

However, breakfast at that time was defined as a plated breakfast with eggs, bacon/sausage, or hot cakes/waffles. Furthermore, the heaviest consumers of breakfast outside the home were middle-aged men, not young families with children who were McDonald's core user at the time. For this core user group, breakfast outside the home was a weekend treat, not a weekday occurrence. (Wow, how things have changed!)

Breakfast at McDonald's did not really take off until the company marketers recognized this difference in the demographic and product usage target of the away-from-home breakfast consumer. They revised their marketing target to the middle age male instead of the family with kids and launched their "Big Breakfast" featuring scrambled eggs and bacon/sausage or Hot Cakes Breakfast. (Ironically, consumers then shifted to handheld breakfast about a decade later and McDonald's once again had to play catch-up.)

Hardee's Thick Burgers

Prior to 2001, Hardee's had one of the most expansive menus in the fast food channel. They had burgers, roast beef and ham sandwiches, fried chicken, hand-dipped ice cream, onion rings, French fries, and about every other food appearing on any quick serve menu. They were trying to appeal to every consumer by offering something for everyone. Their sales continued their long downward spiral.

Hardee's essentially pulled themselves out of financial disaster by retargeting their entire marketing, menuing, and operational programming to the young adult male who is seeking big food. They pursued this target by focusing every essence of their operation to serving the kind of food that this young male target craved and could not get anyplace else. The Thick Burger became their flagship product. The retargeting was highly successful and has resulted in several years of unprecedented volume and share growth.

Targeting drives everything. Defining the precise consumer target and meeting their needs yields tremendous opportunities for growth. Hitting the target wins. Missing or misinterpreting their needs yields a failure.

1.4.2   FACTOR 2: THE FOOD DID NOT PERFORM

It is usually a sign that the food did not perform when a new product initially skyrockets and then crashes. These failures usually manifest themselves in the following seven ways:

1. The appeal of food was not consistent with buyer's expectations.
2. The hedonics did not live up to expectations.
3. The product did not fit the buyer's needs, lifestyle, or shopping and usage regimen.
4. Nutrition delivery did not fit target.
5. Preparation was difficult for the end result.
6. Portion size did not match the usage needs of the target consumer.
7. The product was difficult to use.

The performance of the food in meeting the target's needs will drive repeat purchases and ultimately market success. Conversely, low performance will dampen repeat purchases, usually causing marketplace failure.

Banquet Homestyle Bakes

Banquet's Homestyle Bakes was a 2003 Top Ten new product with sales reaching nearly $100 million. A few years later, the volume dropped dramatically.

The product hit many of the right buttons out of the gate. It was well targeted to a time-starved, purse-stretched consumer seeking a complete meal that could be on-hand and prepared in less than 30 min. It was shelf stable (always available). Furthermore, its ambient temperature enabled it to be faster to prepare than anything frozen or refrigerated. It created a casserole with "freshly baked" bread toppings and fed a family of four. However, it used canned meat (that was included in the kit) that may have offended the sensitivities of many triers. As a result, its repeat purchase levels were fairly low, resulting in a dramatic drop in volume.

Stouffer's Skillet Sensations

In a similar vein as the Banquet product above, Stouffer's introduced Skillet Sensations, a frozen vegetable and meat skillet dinner targeted to a slightly more upscale user than Homestyle Bakes (Figure 1.2).

FIGURE 1.2   Stouffer's Skillet Sensations. (Courtesy of Nestle, U.S.A. With permission.)

Again, Skillet Sensations was well targeted to meet the needs of the consumer and beat out a current line of competitive frozen meal maker kits which required the consumers to add their own meat.

The product was launched with an exceptionally high level of trial and sales. However, the product immediately started to wane. Follow-up research indicated that there was simply not enough meat included in the kit. Adjustments were made to the product, the "more meat" burst added to the package and the product turned around.

Krispy Kreme Donuts

In 2000, Krispy Kreme was the rage of the food world as it exploded from its Southern roots to build its donut factory stores in larger urban communities across the country. The famous "Hot Doughnuts Now" sign brought customers into their units for the delightfully light yeast raised donuts that would melt in your mouth.

The business proposition also called for expanding the output of each factory unit by marketing Krispy Kreme Donuts to local convenience stores and supermarkets. The chain was exceptionally successful in building this distribution and servicing it with daily or near-daily deliveries.

However, a cold Krispy Kreme Donut was not light, did not melt in your mouth, and ate nothing like a fresh hot Krispy Kreme. Purchases from these outside points of distribution fell, leaving the entire chain in a financial bind as unit sales fell significantly. (During interviews, many executives cited the Atkins diet and low carbohydrates as the reason for this lost volume. However, poor food performance away from the Krispy Kreme stores drove overall chain sales to a significantly lower level.)

The performance of the food in meeting the taste, preparation, and portioning needs of the consumer is paramount to generating repeat purchases. "Trial alone will not drive success."

1.4.3   FACTOR 3: THE PACKAGING WAS NOT RIGHT

In the retail channels, the packaging says everything about your product. A product's packaging provides containment, protection, convenience, and communication. In the store, the package communicates what the product is, and what it is not. It is vitally important that the package shape, construction, and graphics properly communicate the product proposition.

And in the home, packaging will often dictate whether the product will be used again. Was it functionally correct? Was it intuitive to use? Did it store well in the pantry, refrigerator, or freezer? Was it easy to get into? Did it hold the right amount for the meal or snack?

Campbell's Souper Combo

Perhaps the granddaddy of all package failures was Campbell's Souper Combo product, launched in the late 1980s. This pioneering microwave product was based on a great food proposition of delivering a convenient way for someone to prepare a grilled cheese sandwich and a bowl of soup in their microwave. The food delivery was pretty good. The sandwich was toasty and cheesy and the Campbell's Soup was piping hot.

But the steps that the consumer had to go through to prepare the soup and sandwich in the microwave was the equivalent of a nightmarish Christmas Eve experience of assembling a tree full of toys. There were no less than four steps for each soup and sandwich. The steps required stopping the microwave at certain points and removing, opening, or adding some part of the package. The product packaging made the product just too difficult for any mortal being seeking lunch. And the product failed.

Deli Pot Roast

Perception is everything in food products. Packaging drives perception. As the consumer's need for convenience continues to grow, many marketers of fresh, value-added meats have tried to provide their customers with fresh meat products that are preseasoned, packaged, and ready to cook.

Unfortunately, few of these products have been truly successful. The major culprit is that the packaging often works against the "fresh delivery" proposition. Most products are "over-packaged" in the view of the consumer, leading to a less than fresh perception. This lack of "freshness" is probably one of the main reasons for the stunted growth of the category.

The National Cattlemen's Beef Association was designing a program to introduce a hot take-home beef product to complement the supermarket deli's highly successful rotisserie chicken program. There were major challenges to overcome. Previous attempts failed due to at least three appearance/packaging issues:

1. Unlike a golden roasted chicken, beef is dark and brown and does not visually stand out under the lights in the "warmer case."
2. The confidence hurdle for beef was much higher than that of chicken in terms of whether the product was freshly prepared rather than dried out and

would taste good. The consumer needed to be reassured that the product was freshly prepared.

3. The density of a piece of beef made it look smaller in the container, thereby making the value proposition lower due to the perceived small portion size.

To overcome these issues, a packaging approach was developed that surrounded the beef entrée (pot roast) with bright and vibrant carrots, celery, and red potatoes. The vegetables' bright colors communicated "fresh" to the consumer, while adding significantly to the portion size. As a result, the deli pot roast was introduced and succeeded in the marketplace (Figure 1.3).

Heinz's Ketchup

As consumers tired of ketchup "anticipation" and sought a means of getting ketchup out of the bottle more quickly and accurately, Heinz, the largest marketer of ketchup, designed a new package.

The first package refinement offered a special squirt lid that allowed the consumer to squeeze the ketchup bottle. However, the ketchup was not consistent in the way it came out of the bottle, due to an uneven amount of ketchup being available to the squirt lid. In some instances, the ketchup produced quite a mess, particularly when little hands were using the bottle (Figure 1.4).

Several years later, the company launched the up-side-down easy squirt package with a wide top and a new "lid side down" center of gravity that allowed the ketchup to fill the squirt mechanism and created a consistent, neater, and more convenient stream of ketchup. This new bottle generated a 25% increase in sales over the 2 years, and it has been on the market by meeting the consumer's need for convenience by all family members (Source: Rountable Magazine, April 2008).

FIGURE 1.3 Deli pot roast. (Courtesy of the National Cattlemen's Beef Association.)

FIGURE 1.4   Heinz's ketchup. (Courtesy of the HJ Heinz Company.)

Earthgrain's Zipper Bread Packages

In the late 1990s, Earthgrains Bakeries (now part of Sara Lee) sought to lead the market with their new zipper closure systems to replace the more standard bread ties and clips. The zipper would add about 5¢–10¢ per package.

Whereas the convenience of the package was terrific, its success required either a significant increase in volume to off-set the higher cost or a higher price point. Increased turns, i.e., more frequent sales, were not significant enough to justify the added package expense, and consumers were reluctant to pay more for the zipper feature. This resulted in scraping the new system and returning to the traditional and lower cost system. The convenience was not worth the added cost.

The right package is a real trial generator. Conversely, bad packaging can reduce both trial and repeat purchases. The consumer looks at the packaging to define the new product. Its construction defines the process and safety of the product and drives the consumer's rational assessment. The graphics and branding drive the consumer's emotional response. Together, the construction and the graphics can provide the product with a major boost.

1.4.4   Factor 4: The Name, Positioning, and Advertising Just Did Not Connect with the Product

There are times when the product, positioning and resulting advertising just do not "play well together." In most cases, the culprit is usually the product itself, which simply does not deliver what it promises, thereby proving the old adage that "nothing will kill a bad product better than good advertising."

But that is not always the case as the two examples below illustrate.

McDonald's Arch Deluxe

McDonald's promised "The Burger with the Grown-Up Taste" when it introduced its Arch Deluxe, a burger that promised zesty flavor that would appeal to adult tastes. Initial trial was significant, but the product just did not deliver the flavor experience promised in the advertising. Somewhere along the way the zesty sauce was reduced to just another condiment.

While it is hard to say that a 100 million dollar (plus) product is a failure, but by McDonald's standards, it did not meet their goals. The product was discontinued after about 12 months.

Kellogg's Ensemble

Kellogg's introduced a new breakthrough product line with very tangible nutritional attributes with its short-lived Ensemble brand. But along the line of promoting the attributes and scientific claims, they failed to translate these into benefits for the consumer.

Like so many functional food marketers, Kellogg's was mesmerized by the magnitude and power of scientific claims of health and wellness benefits. In turn, they revolved their story around the bioactive ingredient and not the benefit which consumers can derive from consuming the product. In fairness, some of this was dictated by Federal laws limiting benefit claims from functional ingredients.

Regardless of the underlying reasons, marketers failed to articulate the product's benefit in a way that is meaningful and clearly understood by the consumer. This left the consumer recognizing the ingredient name but not exactly comprehending what its overall benefit is to the potential user. This was one of the reasons for the failure of Kellogg's Ensemble. This functional food line of nutritionally enhanced cereals, cookies, lasagna, frozen entrees, and baked potato crisps made with cholesterol lowering psyllium husk that was taken off the market a mere 8 months after its launch in 1999.

The product strengths and appealing attributes must be properly positioned in the media and at the shelf. Great care should be taken to ensure the product and the message match, and that they are compelling to the consumer. As a rule of thumb, if the message has more hype than the product deserves, then perhaps the product itself should be revisited and reworked to make its appeal match the message.

1.4.5   FACTOR 5: THE TRADE HAD A DIFFERENT "INTERPRETATION" OF THE PRODUCT

For retail products, the last five yards before scoring lies in the hands of the retailer. There are times when the positive efforts by the trade will create a highly successful product, and other times the actions of the retailer will doom an entire new product proposition.

Trade-driven elements that will affect the success of a new product include

1. Was the product placed in the department of the store where it was intended by the manufacturer/marketer? Placement outside of the intended shelf location can nullify the company's efforts regarding product line definition, competitive set, packaging, positioning, and pricing.
2. Did the trade correctly adopt the defined pricing? In some cases, retailers will take a larger margin than built into the plan, thereby moving the new product outside of strategic price points, again nullifying much of the product proposition.
3. Was the product handled properly or mistreated at store level? Many products are designed to be handled in a specific manner in terms of distribution, stocking, and placement on the shelf. Poor handling will often result in less than ideal presentation on the shelf or cause actual detriment to the food itself.

Hostess Fruit & Grain Cereal Bars

Hostess entered the Breakfast Bar category with a new line of Fruit & Grain Cereal Bars. These products contained an "extra portion" of real fruit (higher than the competition) and was packaged and priced to go on the breakfast aisle next to the other Breakfast bars.

However, because the product had the Hostess name and was store-door delivered, most retail chains placed the product on the snack cake shelf next to the other iconic Hostess Cakes (Twinkies, Cup Cakes, Ding-Dongs, and Ho-Ho's). This placement nullified the nutritional credibility that was built into the product and resulted in less than optimal sales.

Farmland Ground & Browned Ground Beef Crumbles

The pricing that the trade sets for a new product is crucial to its success. If their margin is greater than the norm and results in a significantly higher price point, the results can be disastrous.

In 2000, Farmland National Beef introduced Ground & Browned fully cooked ground beef crumbles. This product was selected as one of the Top 10 New Products of the Year and was the result of preliminary work undertaken by the National Cattlemen's Beef Association and later adopted by Farmland National Beef.

Essentially Ground & Browned was fully cooked browned ground beef, ready for recipes without the time, splatters and grease disposal associated with browning ground beef. It was available in regular Italian and Mexican flavor varieties. A 12-ounce package equaled one pound of fresh ground beef. The product performed well in a number of preliminary home-placement studies and a prelaunch BASES test. With a 40% trade margin (standard value-added category margin), it was intended to be priced at $3.49 with deals at $2.99. This would place the cost of convenience at about one dollar per pound over regular lean ground beef.

However, the trade viewed the product as competitive to the Refrigerated Entrees that retailed for $5.99 and priced Ground & Browned near their significantly higher price point. The results were disastrous because a $2 + up-charge versus fresh ground beef was too high for the target consumer.

It is essential that the new products team thoroughly understand the trade's interpretation of the new product in terms of shelf placement, margins, handling, and pricing. The trade's actions are typically an action by fiat that is seldom shifted and should be clearly understood before the new product proposition is finalized. All products must be properly positioned with the trade so they can extend your product's strengths, not neutralize them.

1.4.6   FACTOR 6: THE PRODUCT WAS INCONSISTENT WITH CORPORATE STRENGTHS OR FINANCIAL GOALS

Defining, developing, and bringing a successful new product to the marketplace is tough enough when all the corporate assets and strengths are aligned. Often the definition of innovation is "moving away from the status quo" into new areas that stretch a company into unknown areas. But, in the end, certain realities can play significant roles.

Certain essential questions must be honestly answered which affect the success of a new venture or new product.

1. Is this product opportunity large enough to support the volume requirement that the company dictates?
2. Does the product line fit the company's manufacturing and distribution assets and expertise?
3. Are the margin levels right for the company to make a profit?
4. Is the sales force sufficiently well staffed and experienced enough to sell this product?
5. Are there other mitigating factors that could hinder the corporation from getting behind the new product?

An honest answer to the foregoing questions will usually yield an important answer regarding whether or not a new venture or product line should be pursued. As you will note from the examples below, there are numerous areas that should be considered before the company should "green-light" a new product or venture.

Simplot MicroMagic

In 1988, the JR Simplot Company decided to expand from their Foodservice roots into the Retail channel. At the time Simplot was the leader in producing frozen

French fries for restaurant chains, most notably McDonald's. Simplot had invented the frozen French fry. Part of the decision to enter the Retail Channel was their invention of a microwavable French fry that they branded MicroMagic. A sales/broker and marketing infrastructure was established and the product line won Product of the Year recognition.

The line was expanded to include MicroMagic Hamburgers and even MicroMagic Shakes. The brand exceeded $80 million in sales, utilized an extensive advertising and promotion schedule. Other retail products generated about another $20 million in sales.

Ultimately, three major factors caused Simplot to pull the plug on the product line and their entire branded Retail venture:

1. The cost accounting for the production of the microwave fries placed incredible pressure on the cost-of-goods because the product line utilized the very high capacity lines generally reserved for the highly profitable foodservice fries. This resulted in the much smaller retail brand incurring significant overhead costs for each pound produced.
2. Simplot was virtually the exclusive producer of fries for McDonald's. Many persons inside and outside of the Simplot organization felt that marketing a retail burger, fries, and shake line of products was an affront to their largest customer.
3. The significant investment spending required in the retail channel for slotting and up-front media was a tough reality given Simplot's corporate culture.

Riceland Rice & Easy

Riceland Foods is a major co-op of rice growers in Arkansas and other southern states. They are the largest producers of rice in the world. Naturally, their mission is to move large quantities of rice for members of the co-op. Riceland's main business at the time of the launch of Riceland Rice & Easy was exporting boat-loads of rice around the world, and selling 5, 10, and 20 pound bags at retail.

Riceland developed and launched Rice & Easy at the time other companies were launching flavored rice kits (e.g., Uncle Ben's Country Inn). The Rice & Easy product line was introduced into strong Riceland retail markets and performed well on the shelf. However, after 1 year, the product line was essentially pulled. The reason was that it simply did not move enough rice for the grower's co-op. Selling a few ounces at a time ran counter to the mission of the cooperative, regardless of the profit potential of the product line.

Ensuring that a new product meets the financial, logistics, and culture of the company is essential for long-term success. Anything less than full commitment will usually result in failure.

1.5 CHALLENGE

The six factors cited above all have a huge impact on the success or failure of a new product or venture. When problems occur, they usually result from compromise made to pacify some concerns about the product or the corporate way of doing things. Thus, in light of the many failures traceable to "ego," "beliefs," and just plain stubbornness, the three key challenges for NPD Teams are

1. How does an organization seek the known and at the same time confront and tame the unknown in order to optimize the success of a new product?
2. How does a company align all team resources to ensure that each element of the new product proposition is addressed, acted upon, confirmed, and optimized?
3. How does the new products team validly answer the truly important questions that must be answered for a new product to be successful in the marketplace?

The Success Equation outlined below and enumerated throughout this book seeks to offer a means for ensuring the right elements are THE PRIORITY in preparing a new product for launch.

1.6 THE SUCCESS EQUATION: AN OBJECTIVE-BASED NEW PRODUCT PROCESS

The Success Equation takes a fundamentally different approach toward defining a food and beverage NPD process. The equation is holistic, i.e., it defines the relationship among the different factors. It has the promise of being quantitative, although still now at the early stages it is more descriptive. The equation was created with three building blocks:

1. The equation takes a Goal and Objective-Based approach rather than an "organizational" or "process" approach. It defines precise goals that are essential to new product success and does not try to dictate a precise set of steps. The substitution of goals for steps means that the equation focuses on the results, not on tasks.
2. The equation is Team Oriented and Interactive so that all members of the team representing precise disciplines are involved in all facets of product development. It recognizes the fluid nature of bringing a successful new product to market versus the linear nature of rigid process steps. The equation places maximum importance on the truly essential elements that drive success.
3. The equation is scalable with its attainable goals. The equation can apply whether the new product is being driven from the back of a garage or a multinational headquarters—a large new branded initiative or a simple line extension or key customer request. The equation also implicitly recognizes that not all new product initiatives are on the same scale. From company size to project size, there must be a basic approach that allows the available resources of a company to be placed behind a new product initiative to guarantee its success. Large food companies may afford many more development dollars and manpower than a mid-size or start-up company. Yet, the market success ultimately depends on whether the end product actually meets a given set of parameters.

The first thing the reader will notice about The Success Equation is that it is not a step-by-step process for defining, developing, and launching a new food and beverage product. Rather it is a set of critical objectives that must be achieved to ensure success. Therefore, the process does not define a series of steps, but rather a series of objectives to insure success. It outlines what should be accomplished to be successful not the steps that should be done to fulfill a process.

This is a very critical difference. Significant time, expense, and talent can be employed to bring information forward to meet a staged progression of tasks, but at the end of the day the process itself brings little to the ultimate outcome unless key issues are addressed and answered.

We define the New Product Success Equation for new food and beverage products as the sequence of five multiplicative factors, shown in Figure 1.5.

Three important notes about the equation itself:

1. Each element is simply stated as a goal that drives success. The element *does not* define *the precise steps on how to get there,* though this book will certainly enumerate each element. Instead, it simply states each important element of success that must be achieved.
2. It is a mathematical equation that uses a multiplication product as its answer. This is extremely important because less than optimal delivery in a specific area of the equation could reduce the level of success, and perhaps even nullify success. If during the development process a decision is made to reduce the new product's appeal in a given area, that decision impacts all elements of the success.

    For instance, a common development problem is keeping the cost of goods in line to make the product financially viable. Often after the food and packaging have been tested, steps are taken to lower the cost of goods. These steps meet the financial goals, but may play havoc with delivering the right food. One part of the equation is enhanced, whereas the other is reduced. Hitting the right balance across these two areas could mean the difference between market success and market failure.
3. It is interactive by not defining specific areas of responsibility. By its nature, the equation ensures that all team members are responsible, even for things that are outside of their immediate discipline.

This equation states that every component of the New Product Proposition must be right (not zero) and maximized in order to have success. At the risk of restating by now the obvious, here are six key considerations and definitions:

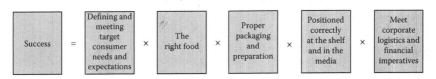

FIGURE 1.5 Definition of the new product success equation.

1. Success is the criteria that your company has established to be a success. These criteria will vary by institution and will relate to quantifiable sales, distribution levels, trial/repurchase rates, profitability, investment, and payback.
2. Defining and Meeting Target Consumer Needs and Expectations is the ability to find an unmet need or a superior way to answer a consumer's need with a new product.
3. With the Right Food that meets the flavor, textural, appearance, portioning performance levels, and overall quality, so it will be purchased again.
4. Proper Packaging and Usage Prep that allows the product to be credible on the shelf and to deliver at optimal performance levels within the realm of reasonable preparation for the consumer.
5. Positioned Correctly at the Shelf, in the Media, and with the Trade to carve out a level of uniqueness that communicates the product's reason-for-being, justifies its price, and stimulates trial.
6. Meet Corporate Logistics and Financial Imperatives to ensure that the business case for the product fits and can be adequately supported by the company's operational and marketing infrastructure, in order to generate overall sales and profitability requirements.

The structure of The Success Equation calls for all elements to be addressed as equals. The nature of multiplication is that the change in any one element affects the outcome. There is no room for "compensation," so that one change can be compensated for by another. Multiplication is associative and distributive, affecting all the variables. Increasing the impact of each element increases the total product of the five components.

Success requires constant collaboration among the disciplines of marketing, consumer research, sales, R&D, operations, financial, and top-management. When integrating the Success Equation into an existing organization in its simplest form,

1. Marketing Disciplines are responsible for creating the elements of the new product proposition that stimulate "trial."
2. Technical and Manufacturing disciplines ensure that the new product fulfill the product promise, resulting in "repeat" purchases.
3. Top Management ensures all disciplines are on-track, prioritized, and meet the financial goals of the company.

And importantly, the optimization of each element is not based on a predicated process, but rather achieving an internal comfort level that

1. The consumer's need represents a viable market for reaching the corporate goals, and that this need is met by the product.
2. The food tastes good, performs well, and is properly portioned for the target user and target occasion.
3. The packaging performs the functions of containment and protection, while meeting the product's specific needs of convenience and communication.
4. The positioning delivers well on the package, with the trade, and in the media.
5. The company can produce the product, market the product, and make money with the product.

1.7   IMPLEMENTATION OF THE SUCCESS EQUATION—USING IT TO QUANTIFY AND BENCHMARK SUCCESS

The remaining question is how does a company optimize the elements of the Success Equation? What role does the equation provide in quantifying this success? Since the Success Equation is goal oriented and does not outline a specific process, the question may remain on how to best ensure that the new product elements are optimized.

For established companies possessing good data on products that succeeded in the marketplace (both internal and competitive), the appropriate scores can be entered into each respective part of the equation. Of course ahead of time, the rules for defining the scores must be established so that across products the scores are commensurate with each other, and can be compared.

Please note: Caution should be used not to enter scores of all products tested to generate an average score, but only the scores of products that have actually achieved marketplace success. The equation is designed to predict success, not just "above average" scores. (Remember if 80+% of all products fail in the marketplace, then above average scores do not have much meaning.)

For companies beginning new programs with little prior experience, a benchmarking approach should be undertaken to help ensure that newly developed products are competitive to products already successful in the market. This is stiff competition, but beating these products will yield a greater chance for success in the marketplace. Again we must emphasize the need to have a common scoring for all products. The scoring criteria will, of course, differ for the different parts of the equation.

1.8   PRODUCTS WHERE THE SUCCESS ELEMENTS WERE RIGHT

There are certain companies that have succeeded in both understanding their target and launching incredible products that have enjoyed success year-after-year. Their success is not only demonstrated in the success of that specific product, but also the fact that the products defined new categories that were later entered by competitive companies. Providing their story, along with the successful revisions made in the examples cited above, may give the reader a more comprehensive view of how to create success in their own new product ventures.

1.8.1   DiGiorno Self-Rising Crust Pizza

The entire Pizza Category took a radical turn when Domino's Pizza created the home delivered market for Pizza. Up until that time, pizza was either enjoyed on premise at a restaurant, carried out from a restaurant or a frozen pizza was heated at home. A few also made their own pizza, but that was a very small market relative to the frozen pizza category.

Category leader Kraft, marketers of Tombstone Pizza and Jack's frozen pizzas, saw a tremendous market in offering the delivery and carryout pizza user with a really delicious frozen pizza, having a restaurant-quality crust. It is important to note that they targeted outside the existing frozen pizza category to a much larger restaurant user who typically spent more for a pizza than did the average buyer of frozen pizza.

Professionals in the "pizza business" understand that "pizza is all about the crust." Kraft understood that frozen pizzas, while immensely popular were known for their "cardboard crusts." Any substitute for a restaurant/pizzeria pizza would have to have a superior crust, no matter the toppings. The crust would make or break their success.

The huge technical expertise at Kraft created the self-rising crust (frozen dough that would rise during the home baking process) that delivered a fresh restaurant crust out of the home oven (Figure 1.6). The product benchmarked well against the major pizza restaurant pizzas.

Kraft also went against the traditional wisdom of NOT trying to introduce a "$1.99" pizza. They went premium, approximating the price points of Domino's and Pizza Hut deals (around $5.99). This took a lot of convincing to the trade and other skeptics who did not believe that a $5 frozen pizza could make it in the supermarket frozen food case. Kraft, as a category leader, had the clout to shift the trade's perceptions that few other companies may have had.

The product was branded using a more authentic quality brand, DiGiorno, rather than one of their current frozen pizza brands. And the product was positioned and used the now iconic "It's Not Delivery, It's DiGiorno" advertising campaign.

Since the late 1990s, the DiGiorno brand and competitive entries (most notably Schwan's Freschetta brand) into the rising crust segment generate annual sales of approximately $1 billion.

FIGURE 1.6 DiGiorno self-rising crust pizza. (DiGiorno is a U.S. registered trademark of Kraft Foods. With permission.)

1.8.2   Hostess Brownie Bites

While the Hostess Brownie Bites story is certainly more modest than DiGiorno, it demonstrates how a company can understand a target, create a concept for meeting that target's need, and commercialize it in a manner that fits corporate assets. And the fact that this product has enjoyed consistent sales for over 15 years demonstrates it longevity.

With the huge growth in the number of convenience stores during the early 1990s, Hostess saw an opportunity to address the precise needs of the heavy C-store customer. They found young male outdoor laborers, technicians, and craftsmen to be consistent shoppers for breakfast and throughout the day. Hostess found that many of these customers would essentially jump-start themselves in the morning with a quart of chocolate milk and some sort of packaged sweet like a brownie or cupcakes. Brownies were their favorite, but found existing products were messy to eat and the package, once opened was impossible to close or even throw away.

To meet the food and package needs of this target, Hostess created a ball shaped brownie that would enable the target to pop the entire brownie, while keeping hands clean (Figure 1.7). Hostess also found that they could not bake and package the round brownie on their regular cake equipment, so they created a process to produce it in their mini-muffin facility. And whereas the "muffin" solution did not create an exact replica of the original round brownie, its appeal was nearly as strong

FIGURE 1.7   Hostess Brownie Bites. (Courtesy of Interstate Bakeries Corporation.)

and commercially viable. As a result, this line extension has become a long-term member of the Hostess product line.

1.9 THE SPIRIT OF THIS BOOK

Essentially, this entire book is committed to helping the reader optimize the entire Success Equation (see Figure 1.5).

Within each section you will find chapters written by both academics and by business-world practitioners from around the world who are experienced professionals in their respective field of food and beverage product development. They provide their insights into optimizing each area for success. Taken together and applied to the Success Equation, the information in our book should help you formulate or perhaps even concretize your own process for optimizing new product success.

REFERENCE

Cooper, R.G. 2001. *Winning at New Products*. 3rd Edn., Perseus Publishing, Cambridge, MA.

Part II

Defining and Meeting Target Consumer Needs and Expectation

| Success | = | Defining and meeting target consumer needs and expectations | × | The right food | × | Proper packaging and preparation | × | Positioned correctly at the shelf and in the media | × | Meet corporate logistics and financial imperatives |

2 Strategic Planning

Todd K. Abraham

CONTENTS

Success	=	Defining and meeting target consumer needs and expectations	×	The right food	×	Proper packaging and preparation	×	Positioned correctly at the shelf and in the media	×	Meet corporate logistics and financial imperatives

If you don't know what port you are going to, no wind is favorable.

Socrates

If you don't know where you're going, any road will take you there.

George Harrison

2.1 INTRODUCTION

Strategic planning is the foundation for everything we do. It is the roadmap to success. However and perhaps more important, strategic planning is the articulation of an organization's objectives. The strategic plan should define what will be pursued and what will not. Deciding what is not going to be done is sometimes more important than what will be done. The overt discussion and debate drive alignment and create clear prioritization that brings focus to the organization.

The plan should address the strategic approach to accomplishing the objectives. The plan should stem from the organization's vision and provide the aspirational goal. John F. Kennedy's aspirational goal of landing a man on the moon was such an objective.

I believe that this nation should commit itself to achieving the goal, before this decade
is out, of landing a man on the moon and returning him safely to the Earth.
— *U.S. President John F. Kennedy, May 25, 1961 (Kennedy, 1961)*

Kennedy's goal provided both a concrete, measurable result, and a definite time
frame. It mobilized the entire American scientific community to unite in common
cause, aligned to a common result. The result was that his vision, essentially a grand
goal, shaped the country's scientific priorities for almost 10 years.

Let us step back for a second to explore Kennedy's vision in more depth, and
see what it can teach us. His was not a fanciful vision aimed at grabbing headlines,
but rather a result of strategic need. The United States had been embarrassed at the
Bay of Pigs. Four years before, in 1957, the Soviet Union had already put a man in
space. This vision was a strategic plan developed in view of the long-term needs of
the organization (the country), shaped by the "competitive environment," and leveraging core competencies with a time frame reflecting the urgency of a competitive
"marketplace." Kennedy's statement and vision was not merely an idle comment or
standard political fodder. The goal was the result of high-level strategic discussions
about what were the capabilities and potential objectives (Kennedy, 1961).

The resulting space program reflected a focused commitment to that goal with
all aspects of the program focused on reaching the objective. Furthermore, the strategic
plan defined how the goal would be accomplished. The complexity of the final
program generated a progression of experiments (missions), each targeted against
different subcomponents of the end objective. First, get a person into space and
back to earth safely. Next, keep that person in space for longer periods of time.
Allow astronauts to work outside the environment of the space capsule. Third, develop
the capability to dock spacecraft. Fourth, figure out how to get to the moon and
return—lunar orbital missions. Orbit the moon. Finally, descend to the moon
and return to the spaceship.

Breaking down the vision and strategic plan to its components allows for better
resource planning, identifies critical hurdles, and progresses relentlessly to the goal.
Importantly, this division into components generates intermediate milestones that
have commercial value along the path. If the space program discovered an unsolvable
problem during the Apollo missions, the United States still had the capability to put
satellites into orbit, clearly an important capability in and of itself.

Another important benefit of having "big hairy audacious goals" (BHAGs) is
the unexpected positive benefits that spin off from the attack on very difficult problems.
The systems utilized today in hospitals to monitor intensive care patients come
from technology developed to monitor astronauts in space. Scratch-proof plastic
lens, which has replaced most breakable glass, is a direct outcome of the early
space work. Antifogging ski goggles came from technology developed to keep
the windows of spacecraft clear before takeoff. The same technology that cools
on board electronics is used today in the artic to ensure that the pipelines running through the tundra remain cold on the outside to protect the environment,
yet warm enough on the inside for oil to flow. These are just a few examples of
spin-off technologies developed from the components of the original strategic intent
(Scientific and Technical Information (STI), 2008).

Without the vision to guide the space effort, we never would have overcome
the setbacks of intermediate failures such as the death of three astronauts early in
the Apollo program. Furthermore, without the overall vision, we would have been
required to define new objectives at every step of the program. The vision and resulting strategic plan provided the roadmap for the nation to accomplish one of America's
biggest BHAGs with incredible speed and efficiency.

The mission itself reflected the success equation.

Success = Meeting the Target Consumer's Needs and Expectations × The Right
Food × The Right Package and Preparation Method × Positioned
Correctly on the Shelf and in the Media × Meet Corporate Logistics
and Financial Imperatives

The need was defined and expectations set—land a man on the moon and safely return him
to earth within the decade. The key components independently and jointly were evaluated
and tested. Coming back to food, for a moment, the project would entail product, process,
package, positioning, price, promotion, consumer target, and distribution capabilities.

For the moon mission some were

1. Launching men into space for an extended period of time
2. Uncoupling spacecrafts and rejoining them
3. Traveling to the moon
4. Leaving and returning to earth orbit
5. Entering and leaving lunar orbit
6. Landing on and leaving from the surface of the moon
7. Safely landing (splashdown)
8. Recovering the astronauts

Each of these eight foregoing components was developed as an independent exercise
or building off a previous milestone, then coupled together to ultimately make a successful program.

2.2 FIRST STEPS TO DEVELOP THE STRATEGIC PLAN: THE SWOT ANALYSIS

Begin with the end in mind. The strategic planning process for the business must
start with the end goal and vision in mind. What is the future state of the business?

The strategic plan must start with an analysis of the organization's strengths, weaknesses, opportunities, and threats (SWOT), a technique credited to Albert Humphrey,
a Stanford researcher (SWOT, 2008). A SWOT starts with the objectives and then
defines the internal attributes that are helpful (strengths) and harmful (weaknesses) to
the effort of attaining the objective. External enablers (opportunities) and conditions
which are barriers to the objective (threats) are defined as well.

The SWOT provides the basis for determining how to leverage the strengths,
overcome the weaknesses, exploit the opportunities, and defend against the threats.
The assessment of capabilities should determine the roles of each business within the

		Positive	Negative
Internal factors		Strengths	Weaknesses
		➤ Technological capabilities	➤ Critical skill gaps
		➤ Brand equity	➤ Undifferentiated brands
		➤ Distribution channels and scale	➤ Disadvantaged distribution
		➤ Customer loyalty/relationship	➤ Low customer retention
		➤ Production network	➤ High levels of consumer complaints
		➤ Scale	➤ Sub-scale
		➤ Management	➤ Management
External factors		Opportunities	Threats
		➤ Changing customer needs and preferences	➤ Changing customer needs and preferences
		➤ Liberalization of geographic markets	➤ Trade barriers
		➤ Technological discontinuities	➤ Technological discontinuities
		➤ Changes in government politics	➤ Changes in government politics
		➤ Favorable policy changes	➤ Unfavorable policy changes
		➤ Change in demographics	➤ Change in demographics
		➤ New distribution channels	➤ New distribution channels
		➤ Broadened market definition	➤ New competitors

FIGURE 2.1 SWOT adapted from material developed by Durham University. (Adapted from Chesbrough, H., *Open Innovation: The New Imperative for Creating and Profiting from Technology*, Harvard Business School Press, Norwood, MA, 2003.)

portfolio. Not all businesses will have the same competitive position. SWOT analysis enables the company to allocate a greater proportion of resources to the business that has the greatest opportunity for achieving its strategic intent.

The SWOT method can also be applied to product lines within businesses or smaller subunits of the enterprise. The result is the same—prioritize resources and focus efforts against the activities that have the greatest chance to succeed, and to achieve the overall business vision.

There are many templates available to facilitate conducting a SWOT. These templates focus on systematically evaluating different areas, functions, and initiatives (SWOT, 2008). Typically these will evaluate people, resources, ideas and innovations, marketing, operations, and finances—a model developed based on work from the Durham University Business School. (See Figure 2.1 for an example.)

2.3 METRICS FOR THE STRATEGIC PLAN

"If it doesn't get measured, it doesn't get done." This is an old adage from many sources that reflects the fact that people and organizations will focus on those things for which they can evaluate results. Creating meaningful metrics for any strategic plan is a critical element to being able to execute the vision.

Metrics play an important role in a number of ways. They break a longer term task into periodic (e.g., quarterly) milestones. Measurement on a periodic basis tracks progress and ensures that the urgency remains in the efforts. Typically, this periodic measurement is best used for complex technical roadmaps that have a number of discrete steps or subprograms that can be tracked on a regular (quarterly) basis. Holding the organization accountable for hitting the milestones will help ensure that the longer term goal is achieved.

A second class of metrics is "rear-view mirror" metrics—evaluating how the organization has performed on the critical components of the plan. Examples of these rear-view metrics might be market share or number of patents issued. Rear-view mirror metrics are valuable for identifying progress and trends, but may be misleading about where the organization is going.

The most difficult to define and measure are predictive measures that will monitor the likelihood of a successful strategic result. Defining these metrics for the strategic plan is critical. An example would be "percentage of R&D supporting new technology development." It alone will not ensure that future programs are successful; however, it is a predictive measure which reflects a higher likelihood of success. Advertising spending for new product launches is another forward looking metric.

The frame of reference for metrics also drives behavior. Market share is a very traditional metric to assess how the organization is performing. Companies like to report that they have the number one market share in key business areas to demonstrate that they are leaders in their field. Whereas useful for the corporate ego, this definition is detrimental to corporate growth. Imagine, as an extreme example, if the world were to be defined strictly as "the company store." Most organizations have 100% share in their company store. There is little opportunity to grow the business with that narrow, biased, and limiting definition.

Organizations will define categories in a way that inflates their success. This may lead to complacency and a lack of innovative new ideas and options. If the business area is "carbonated cola beverages", the programs and strategies are very different from the category that is "all beverage consumption, including tap water."

A better approach to metrics for the strategic plan expands the frame of reference so the organization has a smaller share of a bigger pie. This will lead to new ways to think about categories, opportunities, technologies, and competitors. Going back to the beverage category, the strategic plan might identify how to source volume from sources that was not evaluated before. Would the key beverage companies have a bottled water business if the frame of reference was only "carbonated soft drinks"?

A practical example of this frame-expansion comes from pizza. DiGiorno pizza has been a very successful frozen pizza product for Kraft. By expanding the frame of reference to the broader pizza eating occasions and looking at take-out and home delivery as part of the competitive set, Kraft was able to grow the DiGiorno business even faster. Looking at this challenge from the perspective of the success equation, we can see the logical steps, as follows:

1. Target consumers were defined broadly. The definition became not just frozen pizza users, but rather all in-home pizza eating occasions. The expectations of that consumer group for that eating occasion needed to reflect the new competitive set.
2. The product had to be as good as or better than a home-delivered pizza.
3. This assessment led to the understanding of the importance of the pizza crust in delivering against expectations and the development of a technology program to optimize self-rising crust.
4. The positioning had to reflect the new competitive set—"it's not delivery, it's DiGiorno."

5. The distribution system and financial proposition had to reflect not only the cost structure of the new technology, but also the competitive frame of home delivery. This assessment came from looking at the category as broadly as possible and then narrowing the focus to the areas with the most promise based on the SWOT analysis discussed above.

6. The company's R&D strengths were coupled with an opportunity to make a pizza that better met the consumer's needs for a fresh, hot pizza quickly in the face of a threat from an increasing percentage of pizza eating occasions being served by home delivery. This overcame the historical frozen product weakness because it was hard to engineer the same high crust quality as the consumer would get from a pizzeria.

7. This led to the development of crust technology, consumer messaging, and a broader marketplace view that significantly grew the business.

2.4   PORTFOLIO ANALYSIS AND NEW PRODUCT DEVELOPMENT

Most large companies are portfolios of smaller businesses. One benefit of the diversified company is selection, i.e., the flexibility to allocate resources to the largest opportunity, i.e., using some businesses to fund others. Existing revenue fuels future growth. Inherent in that approach is that all businesses are not treated equally. Certain businesses have greater prospects for growth. Certain technologies have greater chances of success. Some new product programs are better aligned with the business needs.

Another component of the strategic planning process is the ability to define the goal and role for a business in the portfolio. Is it a business that will grow revenue and invest its profit? Is it a business that will be "milked," i.e., a business whose profit grows faster than the revenue, which will be used in order to fund the growth engine?

A commonly used portfolio tool was developed by Bruce Henderson at the Boston Consulting Group (BCG) in the 1970s. We see an example of the portfolio tool in Figure 2.2. The tool provides a good model to start discussions about resource allocation and portfolio focus. The premise is that businesses can be viewed by the market growth potential and relative position in that market (market share). Having high market share in a growing category is a recipe for success and merits a disproportionate share of the company's programs and attention. Conversely, low market share in a stable or shrinking category is an area where there should be fewer resources, less focus, and discussions about how best to deploy any effort against bigger opportunities.

Let us consider the case of Starbucks. Starbucks can look at its portfolio of cafes and retail grocery programs. The grocery coffee category is not growing nearly as quickly as the trend of local coffee shops. Starbucks share in the retail grocery aisle is not nearly as large as its share of "coffee restaurants." Putting a larger share of resources and attention against coffee shops reflects this type of portfolio thinking. Even with this said, it does not mean that Starbucks completely ignored the grocery opportunity. Rather, Starbucks approached the grocery opportunity differently. They entered into an agreement with Kraft to sell and distribute Starbucks coffee in the grocery trade. More comprehensive and complex portfolio management techniques have been developed since the original BCG work and will take into account other considerations.

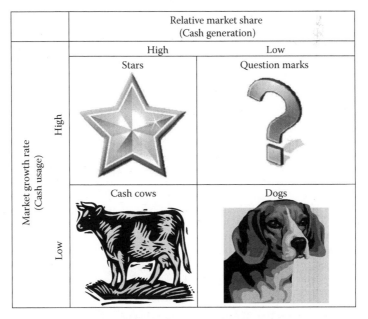

FIGURE 2.2 Portfolio Tool adapted from the material developed by Boston Consulting Group. (Adapted from Huston, L., *Harv. Bus. Rev.*, 84, 58, 2006.)

2.5   THE ROLE OF CRITICAL PROGRAMS
IN STRATEGIC PLANNING

Not only must there be an evaluation of portfolio roles for business, but also a role and evaluation of critical programs. The range of activities should include short-term efforts to support the current year needs, long-term programs focused on defined future growth, and technology development programs to fuel completely new areas.

All programs should be assessed for their risk versus reward. What will it cost, what will it yield, and with what likelihood of success in light of the benefit to be obtained? Some programs are inherently riskier than are others. The ratio of high- and low-risk programs should be managed as well. The balance between them is an important strategic decision and reflects the aggressiveness of the plan.

As an example, imagine two programs. Program A is early in the concept phase and has the potential to generate a large benefit. Yet the technology behind Program A is new and unproven, the market is still not well understood (is it a fad or a trend), and the risk is high. Contrast this with Program B. It will yield a moderate benefit, use known technology, and address a well-recognized source of consumer dissatisfaction, albeit in a unique and novel way. If the company has sufficient resources only for one of the two programs, which program should it choose? Which program is the better investment? The answer is the obvious, clearly it depends! It depends on what else is in the portfolio of programs. It cannot be answered in isolation, but rather needs to be determined in the context of the balance of the activities. What are these other activities? Are they short term or long term? Are the more risky or less risky?

2.6   SCENARIO PLANNING AND STEPS ALONG THE PATH

A strong strategic assessment will lead to the options or scenario planning, a technique that allows the business to assess different strategic alternatives. Each alternative most likely will require a different set of associated tactics associated, and a differ- ent set of required capabilities. Thus there will be gaps that will be identified from the strategic plan. The alternatives should be evaluated for the skill, knowledge, and capability gaps, with plans established for closing those gaps.

The operational plans that flow from this process generally lead to longer term programs targeted at developing new critical capabilities for the organization. Those programs should be designed in a way that breaks them down into meaningful subunits. Ideally, these subunits should be commercialized separately to derive benefit from the intermediate milestones.

Let us look at the series of subunits involved in a strategic opportunity. The oppor- tunity is to design an all-natural zero calorie complex food consisting of fat, protein, and carbohydrates. The program could be attacked through a series of tactics, ultimately leading to the final objective, as follows:

1. Zero calorie fat substitutes (such as sucrose polyester) can be used to move to a zero fat product or for other products that derive all their calories from fat (e.g., cooking oil or margarine). This does not meet the ultimate objective, but is a critical milestone along the path to get there.

2. Replacing carbohydrates with artificial sweeteners may be next.
3. Finding natural alternatives to artificial ingredients would be a key milestone as well.
4. All together, these will achieve the BHAG; individually they are important tools to derive commercial benefit and economic value from the work in progress.

These intermediate milestones and the commercial benefits derived from them are critical to demonstrate the value of the technology programs and to provide rationale for continued funding. In reality, intermediate commercial milestones provide the funding for the longer term big hairy audocious goals (BHAGs).

Another critical component in the strategic planning process is the total enterprise view of the results, and the need for balance among the different businesses in the portfolio. Often there is one dominant business unit or particularly advantaged segment. It would be possible through a fair assessment to focus only on this major business, and ignore the others. The rationale would be easy—the privileged business produces most of the revenue or profits. One could rationally allocate all the resources to that business.

The reality is that the portfolio requires some level of balance among the businesses. A business cannot go completely unresourced lest it wither in the competitive environment. There must be a minimum level of support each business receives, even when designated as a "harvest" business. Strategic planning must have a qualitative component or judgment to look at the *balance* among the various businesses within the enterprise.

In this vein, it is worth mentioning that most systems of metrics tend to bias the organization to either short-term tactical programs or long-term "big bets". Programs cannot be skewed to all "big bet/high risk" programs or short-term tactical efforts that provide little in new capability development. A balance must be established between the long-term efforts and short-term business needs. The portfolio should be evaluated to ensure a stream of innovations for steady growth. Intermediate, commercially viable milestones from the long-term programs should be integrated with the tactical initiatives to ensure that the annual year on year objectives are met. And that the long-term programs are resourced. This balance is often difficult, but critical to the long-term strategic goals of the corporation. Another component of balance is the risk level of the programs. Portfolio planning helps to inform the "what"— in what areas should the company focus and in what areas it should scale back. For any industry, there are considerations that apply to the "where" and the "how." The global economy is requiring food companies to take a broader view of strategic opportunities and identify platforms that can leverage the company's global reach.

Although the business environment is "global," the consumer is all too often "local." Technology may easily cross borders, local taste preferences may not. As projects get implemented, global solutions to technology challenges should be evaluated with a focus on local implementation. Many global products keep their fundamental character around the world, but have local adaptations. An Oreo sandwich cookie will have two chocolate wafers and a crème filling. The chocolate flavor may vary by market to better meet the consumer's preferences, but it will always fit with the basic brand premise.

2.7   ROLE OF INNOVATION

Innovation is the lifeblood of the industry. "Innovate or die" seems to be the mantra of today's consumer product companies. Innovation has often meant finding new product ideas and new market opportunities. Innovation must be more than that. Innovation has become both doing innovative things and doing things innovatively.

One of the key innovations on the innovation process is the broader reach and scope of external involvement and partnerships. Open innovation—reaching outside the corporation for solutions—has become a critical enabler for speed to market. In Henry Chesbrough's book: *Open Innovation, The New Imperative for Creating and Profiting from Technology* (Chesbrough, 2003) he outlines many examples of finding solutions in the most unlikely places, simply because no one has looked there before. There are examples of relationships between historical competitors that helped to bring value to both companies. A good example of such company-to-company cooperation is the Clorox relationship with Procter and Gamble, used to bring P&Gs "press and seal" technology to market in the Glad brand of plastic wrap. Larry Huston discusses this in depth in his seminal Harvard Business Review article "Connect and Develop" (Huston, 2003).

2.8   OVERVIEW

The business strategy process leads to an integrated, cross-functional plan to drive activities in a harmonious way across the entire enterprise. Each functional group should view the finished strategic plan as a framework out of which functional implications and calls to action should arise. The functional input drives the assessment, the alternatives, and the ultimate direction of the strategy. However, the resulting plan requires each functional group to develop its response or implications of the strategy. Without this activity, the alternatives all may be pursued with equal vigor. A consequence is that the programs may reflect individual biases, and the unhappy outcome is the loss of focus.

To sum up, the strategic plan is a process of evaluating alternatives, selecting the best options for the enterprise given an internal and external view, and then aligning behind the chosen alternative. Yet, without the steps of "closing the loop" and determining the implications of the strategic plan in the function, that alignment will be lost.

REFERENCES

Chesbrough, H. 2003. *Open Innovation: The New Imperative for Creating and Profiting from Technology*. Norwood, MA: Harvard Business School Press.

Durham University Business School. http://www.rapidbi.com/created/SWOTanalysis.html (accessed July 2008).

Huston, L. 2006. Connect and develop—P&G's new innovation model. *Harvard Business Review*, 84(3), 58–66.

Scientific and Technical Information (STI). 2008. http://www.sti.nasa.gov/tto/ (accessed June 16, 2008).

Strategic Planning. http://en.wikipedia.org/wiki/Strategic_planning. Wikipedia, Free Encyclopedia (accessed July 2008).

3 Innovation as Science

Martinus A.J.S. van Boekel

CONTENTS

3.1 INTRODUCTION

Production of foods is of prime societal importance. In the Western world, food production has become highly industrialized. On a historical note, things have changed dramatically. At the start of the industrialization of the food production by the end of the nineteenth century, preservation of foods was the main approach, used to overcome seasonal shortages and to prevent spoilage. When that objective was reached, attention shifted. This shift occurred roughly around World War II, with the goal to improve food in the sense of nutritional value and eating quality. Furthermore, processes were optimized and made more efficient. After that, convenience became the goal, starting some time in the 1960s. Now, we are in the era when food and health are the main targets for companies to work on.

The societal impact of this industrialization and the accompanying achievements in delivering high-quality foods have been enormous. Together with the large increase in productivity in agriculture, the advances in food production made it possible for most people not to have to work on the land. Additionally, much less time was needed in the kitchen. This has had a major impact on the participation of employment of women, as at least one consequence. We are now in the stage when food is so abundant that mankind does not seem to be able to handle this situation, at least based on the incidence of obesity, even in developing countries. It should be realized, though, that obesity is not only a consequence of the present food situation but also linked to a major change in lifestyle.

In short, the developments in the food industry exerted a major societal impact. As a result, we see now the opposite happening. Societal trends dictate changes that the food industry must undergo to comply with societal changes. Lifestyle, high incomes, quality awareness of consumers in general, all lead to demand for variety, convenience, attractive and healthy foods, respectively. The food industry is struggling with how to interpret changing consumer needs and how to respond with appealing food products. Clearly, the food industry needs innovative changes.

So, innovation is the keyword, but what does innovation imply and require? In general terms, innovation can be seen as generation of new ideas. But innovation is more. Innovation requires successful exploitation of these ideas so that they lead to new and improved added-value products that are successful in the market. Innovation includes cost efficiency in production, the introduction of new technologies, along with process development and improvement (Watzke and Saguy, 2001). Innovation should be distinguished from invention and it is more than technology. Innovation is a formalized process for identifying, evaluating, and shepherding the best ideas into new product and process concepts. Innovation comprises structured problem solving, orientated toward a solution, with the use of apparently unrelated information. This is different from problem solving using existing or closely related technology.

The involvement of management in the innovation process is essential. Without management's help, the process is bound to fail. Management should address

1. The strategy of the company (what types of products and markets are addressed)
2. Innovation targets (product and process development, technology portfolio)
3. Organizational management (the actual running of the production process)

Management should set the optimal conditions to allow for creative integration of knowledge. However, innovation is definitely not a matter of management alone. Knowledge of technology and the products produced (i.e., food science and technology) are central to the design process.

Production of foods is realized by transforming raw materials coming from agriculture, horticulture, livestock, and fisheries, to foods, followed by distribution and delivery to the consumer. When looking at food products, a division can be made into foods and fabricated foods, meaning that either raw materials are processed directly (milk is only heated, or beans are only canned), or that foods are composed ("fabricated") from various ingredients (bread from wheat, yeast, and salt; and cheese from milk, rennet, and salt).

Nowadays, many foods are fabricated foods. It is also of importance to consider food innovation in relation to the food chain, including new organizational combinations in demand-driven networks and chains (Benner et al., 2003a; Verhallen et al., 2004; Trienekens et al., 2008). This view of food innovation has two important consequences. One is that innovation should be considered across the whole of the food chain, rather than in separate elements. The other is the term "demand-driven." It implies that consumer orientation has become the paradigm, rather than a production-orientated supply-driven chain. This development has been named "chain reversal." Overall, design is becoming increasingly important, not only of products and processes but also of a whole chain.

Many books have been written about the design of products in general. A standard text is the book of Ulrich and Eppinger (2000) who consider the design process for many products but not for foods. A text that is somewhat related to foods is about chemical product design by Cussler and Moggridge (2001) and they give an occasional example of foods, though only marginally. Two textbooks devoted completely to foods are that of Earle and Earle (1999), and of Linnemann and van Boekel (2007). These books have in common that they divide the process of product development generally into four stages (though in different wordings):

1. Identify the needs.
2. Generate the ideas to meet the identified needs.
3. Select the most promising ideas.
4. Manufacture the products, launch, and evaluate.

It is stressed that this is not a sequential process but rather a sequential, interactive, and iterative set of activities involving the four stages.

This chapter presents a global view of the various aspects of food innovation by designing new food products. We will take the so-called success equation as the guiding principle, as shown in Figure 3.1.

This success equation is, of course, qualitative, and it should be seen as a probabilistic equation rather than a deterministic equation—what is the best guess for the chances of success given the aspects mentioned? It is not clear if and how the various parts in the right-hand side can and should have different weights; i.e., whether or not some of the components of the equation are more important than the others. Nevertheless, the equation entails the most important aspects that have a direct or indirect effect on success.

In order to reach that success, it is clear that scientific tools and results (i.e., food science and technology expertise), experience, and creativity should be included (Naes and Nyvold, 2004). At several stages in the design and development process, more or less subjective decisions must be taken, but these decisions should be supported by the outcomes of experimental design. In fact, one could think of the

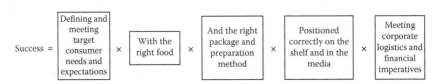

FIGURE 3.1 The success equation.

development of decision support systems, in which the Bayesian approach could be very helpful. The Bayesian approach implies the application of a branch of statistics that allows the use of prior knowledge and it expresses results as probability statements, in terms like, it is 70% likely that this product will be a success. Several existing decision support systems are based on this Bayesian approach (see van Boekel et al., 2004, for more details).

The title of this chapter raises the question "how scientific the innovation process is?" It is proposed that innovation based on science is more successful than an empirical approach "but this does not imply, of course, that innovation is therefore scientific." In the author's view, innovation is a combination of the application of science, creativity, and expertise. Most importantly, it should be a multidisciplinary activity (or even better interdisciplinary), by combining consumer and marketing studies with nutrition, sensory science, food science and technology, and scientific disciplines that cover knowledge about raw materials and ingredients. Specialists should carry out their disciplinary tasks, of course, but in the end, integration is needed to facilitate problem solving and decision making among specialists with different disciplinary backgrounds (Ulrich and Eppinger, 2000). Food design should move away from an *ad hoc* recipe-trying procedure to a systematic knowledge-based design procedure (Earle and Earle, 1999), and in doing so, it becomes a science-based procedure.

3.2 DEFINING AND MEETING TARGET CONSUMER NEEDS AND EXPECTATIONS

As already mentioned, consumer needs and expectations should guide innovation. However, this is more easily said than done. One cannot expect that consumers will be able to articulate their needs for new products in a way that the developer can use as a blueprint. The big problem here is to interpret and to find out what it is that the consumer actually needs and expects. A consumer will express his needs and expectations probably as vague as in the phrase: I want a food that is safe, healthy, attractive, delicious, and by the way has no substantial environmental impact. In addition, aspects such as animal welfare may come into play.

In an attempt to formalize an approach to understanding consumer's needs, Linnemann et al. (1999) considered seven types of consumer images based on an idea of the late marketing professor Meulenberg. These seven images or segments of consumers are

1. Environmental-conscious consumer
2. Hedonistic consumer
3. Nature- and animal-friendly consumer
4. Convenience consumer
5. Price-conscious consumer
6. Variety-seeking consumer
7. Health-conscious consumer

Of course, this is an abstraction of reality. A single real-consumer can belong to more than one segment. Yet such a division helps to guide in developing product assortments, and hence in innovation. Jongen and Meulenberg (1998) summarized this approach in the following steps:

1. Thoroughly analyze the socioeconomic developments in specified markets.
2. Translate consumer preferences and perceptions into consumer categories.
3. Translate consumer categories into product assortments.
4. Group product assortments in product groups at different stages of the food supply chain.
5. Identify processing technologies relevant for specified product groups.
6. Analyze the state of the art in relevant processing technologies.
7. Match specified state-of-the-art processing technologies with future needs.

The foregoing seven steps map the consumer types to product assortment, and from there to technological requirements as discussed below.

The problem remains how to get a sense for consumer's wishes, and more importantly a sense that can be turned into actual products. Several methods are available to get inputs from customers and markets, means-end analysis, laddering techniques, and focus group discussions (Sijtsema et al., 2002; Costa et al., 2004; van Boekel, 2005; Mattsson and Helmersson, 2007). van Kleef et al. (2005) and Linnemann et al. (2006) gave a critical discussion on the techniques available and suggested some guidelines to help product developers select the most appropriate ones.

3.3 GLOBALIZATION

It is obvious that globalization has had a large impact on food choice and eating behavior. On the one hand, companies have access to a large variety of raw materials and ingredients, and always actively seek possibilities to replace one ingredient by another, for cost or consistency reasons. This means that the entire earth is the source of different raw materials, selected for economic reasons.

On the other hand, consumers become much more aware of other food and as a result hitherto unknown foods enter the market. Here, consumer typology may be helpful because not every consumer is thrill-seeking. There is also the phenomenon of neophobia, meaning that some consumers may hesitate to try something new. Both food exploration and neophobia, opposite aspects that need to be taken into account in the innovation process, suffice to reinforce the fact that there is no such single person as "the consumer."

3.4 TRANSLATION PROBLEM

After discovering the consumer wishes and needs, one can then try to realize these wishes by linking them to physical product properties and then to designs that are most appropriate. We see a schematic of this translation issue in Figure 3.2.

One of the few methodologies that are directed toward this problem is quality function deployment (QFD). Its applicability to foods has been reviewed by Benner et al. (2003b) and Costa et al. (2004). The conclusion is that QFD is useful for food design but has its limitations. The usefulness comes from the discipline of forcing a structured approach. Systematics always wins in the long run.

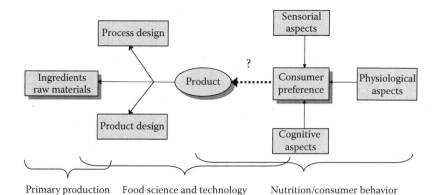

FIGURE 3.2   Translation of consumer wishes into product properties.

The question mark on right-hand size in Figure 3.2 reflects the realization that translation is not straightforward. We do not really know how to transform needs into products from first principles. It is interesting to note that a term has been coined for this—translational science—and in the case of nutritional demands it has been called translational nutrition (Green et al., 2006). It is a plea for integration of nutrition, food science, health science, basic life sciences, and communication to have an integrated view from cell to society. It is also stressed by Green et al. (2006) that such an approach needs to be proactively managed in order to come to deliver nutrition solutions.

On a more general basis, such a translational science can go beyond nutrition alone. Translational science can also extend beyond food science to primary production of raw materials and ingredients, as suggested in Figure 3.2. One of the remaining dangers is that qualitative customer requirements are translated into design parameters, which are seen as independent, measurable quality determinants of the product. The problem here is that these determinants are not really integrated, and therefore may not have the right combinations of the various determinants (Mattsson and Helmersson, 2007). Even though this is not straightforward, attempts are described in literature to overcome this problem (Corney, 2002; Naes and Nyvold, 2004; van Boekel, 2005; Mattsson and Helmersson, 2007).

It all boils down to use structured methodologies rather than so-called cook and look procedures. The ultimate search is for the right food, i.e., the food that has the right combinations of quality determinants. It is essential to involve consumers to evaluate their response to several combinations of quality determinants. It may require that consumer and marketing research be shifted to a later stage in the product development process because real innovative ideas may not be expected from consumers.

Let us first take a closer look at the right food, i.e., what are the possibilities to integrate the right quality determinants.

3.5   THE RIGHT FOOD AND FUNCTIONS OF FOODS: NUTRITIONAL, SENSORIAL, AND SOCIAL HEALTH EFFECTS

Whatever innovation is achieved, it should result in a product of appealing quality. Quality is an elusive concept. It may be helpful to discuss this topic in a little more detail. A very general definition of quality is "to satisfy or exceed the expectation of the consumer." This definition requires us to understand who the consumer is, what does he expect, and how can food properties meet that expectation.

Here, we focus on how quality can be achieved using technology. It should be realized that consumers are not just attracted by a product itself but also by the benefits that are delivered by the product (Spetsidis and Schamel, 2002). Thus, it may be productive to think about the specific functions that a food can deliver. Figure 3.3 lists benefits as well as possible harms delivered by foods in general terms.

On the whole, consumers tend to be rather conservative in their food choice. The reasons are that foods have a large emotional value and that people want to trust what they put in their mouth. If this product they put in their mouth is something unknown, then they will be quite hesitant. The consequence is that radically new food products are not so easily accepted when they are completely unknown.

One practical consequence of the foregoing conservatism is that innovations are put into familiar products. For instance, a margarine that lowers cholesterol content is indeed radically new in a specific function but not in a number of other functions. It is still margarine, which by itself is an imitation of the age-old product butter. When a company develops a successful new ice cream, the key to success is in the choice of ingredients and the combination of new flavors. Yet the product is still the age-old ice

Benefits	Description
Supplying enough energy and adequate amounts of nutrients for growth	Needed for growth and development of the body
Supplying energy and nutrients for maintenance of the body	Replacement of tissues, staying in good health, maintenance of immune system
Supplying energy and nutrients for reproduction	Needed for offspring, embryo development, milk production
Supplying energy and nutrients for physical activities	Moving, thinking, working, sports, in short, being active in life
Giving pleasure	Appealing taste, flavor, texture, satiation, stimulants (alcohol, caffeine)
Social and cultural functions	Eating together, expressing lifestyles
Possible Harms	
Eating too much: imbalance between calorie intake and energy expenditure	Leading to obesity and related diseases
Eating an unbalanced diet, e.g., due to lack of vitamins or minerals	May lead to deficiency diseases and, ultimately, death
Presence of pathogens	May lead to diseases and, ultimately, death
Misuse	Using scarcity of foods in wars and conflicts for political reasons; deliberate poisoning; bioterrorism; hunger strike; boycotting certain products

FIGURE 3.3 Benefits and possible harms delivered by foods. (After van Boekel, M.A.J.S., *Innovation in Agri-Food Systems*, W.M.F. Jongen and M.T.G. Meulenberg (Eds.), Wageningen Academic Publishers, Wageningen, the Netherlands, 2005, 147–172, Chapter 6.)

cream. Almost all novel protein foods that are currently on the market try to approach the characteristics of meat for the same "familiarity" (Aiking et al., 2006).

This conservatism on the consumer part is of importance in the context of innovation. It implies that innovations in the food industry will need to focus more on new combinations of ingredients, new technologies, and semifinished products instead of developing radically new products that are not familiar to the consumer. However, innovations within an existing product are of course possible and could be radically new in terms of the benefits that they deliver. In this respect, the author believes that a methodology such as systematic inventive thinking (SIT) is of special importance to food designers (van Boekel, 2005). SIT is based upon the methodology TRIZ (a Russian acronym for Theory of Inventive Problem Solving), but SIT applies this to existing products and looks critically to functions of the product. It can be characterized as a structured product-orientated way to innovation, by using the so-called creativity templates proposed by Goldenberg and Mazursky (Goldenberg et al., 1999, 2001, 2003; Goldenberg and Mazursky, 2002). The approach as applied to foods is recently reviewed by Stern et al. (2007). In the author's view, it is a nice compromise between listening to the voice of the consumer and listening to the voice of the product (van Kleef et al., 2002). An example of the approach appears in Figure 3.4.

Food quality has many aspects to it and only a very short description is given here. Quality is a multidimensional concept, comprising both subjective and objective elements. Food quality is, furthermore, situation-specific and dynamic in time (Martens and Martens, 2001). A consumer, however, does not analyze all elements of food quality consciously; rather, he gives an integrated response based on complex judgments made in the mind, in which memory and experience play a crucial role. Food quality is therefore a difficult concept to put in operation because it is not only a property of the food but also that of the consumer and his perception of the food (Sijtsema et al., 2002). It is fair to say that there is not yet a common terminology in describing food quality.

In order to make quality more tangible for the food scientist, it is productive to divide the topic into intrinsic quality attributes, i.e., inherent to the product itself, and extrinsic attributes, linked to the product but not a property of the food itself. Extrinsic attributes are, for instance, whether or not a food is acceptable in the perception of the

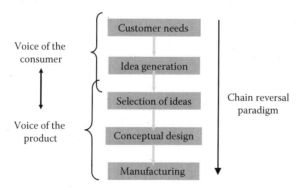

FIGURE 3.4   Product design: listening to the voice of the consumer and that of the product.

consumer for cultural/religious or emotional reasons, or whether the way it is produced is acceptable (use of pesticides, growth hormones, genetically modified, etc.), as well as the price of the food. Extrinsic attributes are therefore not part of the food itself but are definitely related to it, as experienced by the consumer. On the other hand, intrinsic attributes are the chemical composition of the food, its physical structure, the biochemical changes it is subject to, the microbial and chemical condition (hazards from pathogens, microbial spoilage, presence of mycotoxins, heavy metals, pesticides, etc.), its nutritional value and shelf life, the way packaging interacts with the food, etc.

A hypothetical quality function Q has been proposed for chemical products (Kind, 1999). We can do more or less the same for a food product:

$$Q = f(Q_{int}, Q_{ext}) \qquad (3.1)$$

In words, this equation states that quality is a function of intrinsic and extrinsic quality attributes and can, in fact, be decomposed into intrinsic and extrinsic quality functions Q_{int} and Q_{ext}. The nature of this function remains as yet obscure. For instance, we do not know whether we are allowed to sum intrinsic and extrinsic quality attributes, or multiply them, etc. We do not know how to "operate" on these terms to generate the final quality value. Molnár (1995) proposed to sum the weighted individual quality parameters as judged by experts, but the relevance for consumer evaluation is unclear. In any case, it helps the technologist to disentangle intrinsic and extrinsic quality attributes to make clear, which factors are controllable by a technologist.

A further decomposition of the intrinsic quality function, Q_{int}, into its intrinsic quality attributes I_i is possible:

$$Q_{int} = f(I_1, I_2, ..., I_n) \qquad (3.2)$$

Analogously, the external quality function can be decomposed into its extrinsic quality attributes E_i:

$$Q_{ext} = f(E_1, E_2, ..., E_n) \qquad (3.3)$$

The reason we focus here on intrinsic quality attributes is that those can be controlled via technology. $I_1, ..., I_n$ are thus intrinsic quality attributes that can be objectified, in principle at least. I_1 could be color, I_2 the microbial status, etc. The technological challenge is to find the relations between these attributes and product and process engineering. One way of doing that is by establishing quantitative relations between a quality attribute and product composition, i.e., by deriving mathematical models. These models should take into account also interactions between quality attributes. Such quality change models provide information in the form of rate constants, activation energies, etc., which can be used in cooperation with the process engineer to design appropriate processes to realize the quality function Q (van Boekel, 2008, 2009).

In developing right food, one needs to consider how to develop them at minimal cost and time. The resulting products should, of course, be reliable, that is to say that they should be able to deliver what they are supposed to do. For foods, this implies that they should be safe and that they do not lose important quality characteristics within their prescribed shelf life. Although not developed for foods, a method exists for reliability estimation during early product development stages (Yadav et al., 2003). It uses a Bayesian approach as mentioned in the introduction. The approach considers each step in the new product development process as an interim milestone, with which one can assess product reliability to make further decisions in a so-called input–output model. The inputs are various types of information, and outputs are reliability estimates.

Once the most important quality attributes are known, the question is how to exploit this knowledge. For instance, if consumers become bored with flavor of a certain food very quickly, then the food is going to fail in the market because those who buy and "try" the food stop consuming it because of the boredom. So, insight in how to influence quality attributes is essential. The author's group has proposed to apply a concept called quality analysis critical control points (QACCP), inspired of course by hazard analysis critical control points (HACCP), but now applied to quality (van Boekel, 2005). It implies that one should be able to control the factors that have a critical impact on the major quality attributes. Figure 3.5 shows a schematic picture of this approach.

In order to apply this approach, it is necessary to consider the major quality attributes for the particular food under study. These quality attributes may be further decomposed into performance indicators. For instance, red color (a specific quality attribute) may be indicated by absorption at a certain wavelength (the performance indicator). It also requires expert knowledge about what can happen to the food and

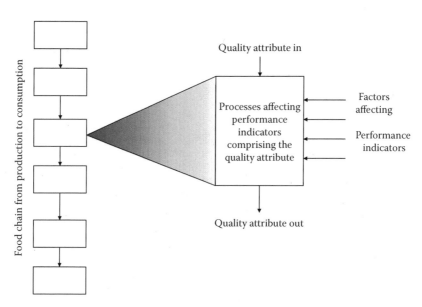

FIGURE 3.5 Schematic picture of the QACCP approach.

specifically how the quality attribute changes as the product passes through the distribution chain. By connecting the various elements, one should begin to understand the interrelations among the main quality attributes, and how to control them in an operationally meaningful way.

3.6   RIGHT PACKAGE AND PREPARATION METHOD

Packaging has many functions, but the three basic ones are

1. To protect and to stabilize the food that it contains
2. As an aid in distributing the food over the food chain to the consumer
3. To transfer information about the product via labeling and the like

Food design and production require an integrative approach (Linnemann and van Boekel, 2007). This approach evolves by combining knowledge of the product, its production process and the packaging, and most importantly, how these affect one another (Hart, 1997). Nowadays, packaging has become an integral part of food design. One can influence the quality of fresh produce by applying concepts such as controlled atmosphere packaging and modified atmosphere packaging. New developments are emerging, such as active packaging and intelligent packaging in which the package actually interacts with the food in the package. The use of microsensors and nanotechnology will become important as providers of feedback on the quality status of the food, and permit interference when necessary. In the author's view, this is going to be one of the relevant innovations in the food industry in the near future.

The right preparation method depends very much on the product. Nowadays, convenience has become a major topic and opportunity for the food industry. Convenience has led to and will further lead to semifinished products that can be easily prepared in the microwave. Examples are ready-to-eat meals that only need to be heated at home. The challenge here is that the important quality attributes remain at a high level, and that the products deliver what the consumer expects (see Figure 3.3). Innovations that will further improve the convenience with which foods can be prepared no doubt will be successful.

3.7   "POSITIONING" ON THE SHELF AND IN THE MEDIA

As argued by Earle and Earle (1999), it is important to consider the diffusion of innovation. Specifically, how innovation will be spread through the company itself and find its way to markets and society as a whole. This is not a simple issue to address. There is clearly a role for advertising and creating an image of a newly developed food. There are also ethical issues at stake. It is quite easy to create illusions by making certain claims (health claims, for instance), but if these claims are not justified, or only partly justified, then the public will become very hesitant in the long run to accept such claims. So, the food industry should pay serious attention to make sure that claims are really substantiated and it actually becomes a big investment to do that convincingly.

There is nowadays also a big influence of retail. The retailer has become actually very powerful in determining what can come to the market and what cannot. The position on the shelf is no longer the decision of the food producing company but of the retailer. When the retailer is not willing to accept a new concept or innovation, then there is a structural problem with the new product.

When it comes to food advertising, further interesting issues are emerging. Although food production is very high-tech, the public wants to hear about "tradition," the way things were. This trend for the traditional, for the artisanal, embodied in the so-called slow food movement, strongly appeals to the consumer, or at least to some of the consumer types mentioned above. This has to do with romanticizing on the part of the consumer and a longing for the good old days when everything seemed to be clear. The author considers that "false romanticism" because all were not well in the good old days when it concerned food. Nevertheless, it is a sentiment that is present and represents a fear for new technology for the consumer. It is questionable whether the food industry should go along with this sentiment by supporting the artisanal image. It is definitely a barrier for food innovation. It seems that this desire for the artisanal is truer for the European consumer than the American consumer. One striking example is the attitude toward genetically modified raw materials.

It is the author's view that honest communication about what goes on in the food industry is of utmost importance in creating a better image with the consumer. Much of the consumers' fears come from the reality that they cannot see what goes on with their food in the production stage. More openness about development and production would help here. Otherwise, innovations may be very well conceived technically but nevertheless the consumer will not be able to see the benefit.

3.8 MEETING CORPORATE LOGISTICS AND FINANCIAL IMPERATIVES

A corporation's strategy is important to consider in the process of innovation. The two extremes are to go for either market-pull or technology push. A combination is of course also possible. Questions to be asked are: Is it possible to do this in the company, should we do it, and are we allowed to do it? There are at least four criteria to be met:

1. Financial criteria: These criteria comprise questions about return on investment, the time allowed to reach a breakeven point, and the willingness to invest.
2. Marketing criteria: Questions to answered here are about which market segments should be considered, about marketing channels, and about the desired market share.
3. Organizational criteria: How much manpower should be involved, what will be the management style, and what will be the level of control?
4. Technological criteria: What are the core competences, will it be a high-tech achievement, and how much R&D should be involved? Within the R&D, the issue should be how much background research is needed and how much development.

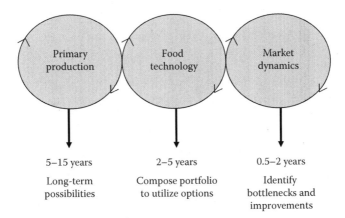

FIGURE 3.6 Different timescales in primary production, food technology, and in the market.

With respect to organizing the innovation process, two extremes can be thought of (Cussler and Moggridge, 2001). The first is organization by function, meaning that different divisions (marketing, R&D, engineering, and legal matters) have different responsibilities and the organization is sequential. When one division is finished, the project is handed over to the next division. The second is organization by project, meaning that a core team is formed from different divisions, and the core team members have complete responsibility. The aim is to create a synergy between the various functions. This second way to organize is more complex from a managerial point of view but obtains results more quickly. It is also more popular in practice.

A major challenge when innovation is considered in the food chain is the fact that the timescales at various parts of the chain are so different, as we see in Figure 3.6. This has practical implications. For example, when one wants to change something in the raw materials (for instance, a higher amount of lycopene in tomatoes), then this objective will take considerable time via breeding. If one wants to use a new technology (for instance, high pressure), then this objective requires a large investment in equipment as well as in expertise. Finally, if one merely wants to change a product only in a small detail, this is the easiest way but it leads to me-too products and line extensions that cannot be really called innovative. Real innovation, not just line extensions to a different flavor, demands important decisions at the corporate management levels. Such decisions may have far-reaching consequences.

3.9 OPEN INNOVATION

Traditionally, innovations come from an internal process. Though not really new in general, an emerging trend for companies is to work together to realize new concepts. For the food industry, these may be both food and nonfood companies. By opening up doors and sharing knowledge and expertise with others from outside the company, new unconventional ideas may be discovered (Rudolph, 2004; Erickson, 2008). It is, of course, not something that should be done lightheartedly. There is the risk of partnering with an incompetent partner; there may be trouble with regards to IP rights, etc. One has to invest into partnership and trust is essential.

Once the right partner is found, however, the partnership may allow a greater level of experimentation and a lower lever of risk. Erickson (2008) gives a list of do's and don'ts. A very successful example from the Netherlands is the coffee machine Senseo, developed by a coffee selling company and a company that produces coffee machines. The basic idea was to make it very easy for a consumer to produce quickly a cup of coffee using "coffee pods" or little plastic containers of coffee, in a machine designed specially for this purpose. The big advantage of such open innovation is that the companies working together are not a threat to each other, and therefore they do not feel inhibited in sharing information.

3.10 CONCLUSION

Critical success factors in innovation are efficiency, responsiveness, quality, and flexibility (Trienekens et al., 2008). It is clear that innovation is not to be taken too lightly. Innovation requires a strategy that is supported throughout the company, management that stimulates efforts, a cadre of capable and knowledgeable people who know how to use technology applying scientific principles, and finally a good knowledge of the customers and their latent needs. Looking back at the success equation, every element appears to be important, and the elements depend on each other. Most importantly, an integrative approach and broad view are the *sine qua non* for innovation. Innovation as science is the art that applies various scientific disciplines to the problem at hand, and integrates their outcomes into the solution.

REFERENCES

Aiking, H., De Boer, J., and Vereijken, J. (2006). *Sustainable Protein Production and Consumption: Pigs or Peas?* Drodrecht, the Netherlands, Springer, 226 p.

Benner, M., Geerts, R.F.R., Linnemann, A.R., Jongen, W.M.F., Folstar, P., and Cnossen, H.J. (2003a). A chain information model for structured knowledge management: Towards effective and efficient food product improvement. *Trends in Food Science and Technology* 14: 469–477.

Benner, M., Linnemann, A.R., Jongen, W.M.F., and Folstar, P. (2003b). Quality function deployment (QFD)—Can it be used to develop food products? *Food Quality and Preference* 14: 327–339.

Corney, D. (2002). Food bytes: Intelligent systems in the food industry. *British Food Journal* 104: 787–805.

Costa, A.I.A., Dekker, M., and Jongen, W.M.F. (2004). An overview of means-end theory: Potential application in consumer-oriented food product design. *Trends in Food Science and Technology* 15: 403–415.

Cussler, E.L. and Moggridge, G.D. (2001). *Chemical Product Design*. Cambridge, UK, Cambridge University Press.

Earle, M.D. and Earle, R. (1999). *Creating New Foods. The Product Developer's Guide*. Oxford, Chandos Publishing.

Erickson, P. (2008). Partnering for innovation. *Food Technology* 62(1): 32–37.

Goldenberg, J., Mazursky, D., and Solomon, S. (1999). Toward identifying the inventive templates of new products: A channeled ideation approach. *Journal of Marketing Research* 36(May): 199–210.

Goldenberg, J., Lehmann, D.R., and Mazursky, D. (2001). The idea itself and the circumstances of its emergence as predictors of new product success. *Management Science*
47(1): 69–84.

Goldenberg, J. and Mazursky, D. (2002). *Creativity in Product Innovation.* Cambridge, UK,
Cambridge University Press.

Goldenberg, J., Horowitz, R., Levav, A., and Mazursky, D. (2003). Finding your innovation
sweet spot. *Harvard Business Review* 81(3): 3–11.

Green, J.H., van Bladeren P.J., and German, J.B. (2006). Translating nutrition innovation
in practice. *Food Technology* 60(5): 26–33.

Hart, B. (1997). Technology and food production. *Nutrition & Food Science* 2(March/April):
53–57.

Jongen, W.M.F. and Meulenberg, M.T.G. (1998). Summary and future prospects. In: *Innovation
of Food Production Systems. Product Quality and Consumer Acceptance.* W.M.F. Jongen
and M.T.G. Meulenberg (Eds.). Wageningen, the Netherlands, Wageningen Pers.

Kind, M. (1999). Product engineering. *Chemical Engineering and Processing* 38: 405–410.

Linnemann, A.R., Meerdink, G., Meulenberg, M.T.G., and Jongen, W.M.F. (1999). Consumer-
oriented technology development. *Trends in Food Science and Technology* 9: 409–414.

Linnemann, A.R., Benner, R., Verkerk, R., and van Boekel, M.A.J.S. (2006). Consumer-driven
food product development. *Trends in Food Science and Technology* 17: 184–190.

Linnemann, A.R. and van Boekel, M.A.J.S. (2007). *Food Product Design. An Integrated
Approach.* Wageningen, the Netherlands, Wageningen Academic Publishers.

Martens, H. and Martens, M. (2001). *Multivariate Analysis of Quality.* Chichester, U.K.,
John Wiley & Sons.

Mattsson, J. and Helmersson, H. (2007). Food product development. *British Food Journal*
109(3): 246–259.

Molnar, P.J. (1995). A model for overall description of food quality. *Food Quality and
Preference* 6: 185–190.

Naes, T. and Nyvold, T.E. (2004). Creative design—An efficient tool for product development.
Food Quality and Preference 15: 97–104.

Rudolph, M.J. (2004). Cross-industry technology transfer. *Food Technology* 58(1): 32–34,41.

Sijtsema, S., Linnemann, A.R., van Gaasbeek, T., Dagevos, H., and Jongen, W.M.F. (2002).
Variables influencing food perception reviewed for consumer-oriented product development. *Critical Reviews in Food Science and Nutrition* 42(6): 565–581.

Spetsidis, N.M. and Schamel, G. (2002). A consumer-based approach towards new product
development through biotechnology in the agro-food sector. In: *Market Development
for Genetically Modified Foods.* V. Santaniello, R.E. Evenson, and D. Zilberman (Eds.).
Oxford, U.K., Oxford University Press, pp. 63–79.

Stern, Y., Taragin, R., and Larry, S. (2007). New thought for food. *Food Technology* 61(7):
34–40.

Trienekens, J., van Uffelen, R., Debaire, J., and Omta, O. (2008). Assessment of innovation
and performance in the fruit chain. *British Food Journal* 110(1): 98–127.

Ulrich, K.T. and Eppinger, S.D. (2000). *Product Design and Development.* 2nd Edition.
Boston, MA, Irwin McGraw-Hill.

van Boekel, M.A.J.S., Stein, A., and van Bruggen, A. (2004). *Bayesian Statistics and Quality
Modelling in the Agro Food Production Chain,* Dordrecht, the Netherlands, Kluwer
Academic Press.

van Boekel, M.A.J.S. (2005). Technological innovation in the food industry: Product
design. In: *Innovation in Agri-Food Systems.* W.M.F. Jongen and M.T.G. Meulenberg
(Eds.). Wageningen, the Netherlands, Wageningen Academic Publishers, pp. 147–172
(Chapter 6).

van Boekel, M.A.J.S. (2008). Kinetic modeling of food quality. A critical review. *Comprehensive
Reviews in Food Science and Food Safety* 7: 144–157.

van Boekel, M.A.J.S. (2009). *Kinetic Modeling of Reactions in Foods*. Boca Raton, FL, CRC/Taylor & Francis.

van Kleef, E., van Trijp H.C.M., Luning, P.A., and Jongen, W.M.F. (2002). Consumer-oriented functional food development: How well do functional disciplines reflect the 'voice of the consumer'? *Trends in Food Science and Technology* 13: 93–101.

van Kleef, E., van Trijp, H.C.M., and Luning, P.A. (2005). Consumer research in the early stages of new product development: A critical review of methods and techniques. *Food Quality and Preference* 16: 181–201.

Verhallen, T., Gakeer, C., and Wiegerinck, V. (Eds.) (2004). *Demand Driven Chains and Networks*, the Hague, Reed Business Information.

Watzke, H.J. and Saguy, I.S. (2001). Innovating R&D innovation. *Food Technology* 55(5): 174–188.

Yadav, O.P., Singh, N., Goel, P.S., and Itabashi-Campbell, R. (2003). A framework for reliability prediction during product development process incorporating engineering judgments. *Quality Engineering* 15: 649–662.

4 Innovation: Integrated and Profitable

Paul Chaudury

CONTENTS

Success	=	Defining and meeting target consumer needs and expectations	×	The right food	×	Proper packaging and preparation	×	Positioned correctly at the shelf and in the media	×	Meet corporate logistics and financial imperatives

4.1 INTRODUCTION—DEFINING INNOVATION

Innovation comprises the delivery of added value of products and services to the consumers and/or customers in a way that drives profitable top line growth. When an organization delivers innovation consistently, consumers, customers, financial community, and the media recognize and reward it appropriately. With that in mind, let us look at some of the different aspects of innovation as practiced by a large consumer food products company.

4.2 TRUE INNOVATION WILL IMPROVE TOP LINE, BOTTOM LINE, AND BRAND IMAGE

Organizations with multiple business units and/or brands typically can find the right people resources and muster the appropriate funds. Yet, year after year these organizations continue to struggle in creating a steady pipeline of new products and consistent launches in order to meet their desired financial goals. With the pressure on bottom line profit, the challenge is not about finding resources, but rather about developing and successfully implementing a process to optimize resource usage and allocation. This would maximize output and minimize waste and risk.

This chapter addresses how to develop a holistic approach from day one that invests time and scarce resource at every step of the innovation process and along the way achieves financial goals. There are often no shortages of creativity and financial acumen in the different functional areas of any organization. Marrying such creativity and financial acumen into one resource pool sparks profitable innovation. The goal is to foster an environment where creativity with business savvy is encouraged.

The new approach to innovation brings three critical elements that can be leveraged for success. They are

1. From product to platform approach
2. From "black box" to transparency
3. From separate resources to one resource pool and process

4.2.1 FROM PRODUCT TO PLATFORM

The innovation process often starts with product ideas (see Figure 4.1). In corporations, employees are socialized into a specific way for creativity. Typically socialization teaches the employees to ideate in groups, rank them, evaluate them, and work as a team to move forward the best ideas. We feel good about the process because a team assesses feasibility and invests funds to build the better ideas in the development phase. As the teams begin creating business propositions, it becomes easier for senior management to evaluate these propositions, because considerable data will have been amassed using the people involved and the expert guidance of those who are adept at financial analysis and planning.

As idyllic as the foregoing description may be, it is fraught with several pitfalls. Today, experts tend to wait to screen many projects in the development phase rather than critically evaluate business value earlier, before the company deploys its development resources. In general, only a few of the projects actually meet the

Ideas	Concepts	Feasibility	Product develop-ment	Production scale-up	Implemen-tation	L a u n c h	Postlaunch tracking

FIGURE 4.1 Schematic of the overall idea-driven innovation process. (Courtesy of Sara Lee. With permission.)

targeted business criteria, such as brand equity fit and financial return. This later assessment allows true evaluation, but only after significant resources have been utilized due to the high cost for each project in the development phase. Finally, many projects seem good individually. Yet they fail to excite the corporation, and may not fit as part of the more exciting, larger strategy on which the corporation focuses.

The ideal situation is to assess a large number of ideas and concepts as a group, along with existing strategic information, well before the development phase and not after significant investments have been made.

This process suggests a platform approach where significant insights-driven learning up-front creates the basis for innovation. This platform is different from the more conventional approach that starts with ideas. The earlier assessment allows the most important criteria for concept evaluation to emerge at this earlier stage. All learning at this early stage of innovation process focuses on improving knowledge of the target market, frame of reference, and point of difference for ideas and business propositions. This knowledge will build stronger concepts.

This approach starts with platform development (see Figure 4.2). The shape of this process also suggests filling the funnel with a large number of ideas and concepts and assessing them together more cost effectively, using existing strategic information as an aid in the screening process. The process moves through the tunnel only those specific ideas that are highly likely to go the market and also likely to meet targeted business objectives. This approach substantially reduces the number of projects that might be killed after investments have been wasted in the different phases embodied by the tunnel. The tunnel shape suggests that ideally no project will be killed since the considerable up-front work will have created stronger projects.

4.2.1.1 Platform Definition and Characteristics

A platform identifies a broader and motivating consumer need that inspires innovation of multiple ideas. The platform allows competitive advantages to be created. The platform drives the strategy. Seven key characteristics of a platform are

1. Grounded in deep insights on consumers, market structures, and brand positioning
2. Connecting many business dimensions and inputs
3. Part of a long-range plan (LRP) and should have longevity

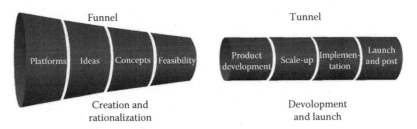

FIGURE 4.2 Platform-driven innovation process. (Courtesy of Sara Lee. With permission.)

4. Evolves every few years and continuously improves
5. Created with inputs from multiple functions, and aligned with senior management
6. Goes beyond a consumer segment, a product attribute, technology, a key usage occasion, or an unmet need
7. A business unit should have about three–five platforms to drive innovation

Four examples of platforms are

1. Today's meals for busy nights
2. Effortless traditional family meals
3. Beverages for aspiring athletes
4. Body care for teen and young males to impress the other sex

4.2.1.2 Developing Successful Platforms

Various approaches, from both the world of research and the world of management, can help to develop meaningful platforms. The following list is not complete, but it should give the reader a sense of the wide range of different inputs that play a role in platform development.

4.2.1.2.1 Consumer Segmentation

Consumer segmentation based on consumer attitudes and purchase behavior will help develop very specific target markets for the different brands in a company's portfolio. All too often target markets for individual brands are developed in isolation and are too broad. The unhappy result is that these brands present similar characteristics to a limited set of target consumers and are less differentiated. Thus, the different brands essentially compete with each other. A well thought out portfolio approach maximizes the differences among the target markets, minimizes cannibalization, and thus, when well executed, allows broader reach and greater volume potential. Figure 4.3 gives a sense of such portfolios for meat and meal brands for the Sara Lee portfolio.

4.2.1.2.2 Category Segmentation

Many food and household product categories overlap with many adjacent and other related categories that consumers quickly consider before selecting a product. For example, there are over 200 defined categories of packaged goods that can be considered before buying a packaged and branded food product for a dinner occasion. Creating the frame of reference and identifying the competitive set of products are essential for discovering the "playing field" that will match the core competency of an organization.

Given the need for clarity about where a brand competes, category segmentation helps develop the frame of reference. This frame of reference is also known as the competitive set of products and brands for the consumers. The so-called market structure approach reveals how the consumer perceives the different products in the category. However, commercial analyses do not necessarily line up with actual consumer behavior. Indeed, the commercial definitions of a product "category" from

FIGURE 4.3 Examples of target markets for a portfolio of meat and meal brands. (Courtesy of Sara Lee. With permission.)

data vendors like Information Resources Inc. (IRI) or A.C. Nielsen almost always differ from the way consumers look at today's offerings and shop for them.

The frame of reference from market structure is not sufficient. There needs to be another vantage point, called the "need state." Consumers have needs and wants. Need-state research identifies the language that will connect in a relevant way to consumer needs. In contrast, the market structure approach identifies specific product categories and attributes. Need states are generally higher order consumer needs. These must exist in the mind of the consumer and influence the way the consumer thinks about product attributes and benefits. The more systematic knowledge one can gather about the mind of a consumer (need state) and the way the consumer mentally organizes the products available (market structure), the better will be the quality of the platform. Figure 4.4 shows an example of this stronger frame of reference,

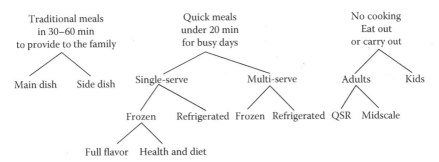

FIGURE 4.4 Example of frame of reference that combines market structure with need state. (Courtesy of Sara Lee. With permission.)

which combines market structure with a need-state analysis. Although the figure looks complicated, consumers process this type of decision tree quite quickly, often unconsciously, before they buy a product.

4.2.1.2.3 Unmet Needs and Trends

Once the playing field is identified with the help of the market structure and need states, consumer-need gaps can be uncovered within these focused areas. For broad food categories, research typically identifies attributes and benefits that fall into three areas: taste, convenience, and nutrition. The richness comes from focusing, not from broadening, however. These attributes and benefits become more detailed, richer, and far clearer when they are considered within a narrower, more focused field.

Beyond the unmet needs are trends. General trends can also be focused to the specific area in which innovation will take place. Combining need states, unmet needs, and trends in broad categories, along with the understanding that comes from market structure creates a knowledge base and leads to insights and eventually to better innovation.

4.2.1.2.4 Technology and Core Competency Assessments

We move now from the consumer to the company. Core competency assessment requires understanding an organization's competencies in several contexts. These include insights and opportunities from the consumer and category segmentations along with the strengths and weaknesses of specific key competitive brands. The competency area includes manufacturing, R&D, sales, and distribution capabilities.

Such a disciplined assessment is essential before the company proceeds too far in the innovation process. Technology assessment will help create competitive sustainability of concepts. In turn, understanding one's core competency will enable the organization to take advantage of its strengths and of its competitive advantage. The reality is that segmentation and need-gap research will always identify a large number of gaps, i.e., opportunity areas. However, only those that are relevant and that can be properly addressed with a sustainable competitive advantage, are worth considering.

4.2.1.2.5 Portfolio Brand Positioning

Once again, although there are always a large number of opportunity areas, only those that fit the brand positioning should be explored. This assessment is often addressed after several phases of the innovation process have been completed. Quite often, and unhappily so, after considerable investments, the participants begin to realize that the business opportunity simply does not fit the brand positioning. Up-front assessments would have saved considerable resources.

4.2.1.2.6 Cost and Feasibility Issues

Whereas strategic up-front learning requires investments in segmentation, market structure, and need-gap research, typically these are all one-time expenses, probably not necessary again for several years. With this one-time cost, large numbers of ideas and concepts can be strategically assessed with minimal investment. The output is a

Product approach	Platform approach
• Easier but costs a lot	• Difficult but costs less in the long run
• Tactical solutions	• Strategic and holistic approach
• Stops after investments	• Management buy-in from early stage
• Less efficient prioritization	• Minimal waste
	• Early, efficient, and strategic prioritization
	• Bigger ideas versus many smaller ideas
	• Higher concept scores with better inputs
	• Portfolio research costs less than many individual research projects
	• Avoids cost of killing projects in tunnel phase

FIGURE 4.5 Advantages of the platform approach versus the product approach. (Courtesy of Sara Lee. With permission.)

rich up-front knowledge that gives depth and richness to the otherwise simpler concept scores. Concept scores alone are not true indicators of success.

Furthermore, the cost is eventually less of a burden. The portfolio research information is used by all brand groups or business units instead, which then need not conduct their own individual research. With organizational alignments at all levels coming out of the focused platform approach, the teams now can move to the ideas and concepts phase with minimum supervision. Figure 4.5 shows the advantages of the platform over the product approach.

4.2.2 FROM "BLACK BOX" TO TRANSPARENCY—OPENING UP THE PROCESS TO INCREASE SUCCESS

Leadership roles in the process of innovation often reside in the hands of selected functions: marketing, R&D, and business operations. Yet a typical project team includes individuals from many other functions, each of which must make a contribution. During the course of innovation, various criteria for moving forward are used. Too often these criteria may not be explained to all of the participants, who are expected to go along without understanding. For example, several key decision steps in innovation such as criteria for entering the innovation process along with moving between "stage gates" are not openly available or clearly articulated to the entire project team. Furthermore, a large portion of an entire organization does not participate in the innovation process due to lacking knowledge of it. Transparency in the following areas can help educate and motivate an entire organization to participate in the innovation process with superior knowledge. The following are just a few.

4.2.2.1 Financial Goals for Innovation Linked with LRP and AOP

Develop financial goals for innovation for specific business units or brands, align with long range plan (LRP) and annual operating plan (AOP), and then share these goals with all employees. The goals should be included in the annual objectives set forth by senior management, and then presented to all employees.

Deliverables

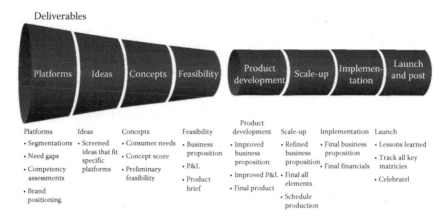

Platforms	Ideas	Concepts	Feasibility	Product development	Scale-up	Implementation	Launch
• Segmentations	• Screened ideas that fit specific platforms	• Consumer needs	• Business proposition	• Improved business proposition	• Refined business proposition	• Final business proposition	• Lessons learned
• Need gaps		• Concept score	• P&L	• Improved P&L	• Final all elements	• Final financials	• Track all key matricies
• Competency assessments		• Preliminary feasibility	• Product brief	• Final product	• Schedule production		• Celebratel
• Brand positioning							

FIGURE 4.6 Example of deliverables at each phase of innovation. (Courtesy of Sara Lee. With permission.)

4.2.2.2 Clearly Articulate Objective Criteria or Hurdles for Innovation in Each Phase

Develop quantitative or objective deliverables and hurdles for any idea or project that will enter and move through the innovation stage gate process. Apply these criteria as metrics consistently to all product ideas. Then include these criteria clearly as part of a document that outlines the specific innovation process being worked on. This will ensure that everyone understands what makes a project move forward in the process and how it compares to others when choices have to be made. The project teams progress together with the help of an aligned business proposition document supported by everyone, instead of each person supporting only the information from his or her own function. Figure 4.6 shows an example of these specific deliverables.

4.2.2.3 P&L for Dummies

Financials are key drivers of projects but often are neither accessible nor understandable to the broad group involved with innovation. Whereas nonfinancial matrices are often subjective, financial measurements, once simplified, easily align project teams to a common goal. Create easily understandable profit and loss (P&L) financials for all projects in a common format. Allow access to the P&L to all involved with innovation so that each person can understand and align with the financial rationale for project decisions. We see examples of P&L sheets in Figure 4.7.

The P&L need not be complex. One can create a comprehensive, 5 year P&L by knowing just a few of the inputs, such as selling price, cost of goods, level of distribution, selling rate, etc. The project leader can create this type of P&L. Using Excel® or other spreadsheets allows the project leader or indeed anyone involved to determine what to do for different inputs (costs, price points, distribution levels) in order to improve their own top line and bottom line, as well as run "what-if" scenarios.

The P&L does not have to stand alone. It can be linked with the total portfolio innovation P&L of the organization to show how one element in a project impacts the

Innovation P&L Samples

Assumptions and inputs to P&L Proforma statement of a profect

Gross sales rate calculation		
Suggested retail price (to consumer)	$	—
Retailer commission		0.0%
Gross sales per unit	$	—

Volume calculations		Year 1	Year 2	Year 3	Year 4	Year 5
ACV %		0%	0%	0%	0%	0%
Units/store/week		0.00	0.00	0.00	0.00	0.00
Number of SKUs		0.0	0.0	0.0	0.0	0.0
Total number of grocery stores (nationwide)		0	0	0	0	0

Cost of Goods calculations			Year 1	Year 2	Year 3	Year 4	Year 5
Cost of goods per unit	$	—	0	0	0	0	0
Marketing spending	$	—	0	0	0	0	0

5 Year P&L

000's	Year 1		Year 2		Year 3		Year 4		Year 5	
Sales ($)	0.0	0.0	0.0	0.0	0.0	0.0	0.0	0.0	0.0	0.0
Cost of goods sold ($)	0.0	0.0	0.0	0.0	0.0	0.0	0.0	0.0	0.0	0.0
Gross margin ($/%)	0.0	0.0	0.0	0.0	0.0	0.0	0.0	0.0	0.0	0.0
Total marketing spending ($)	0.0	0.0	0.0	0.0	0.0	0.0	0.0	0.0	0.0	0.0
Marketing contribution	0.0	0.0	0.0	0.0	0.0	0.0	0.0	0.0	0.0	0.0
SG&A ($)	0.0	0.0	0.0	0.0	0.0	0.0	0.0	0.0	0.0	0.0
Profit ($/%)	0.0	0.0	0.0	0.0	0.0	0.0	0.0	0.0	0.0	0.0

Portfolio Summary P&L for a Division or Entire Organization

	Project 1	Project 2	Project 3	Project 4	Project 5	Project 6	Project 7	Project 8	Grand total all projects
Sales ($)	0.0	0.0	0.0	0.0	0.0	0.0	0.0	0.0	0.0
Cost of goods sold ($)	0.0	0.0	0.0	0.0	0.0	0.0	0.0	0.0	0.0
Gross margin ($/%)	0.0	0.0	0.0	0.0	0.0	0.0	0.0	0.0	0.0
Marketing contribution ($)	0.0	0.0	0.0	0.0	0.0	0.0	0.0	0.0	0.0
SG&A ($)	0.0	0.0	0.0	0.0	0.0	0.0	0.0	0.0	0.0
Profit ($/%)	0.0	0.0	0.0	0.0	0.0	0.0	0.0	0.0	0.0

The above includes key elements of a typical P&L. A real P&L includes several more lines.

FIGURE 4.7   Examples of P&Ls on a common drive with access to those involved with the project. (Courtesy of Sara Lee. With permission.)

portfolio of the organization. Nothing else motivates a lower level member on a team as the ability to see his or her impact on the financials of the entire organization. The P&L, properly created, does that job. In fact, Figure 4.8 summarizes the advantages of this transparency over the black box approach.

Black box approach	Transparency approach
• Elite functions driven	• Opens the door for everyone
• Emotions and subjectivity	• Built from bottom up
• Lack of buy-in by all functions	• Objective numbers
• Less efficient prioritization	• Multifunctionally aligned
	• Early and efficient prioritization
	• Ongoing project quality improvement
	• Broadens innovation participation
	• Improves motivation to participate
	• Financials play key driver role

FIGURE 4.8 Summary of the advantages of transparency over the black box approach. (Courtesy of Sara Lee. With permission.)

4.2.3 FROM SEPARATE RESOURCES AND SILOS TO ONE RESOURCE AND AN INTEGRATED PROCESS

Separate process and resource pools hinder in evaluation consistency and optimal resource utilization. For example, there are different innovation processes, financial formats, and resource pools for the following three areas:

1. Different brand groups or business units within an organization or a division
2. Types of innovations—breakthrough ideas, new platform or category creation ideas, and line or flavor extensions
3. Different channels—grocery, mass merchandisers, wholesale clubs, and other channels

A common innovation process applied consistently in each of the three aforementioned situations allows the management of an organization to understand, evaluate, and approve all opportunities on equal terms and maximize the resource utilization. A portfolio approach to resource allocation at the highest level will identify and rank all opportunities at the same time to maximize new business value creations.

Different divisions of an organization can inadvertently create ideas that compete for the same need states in the minds of consumers. In such situations, a portfolio approach to maximize the potential of all brands with minimum overlap is difficult, and therefore, full potential is not achieved. Realigning the capabilities of an organization to fit the consumer and market needs allows the creation of and the commercialization of greater business opportunities that otherwise might be missed by an existing organization structure. A realigned organization without overlap to pursue focused consumer groups, need states, or business opportunities will also create one team, one vision, and common goals for all the functions (see Figure 4.9).

Once all ideas and projects are strategically aligned with platforms and business objectives, innovation information, especially financials, are transparent; resources are efficient with one common and consistent process; and more employees can

Separate resources	One resource/one process
• Traditional with direct line structure	• Costs less in the long run
• Expensive with duplications	• Efficient
• Inefficient due to absence of center of excellence	• Aligned with a strategic portfolio
• Contradictory projects in the portfolio	• Brings synergy
• Nonaligned when assessed in a portfolio	• Early and efficient prioritization
• Longer communication and reporting with cross-functional involvement	• Creates a strategic portfolio
	• Takes full advantage of marketplace gaps and trends at a higher level

FIGURE 4.9 Summary advantages of one resource/process versus separate resources. (Courtesy of Sara Lee. With permission.)

participate in the innovation process. The more deeply employees understand business needs and find it easy to participate, the more they become motivated to be a part of the growth plan. This will be the first step to ignite the entire organization toward common growth goals.

Finally, the importance of the people element and the entrepreneurial spirit needs to be acknowledged in order to lead the process with profitability as a key goal. Recruiting or identifying the right background and experience is important for innovation jobs at any level and with any function. The qualifications needed for those in the creation phase of innovation differ somewhat from the qualifications needed in the commercialization phase. A well-rounded marketing background is essential for marketers in both the creation and commercialization phases. However, the ability to deal with ambiguity is needed in the early stage. In contrast, a more disciplined adherence to the process helps in the commercialization. Innovation is part science and part art. Whereas a disciplined process is essential for successful innovation, appropriate flexibility is needed in order to know when one should call on the intuition and judgment of the team members. Finally, consumer research in every step is important. Yet, depending completely on research in order to find the options should not be encouraged. Instead, team members' collective judgment, research, and keen trend observation will generally provide the best direction.

4.3 DELIVERABLES FROM AN INTEGRATED PROCESS—EXAMPLES FROM SARA LEE

The proof of the pudding, so to speak, is of course in the eating. And so, to end this chapter, let us see what an integrated process with many of the elements described above has delivered to Sara Lee. We get a sense of the output of such an integrated innovation process from Figure 4.10. It shows some of the many new products the Sara Lee organization has launched recently.

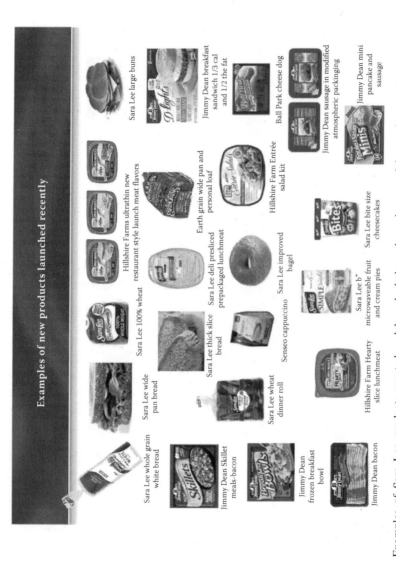

FIGURE 4.10 Examples of Sara Lee products created and launched through an integrated innovation process. (Courtesy of Sara Lee. With permission.)

5 Innovation Partnerships as a Vehicle toward Open Innovation and Open Business

Helmut Traitler

CONTENTS

5.1 INTRODUCTION: INNOVATION PARTNERSHIPS AS A VEHICLE FOR OPEN INNOVATION SUCCESSES

5.1.1 WHAT TODAY IS UNDERSTOOD AS OPEN INNOVATION?

In today's competitive environment innovation is a must. It is the new mantra. *Innovate* or *die* could be the leitmotiv for all companies, and although it sounds crucial, it is all so true. Going hand in hand with this mantra it becomes clear that all innovation emanates from top-class new product development. Such new product development is critical for success, and must be the DNA and genes of every successful company today.

A few years ago, the Nestlé Company made an important strategic decision. Open innovation was going to be the new role model for accelerating and optimizing their strive for innovation on the road to become the number one consumer goods company for Nutrition, Health & Wellness. That meant that Nestlé had to dramatically change its ways of thinking and working in all areas of technological, consumer centric, and R&D driven developments.

Nestlé's history shows three remarkable facts: First, the company has been in existence for more than 140 years. Second, Nestlé has been exceptionally successful, particularly during its most recent history of the last 10 years or so. Third, almost all of its successful innovations came from the inside.

One can imagine that in such a situation and business environment, it was not an easy task to explain everyone in the company that these three factors were also the most important reasons for the need for change, especially when it comes to the source and origin of innovation.

Back in the middle of 2005, a very small team comprising just a few individuals started to discuss, model, and shape Nestlé's new way of innovating and also conducting business with our partners. Subsequently, in October 2006 it became what we call Innovation Partnerships and it drives the journey toward the already elsewhere largely described Open Innovation[1] and Open Business.[2]

If we were to give a stab at the definition of Open Innovation we can truly say that

1. Open Innovation is a new way to collaborate in all areas of discovery and development with external partners who can bring competence, commitment, and speed to the relationship.
2. The key way to achieve this major goal is to accelerate necessary innovation and reduce the burden of resources and time pressure off the shoulders of one partner alone.

Perhaps now is the moment to describe and discuss the concept of innovation in a more general way. It is not really so important to distinguish between incremental (i.e., evolutionary) and disruptive (i.e., revolutionary) innovation. In the author's opinion, these terms are totally artificial and do not really describe the reality. They fit in textbooks to aid thinking. Moreover, they are at best a moving target. Some people say that "… you recognize innovation when you see it." However, although this truism is not wrong at all, it simply fails to reach the heart of the matter. The definition that one would find in dictionaries is "the introduction of something new." This maybe too large of a definition and an example for this would have been the introduction of the Euro as an overarching European currency in January 2002. But can it be called innovation? We should leave this for others to judge. Some say that innovation is "… delivering expected and unexpected solutions to people in new ways." The creation of the iPod is a great example for that.

In the fast moving consumer goods (FMCG) industry, innovation is a combination of a product's inherent values to the consumer, combined with elements of service, well-being, good-for-me, and convenience. In general, well-executed innovation with value for the consumers always comes from challenging conventional wisdom and searching, finding, and implementing new solutions to old or new problems.

5.1.2 THREE STAGES OF INNOVATION: EARLY–MID–MATURE

In order to fully understand where and how innovation is created and how it can flow into an organization, we have established a simple basic model that distinguishes between the generic stages of innovation: Early–Mid–Mature. Each of these stages, in the context of open innovation requires a different focus.

In the case of the model of the Nestlé Company we defined two main requirements (Figure 5.1):

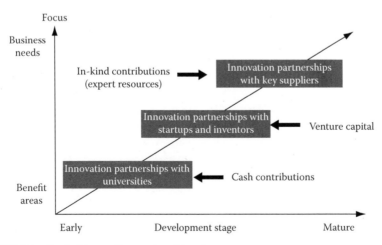

FIGURE 5.1 Typical innovation partnerships stages.

1. Any early stage idea should at least fit into the concept of the predefined "Benefit Areas." These areas were internally defined and the present example represents 13 areas altogether. These comprise eight related to different aspects of Nutrition, Health & Wellness, three related to compliance, i.e., quality and safety, and two related to taste. Among the Nutrition, Health & Wellness areas are topics such as digestive comfort, healthy aging, growth and development, performance (physical as well as mental), weight control, skin health and beauty, protection, and healthy recovery.

2. In a more mature stage, innovation needs a clear business focus and business goals, which are to be established by the business at large, defined in a collaborative fashion between technical and commercial representatives of the business.

5.1.3 REQUIREMENT FRAMEWORKS: BENEFIT AREAS–BUSINESS FOCUS

As a general rule, if one person or group is to collaborate with someone who might have the right competences and resources, it is important to clearly define one's own needs. The solution requester has to let the proposed solution provider know precisely one's needs and wants. Gone are the days of meeting with solution(s) savvy people, where the operative procedure was to 'let them parade on the innovation catwalk and show us what they have until we say *yes, that's it*'.

Today, we have to define such needs and requirements. Furthermore, it is imperative to share these needs/requirements in an open and clear way with the proposed partner. It is clear that such information is of strategic importance. Hence, confidential nature and a proper protection framework must to be established in the appropriate way prior to such sharing.

5.1.4 HOW THIS CAN OPTIMALLY WORK IN THE FMCG INDUSTRY

The FMCG industry has, as do many other industries, a strong focus on intellectual property (IP) as an important financial asset. It is therefore not surprising that the FMCG industry tries to protect, potentially patent and trademark as large areas as possible in order to increase its inherent value. Patents are the most prominent vehicle. Many FMCG industries attempt to own patents as much as possible and to cover as much ground as possible in order to shut out competition from its own heartland.

In recent years, this process has become more strategic. As a consequence, companies constructed the so-called IP fortresses. As important as the IP fortress may appear on the surface, it could also be counterproductive when forging strong alliances with potential innovation partners. This is one of the reasons why even large corporations who typically work with partners in the area of more mature stages of innovations, and increasingly turn to early stage cocreation and can build the IP fortress jointly and in a more natural and harmonious way.

It is important to bring in a word of caution here, pertaining to relationships of any kind and size. It is vital to ensure that none of what is done jointly with an innovation partner can be seen in any antitrust light. Everything must be presentable as having a clear goal in mind—innovation. It is paramount that everything that is done be transparent and withstand any scrutiny of any trade commission of any country, especially of course in the area of operations.

As a general rule, when one works with a partner who is on a different level of the value chain, in principle there should not be a problem. Working with a competitor, however, is a completely different story and occasionally much more difficult.

5.2 BASIC RATIONALE FOR PARTNERSHIPS

5.2.1 What Are Partnerships?

In our approach, partnerships are, what the name suggests, a true teaming up between two or more partners of different need states. There always is a requesting end and a responding one who team up together in order to solve a problem, fill a gap, or find an answer more effectively and overall more quickly. Effectiveness and speed are the operative and overriding principles of any innovation partnership.

We have created a very simple motto for such partnership: "Sharing is Winning." This definition really describes the spirit of any partnership. It does not suggest a naïve approach that either of the partners is giving up any proprietary territory, but rather it expresses the deep respect that partners have for each other once they enter into a codevelopment.

On the journey toward the perfect innovation partnerships, one passes through the three already mentioned essential stages. The first stage is to establish trust. The second stage is to build goodwill. The third and of course most relevant stage is to create value. By teaming up with world-class innovation partners of all sorts, the value creation becomes stronger and, more importantly, happens faster.

5.2.2 Why Partnerships?

Partnerships imply an inside and outside innovation. In turn, there are several approaches to the innovation that comes from the outside. All of the approaches share the common first step, which defines the gaps, the needs, and requirements as we previously called them. Then depending upon the corporate preferences or modes of operation, there is a path to partner. One might be procuring the innovation for some defined sum. A second might be codevelopment through combining expert resources. Only in the case of codevelopment can we say that a true partnership has been established.

Partnerships also constitute the less risky approach. For example, at the onset of any codevelopment no investment is needed other than the existing expert resources. It is well understood that existing resources cost money too and that many internal branches of any given organization compete for such resources. However, from experience, it is much easier to ascertain such resources for purposes of new product development and innovative codevelopments than it is to secure necessary funds to pay for such product development or acquire solutions. Moreover, resources can be redeployed to other projects if a codevelopment would fail. What also becomes apparent from this is that in any case, world-class project management in new product development is a paramount key.

Each partner, namely the competence and solution provider as well as the partner whose requirements are potentially fulfilled, enter into a real partnership. Upfront, it is vital to define and specify clear goals, resources, timelines, milestones as well as assigning of IP and value sharing solutions are defined upfront. Overall, this is really the more powerful and sustainable model, because the risk of too early financial commitments into projects is kept extremely low.

5.2.3   WHO TO PARTNER UP WITHIN THE FMCG INDUSTRY: UPSTREAM AND DOWNSTREAM

The approach just outlined enables partnerships with innovators of any possible kind. There are two major types of partnerships. First, we have the "Upstream" partnerships, comprising partners who represent early, mid, and mature stage of innovation. Oftentimes these partners come from universities, start-up companies, and inventors and, last but not least with large industrial partners, also known as suppliers. "Downstream" partnerships occur with a select group of large customers, with the goal of identifying innovation based on shopper insights, and having strong consumer relevance.

This two-pronged character of partnerships, upstream and downstream, often leads to new product development which is clearly shopper-insight driven, potentially resulting in services and products that "fly off the shelf." The underlying principles are always the same: Bring together needs and requirements from one group with the competences and solutions from the other. For reasons of simplicity, the rest of this chapter focuses on innovation partnering tools and elements in the framework of industrial partners of any kind. The process can equally be applied to relationships with academic institutions, start ups, and small technology and know-how providers.

5.2.4   THREE STAGES OF PARTNERING

As mentioned earlier, all partnerships evolve through three essential stages (Figure 5.2). These three stages are trust, goodwill, and value. It is really the value creation at the

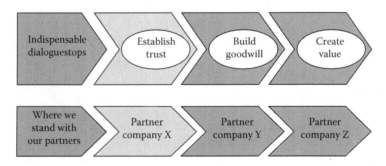

FIGURE 5.2   Relationship stages of innovation partnerships.

end that is the ultimate goal for any partnership. We have to emphasize that without such value creation the whole concept would remain hollow, without worth to both innovation partners.

In order to drive value creation we need to apply a disciplined approach to the innovation process. There are a variety of elements that are of crucial importance in this process. We will describe them in more detail later on. It is important to emphasize here, once again, that no matter what the nature of the innovation process, well-understood project management tasks and excellent management itself are key for continuing success. Without such project management, which includes clear briefs and good understanding of the goals and manages expectations, many projects are doomed to failure, whether these are run internally or in a partnership model. In the case of partnerships matters become even more difficult when the project fails. Such failures by a partnership affect resources from both parties. Furthermore, they demand explanations about why the project fails. These explanations have to be given to two management teams. This makes the whole concept of partnering even more complex and difficult. Hence, flawless planning is even more important.

It is difficult to pinpoint with any degree of accuracy the specific stage of the relationship, as it describes a particular partner company. Evolution along the stages of the relationship is fluid, episodic, and hard to pin down. Yet the evolution exists. Most important is the acceptance by both parties that there is such an evolution, even if it cannot be accurately described, or subjected to metrics, as companies like to do. Finally, it is vital to recognize and accept that it is necessary for both parties to "sense" where they stand in this evolution to manage the mutual expectations that such partnerships bring.

Given the equality or disparity of sizes of the partners, it can easily happen that with one and the same partner one could be "at the doors of the courthouse" in an area of collaboration, and paradoxically be enjoying a full-fledged "love-in" in another. Such contradictory situations with the same partner are, of course, not ideal, but they happen time and again. They should not be used as an excuse to stop working with the partner. Rather, both parties should attempt, in their spirit of collaboration, to resolve the issues that divide them.

5.3 TRUST-BUILDING FRAMEWORK

5.3.1 SECRECY AGREEMENTS, NONDISCLOSURE AGREEMENTS

One important element in any type of collaboration is first understanding and then further defining the needs of the partners involved. The partner who initiates the relationship because of a need must know which specific competencies and innovative solutions are being offered, and how these competencies and solutions will actually help the innovation. In the case of Nestlé specifically, and only for upstream partnerships, we have compiled our company's future needs and requirements for all the businesses and individual business units, in order to be able to share them with our innovation partners.

Moving to downstream partnering, one needs to do the opposite. This time, it is vital to match customer and consumer needs revealed by shopper insights with the company's specific competencies, and what the company can "bring to the party."

Returning to the upstream, one has also to consider legal issues. Before the company shares its needs list, it must establish a legal framework in which to conduct the relation. At Nestlé we first used existing agreements, such as confidentiality agreements, under which terms we were able to share strategic and confidential information. However, as we followed the path of shared innovation, we found that the specific secrecy or nondisclosure agreements were inadequate. The drawback is simply that for the most part, they cover mostly only one specific topic, e.g., sharing information embedded in a document. When one gets into the goodwill phase, i.e., opening up and putting ideas openly on the table between the two partners, then sharing becomes very tricky and far more complicated. New ideas are not necessarily covered by such an agreement anymore, unless one specifies the topic in new agreements each time.

5.3.2 MASTER JOINT DEVELOPMENT AGREEMENTS

In order to avoid the above obstacles and to make our lives much easier, we developed and established what we called master joint development agreements (MJDAs). Such an MJDA comprises two parts:

1. The first part covers all terms of confidentiality between the two parties, as well as covering possible affiliates of each partner. In some cases, this first part may also contain a definition of the potential ownership of the jointly created innovation solutions.
2. The second part comprises basically a detailed description of a project resulting from one or more joint ideation meetings or discussions between the two partners. Typically, the main body of a MJDA really describes in all details the project, expectations, resources, timelines, and all other elements necessary for best practices in projects management.

As a rule of thumb, the most straightforward way to share IP applies a simple rule. The rule states that every physical solution such as ingredients or even technologies is owned by the competence-providing partner. In turn, the "smart applications" of this solution are owned by the receiving party, which in our case is the Nestlé Company. Our requirements, our partners' competences, and the legal frameworks helped us to quickly reach the phase of building the goodwill mentioned earlier.

What is really meant by this, is establishing a climate of openness. Within this open and secure relation, every member of the innovation partners can speak and ideate freely with every other member, without having to revert to yet another secrecy agreement for any new topic that innovation partners would like to come up.

This process is largely built on understanding and then merging the needs and requirements of one innovation partner (Nestlé in the case of the upstream model, the retail customer in the case of the downstream model), with the competences of the respective other partner.

The MJDA was the crucial element to break the ice and speed up the entire process. Once Nestlé established these MJDAs with a select number of Innovation Partners, we could enter into the next phase, which we called "Discover the Opportunities."

5.3.3 IP Issues: Share IP or Own IP?

There are some best practices one can follow when it comes to sharing IP of jointly created (or "cocreated") value. To reiterate what Nestlé developed, the principle in our case is that the competence-bringer owns the solution(s) and the need requester owns the smart application(s) in specified areas. In this way, the partnership can avoid many unnecessary, unproductive discussions and often unpleasant, relationship-damaging conflicts. Obviously, it is not always as easy as this and there may arise many complex situations in which, for tactical or strategic reasons, this simple sharing of IP pattern does not apply anymore. One such example could be that the solution and enabling technologies or ingredients sit right in the heartland of the requester and therefore should remain with the requester.

Good upfront negotiations are crucial, not only involving IP experts but also commercial representatives. Such negotiations should be fact-based, nonemotional, and done expediently and with good will. Nothing can be more counterproductive to the success of a joint project than overly lengthy and bitter discussions that wrangle over IP and who owns what and why.

Often, rights to IP are mistakenly linked only to the specific question of who owns the patent. Patent ownership is too limiting a view. There are many more elements in the field of IP such as trademarks, manufacturing secrets, know-how, exclusive access to necessary supply chains, and potentially many more. It was precisely due to the recognition that "proprieties" could be defined in a new way and resulted in the above-cited best practice of sharing IP. As long as the requester has exclusive access and right of usage, it is no longer important to argue over, for instance, a patent. Ideally, any patent should only be owned by one partner, because shared patents can sometimes be tricky, especially when it comes to the long-term vision on such a patent. Interests of either partner can change and/or evolve over time. In turn, with such change, the continued management of such a shared patent becomes complex.

5.3.4 Commercial Issues: Involvement of Procurement

In principle, successful Open Innovation in the partnership model always should lead to cocreated values. The financial investment that happens at the beginning of a simple insourcing approach (i.e., solutions for money) are delayed in the partnership until such time that a value proposition has been established by successfully accomplishing a joint project. Until then, the real investment is the use of expert in-kind resources from both partners into the project. Once the cocreated innovation can be implemented, solutions have to be found concerning how it would be best to share the commercial values of the innovation. The solution can involve the traditional "goods and services versus cash" model with the partner's cocreation costs built into the delivery of the innovation. Sometimes the solution can involve other ways to share commercial value, such as open-book and royalty payments.

It should become increasingly clear that commercial considerations must be part of the cocreation process, and that these commercial considerations should take place at the very onset of any joint project. From the practical point of view, commercial representatives from both partners, e.g., from procurement or sales, depending on the partner's status, have to be members of the project team, although they should play more of an observational and advising role.

5.4 FROM OPEN INNOVATION TO OPEN BUSINESS

5.4.1 JOINT VALUE CREATION AND INNOVATION

The concept of joint value creation was already described above, and attributed to the process of cocreation. It is clear that cocreation does not come by itself. Rather, the partners have to put in place a framework and an environment conducive to ideation and its execution, the real innovation. Creating value within one well-structured and focused organization is already quite difficult. One need only imagine that, with the cultural and commercial differences that characterize partners, such creation becomes more complicated and complex when innovation has to be cocreated and executed in a partnership way.

In the following section, we will discuss proven and successful pathways focusing on how innovation starts from a structured and disciplined approach, how to involve many players, and finally how to develop the ideal platform for joint ideation as a first, yet crucial step in the innovation process.

5.4.2 NEW WAYS OF VALUE SHARING: THE OPEN BUSINESS MODEL

As the title suggests, we are beginning to look into new ways of value sharing that do not necessarily do away with the existing, traditional ones but rather better reflect our striving toward the next important step in the journey, the one toward an open business model. What does open business really mean? In a similar fashion as we join forces with innovation partners in the cocreation of innovative value, we also will have to join forces when it comes to sharing such cocreated values.

We still have a long way to go but the direction is becoming clear. If we want open innovation to be a sustainable form of development, then the next step is inevitable. We need to add new ways of sharing commercial value to the predominant one, which is the earlier-mentioned goods and services versus cash model.

People dealing with this every day will have to learn to respect an innovation partner in a different fashion than a simple supplier or contractor. These two words, contractor and supplier, set a specific tone for a very different relationship to an innovation partner. Again, if we deal with innovation partners in a commercial sense the same way with this mind-set, then the relationship is not really sustainable. New ways of commercial value sharing will have to be developed. Elaborating this topic further, however, goes beyond the scope and the intention of this chapter.

5.5 CREATIVE PROBLEM SOLVING

5.5.1 INTRODUCTION

As described earlier, innovation can have many faces and can come from many sources. There are, however, several principles. The multiplicity of principles is a common pattern when we try to define the origin, the generation of innovation. The source of any innovation really lies at the fracture lines of multiple disciplines. It is always about bringing unity to diversity and bringing people of many different backgrounds and disciplines together.

The environment in which this multiplicity best thrives should balance degrees of freedom with disciplined and focused spirit. Surprisingly, the overabundance of resources is counterproductive to real innovation. In such situations, solutions are enabled too easily. The reality is that often nothing really new and surprising can come from such an environment. Lastly, it can be said that true innovation is really generated in a spirit of deep understanding of people's needs, dreams, desires, and hopes.

With this in mind, we come to the concept of creative problem solving tools as they are, for instance, expressed in the IdeaStore™. The IdeaStore is one variant of a brainstorming-based tool with three main characteristics, namely

- Facilitation by visualizers
- Large diversity of participants
- Disciplined and stringent selection and prioritization process

Some of these tools of creative problem solving which is an important part of the cocreation process are discussed below.

5.5.2 CREATIVE PROBLEM SOLVING TOOLS: IDEASTORE

It was mentioned in an earlier section that in order to jumpstart innovation partnerships we need to introduce new, yet proven ways of joint creative problem solving tools. The principles of these tools have extensively been described by VanGundy.[3,4] IdeaStore is one of such tools and was developed by us in collaboration with design experts. Visualization plays a crucial role in this process. The IdeaStore uses well-known brainstorming techniques, combining them with clear goal setting, diversity, divergent and convergent idea flows, stringent and fact-based selection, as well as the thorough elaboration and then prioritization of selected ideas.

The main thrust of the IdeaStore approach is to discover opportunities, creating these opportunities by matching of needs and requirements with relevant competences. An IdeaStore session typically takes place over 2 days. The session brings together an approximate equal number of representatives of both partners together. The workshop for practical reasons is led by two facilitators who have a design background as they also act as visualizers. They can make the sticky images which help better explain the potential concept and give a potential project a face.

From experience, the total number of participants should not exceed 12–14. This optimum number enables every participant to creatively contribute to the brainstorming, selection, elaboration, and prioritization process as crucial elements of the IdeaStore process. The background of the participants in the IdeaStore should be as diverse as possible. From our experience and being a company that has put Nutrition, Health & Wellness as its prime target, we always involve a nutrition-savvy person. Furthermore, we always involve a person from procurement, so that even from the beginning commercial thinking can flow into the process of the ideation. Finally, we always involve people who can make decisions when the time comes to dispatch necessary resources into projects.

Every IdeaStore process begins with the definition of the wish statement. This wish statement matches the needs and requirements of one partner with the competences of the other. The wish statement typically might read this way: "It would be great if we could. ..." The statement will comprise short-term versus long-term concepts. When formulating the statement, one must consider some factors such as value of concept, whether or not the concept cover a large enough business opportunity, can resources be put behind, realistic timelines, etc. The statement might also deal with the more soft targets such as building a team spirit between the two partners.

In order to improve chances of successful follow up of any of the concepts developed during the IdeaStore, it is advisable to limit the number of projects in the selection process to not more than 10. Once such potential concepts have been elaborated and brought into the format of a concept sheet, they will be submitted to the relevant businesses and our R&D for validation.

Typical success rate of this validation is around 30%, i.e., something like three projects are being further elaborated on and worked on jointly between the two companies. The pattern that emerges from such workshops can be described as

$$200 \rightarrow 10 \rightarrow 3 \text{, or from 200 ideas to 10 concepts to 3 projects.}$$

IdeaStore workshops also create an intangible product—team building. Once the groups interact, they know each other better, they understand what each group brings to the relationship, and the interaction opens new pathways to direct cocreation. Quite often, the IdeaStore leads to other innovation activities that involve the partnership. The new innovation activities often transcend the original limited goals and outcomes of the IdeaStore project.

The Stanford Research Institute (SRI) International in Palo Alto, California, uses a similar approach to the IdeaStore but goes even further. After a first round of "200 → 10 → 3" SRI organizes a second workshop on the "3" to drill it down to "1," which ultimately can be further elaborated in drilled down in a third workshop.[5]

5.5.3 CREATIVE PROBLEM SOLVING TOOLS: OTHER TOOLS

At the Nestlé Company we use also other tools, one of which is applied for packaging innovation. We call this too FastPack™. FastPack has the same roots and elements as the IdeaStore. The key difference is that FastPack is exclusively designed to generate a limited number of packaging concepts. The timeline for FastPack is accelerated. The concepts can accelerate toward to rapid prototyping and consumer research and potential feedback within a few weeks only.[6] As for the case of the IdeaStore, efficiency, focus, and speed are of the essence.

FastPack is a well-proven tool. Nestlé has successfully applied it for many years. Many successful Nestlé packaging innovations have come out of FastPack, such as the Moevenpick iconic ice cream tub, new Nescafé containers, coffee creamer containers, new baby food containers, and also shelf-ready packaging solutions, the so-called pdq's ("pretty darn quick") for large retailers such as Wal-Mart.

5.6 OUTLOOK

5.6.1 FROM OPEN INNOVATION TO OPEN BUSINESS TO OPEN SOCIAL

In today's business environment, "open social networks" are emerging as increasingly important for people and for business as venues for commerce and of course innovation. The examples of Google, Facebook, and My Space, etc., are just a few and have one major element in common: they all play a role in the "virtual world" on the Internet and Web-based platforms. It is, however, becoming increasingly clear that such open social networks can and will play a role, yet in a different way. It is quite likely that these networks will also play a significant role in FMCG as well.

5.6.2 WHAT DO OPEN SOCIAL NETWORKS MEAN FOR AN FMCG COMPANY?

A recent article in the McKinsey Quarterly[7] describes the opportunity of such open social networks in more detail. The article points out that the open innovation paradigm is really the open social component. It is also clear that it will still take a while until we will get there. Yet the way forward appears to be clear. Open innovation is just the first step in this journey.

With the foregoing paragraph as prologue, let us try to answer the question about the meaning of the open social network specifically for an FMCG company. We must realize that the answer to this question is really highly speculative just right now. The FMCG company, especially those executives entrenched in today's best practices, will probably be asked to change some of their mind-sets, even about methods and ways of thinking that have been successful to date. Companies will have to transform their rather slow, measured, studied, and occasionally ponderous approaches to innovation and execution into very fast ones, not always perfect, but with major opportunities as well. The real protection of innovation will come from speed rather than from patents.

It also means to search, find, and implement just about any potential solution to an opportunity from just about any source, and become the best and fastest in translating, adapting, and executing such solutions. It means that we may have to define the field of innovation in a new way, at least within the realm of a packaged goods company, to become a packaged goods and services to the Consumer Company, making lives more pleasant, more convenient, and bringing affordable and nutritionally sound goods and services to our consumers. This might be the real answer to the open social network in the FMCG industry. Needless to add that this paradigm shift in innovation will have a pronounce impact on future new product development practices. Hopefully this will also bring a much better understanding of the marketplace and the consumers so the success rate of NPD will be improved dramatically.

5.6.3 Mind-Set Change into Sharing-Is-Winning

Finally, we come to the real conclusion of this chapter. It is all about mind-set change, as so often proposed and promoted, yet this time apparently for real. There is no open innovation and no open business model without clearly changing the mind-set from "attempting to do everything within" to "searching out the most appropriate partners for success."

This mind-set change can most easily be described as sharing-is-winning. This descriptive motto does not mean that we naïvely give away our secrets. Rather, as described above, we build the framework by appropriate agreements, we share our needs and requirements, searching for the appropriate partners with the appropriate competencies. Jointly we discover the opportunities, we enter into the trust to good-will to cocreate value cycle and we stubbornly pursue opportunities professionally supported through world-class expert resources and top-class project management.

At the end, everything that we do in a company should make sense in terms of value creation. If we do not create value, then we will have failed. Yet, by doing things right, and doing the right things, it is likely that so much transformation value will be created that there will not be the nagging question about "what is the right way," "us versus them," and "we can't let go." Rather, sharing-is-winning will become the overarching principle. This new mind-set will also affect all aspects of NPD. Undoubtedly, new practices will emerge to meet consumers' expectations and wishes. The end result will be positive—significantly enhanced success rates in the ever increasingly competitive marketplace that confronts all of us.

REFERENCES

1. Chesbrough, H. 2003. *Open Innovation, the New Imperative for Creating and Profiting from Technology*. Boston, MA: Harvard Business School Press.
2. Chesbrough, H. 2006. *Open Business Models, How to Thrive in the New Innovation Landscape*. Boston, MA: Harvard Business School Press.
3. VanGundy, A.B. 1992. *Idea Power: Techniques and Resources to Unleash the Creativity in Your Organization*. New York: AMACOM.
4. VanGundy, A.B. 1988. *Techniques of Structured Problem Solving*. New York: Wiley, John & Sons, Inc.
5. Carlson, C.R. and W.W. Wilmot, 2005. *Innovation, the 5 Disciplines for Creating What Customers Want*. New York: Crown Business.
6. Lane, G. 2002. The Nestlé Company, private communication.
7. Bughin, J., M. Chui, and B. Johnson, 2008. The next step in open innovation. *McKinsey Quarterly*, June.

6 Trend Monitoring: Sorting Fad from Long-Lasting Profit-Building Trends

David Christopher Wolf

CONTENTS

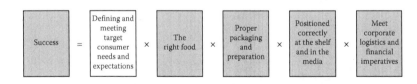

6.1 TRENDS VERSUS FADS: DEFINITIONS AND PERSPECTIVES

When it comes to distinguishing between "trends" and "fads," marketers have long been interested in schooling on the art and science of determining what is a "flash in the pan" versus what has staying power for the industry. Presumably, this interest is reality-based. This is because there is nothing costlier than investing millions of dollars behind a new product effort (which can take 2–3 years in large companies from initiation to launch) designed to address a growing "trend," only to have these products hit the supermarket shelves precisely at the same time pundits and consumers have declared the so-called trend to be finished. Not just flavor trends, but also a number of health-related trends have had their share of peaks and valleys in recent years. These "roller-coaster" rides have given marketers pause as to the wisdom of addressing a costly and speculative opportunity.

The rise and fall of the interest in low carb diets around 2004 is a prime example of a consumer movement that spurred hundreds of companies—large and small—to develop, launch, and merchandise hundreds of products with low-carbohydrate profiles, only to watch them languish and fail as quickly as they had come out (Figure 6.1). The success of early entrants into this category, primarily small start-up companies, inspired larger and larger companies to enter the market, spurred on by consumer studies showing a rise in obesity as well as interest in Atkins, South Beach, and other high-protein, low-carbohydrate diets. In hindsight, many observers and journalists call interest in low-carbohydrate diets and products a passing fad that began and ended in a year's time. Others would argue it was a trend that had been growing incrementally for many years before growing exponentially over a few months and peaking. Still others would claim that the trend simply splintered into other trends (complex carbohydrates, low-glycemic index foods, whole grain foods, high protein, etc.) and is still an important dietary influence today. What is the correct verdict?

To help sort out trend from fad, first we must examine the definition of these concepts in the two contexts of consumer trend analysis and new food product development. At its core, a "trend" is simply defined in most lexicons as "the general direction in which something tends to move." A "fad," on the other hand, is often defined

Categories	No. of new products
Baby Food	187
Beverages	3664
Breakfast cereals	417
Confectionery	2531
Sauces & seasonings	1933
Desserts & ice cream	807

SOURCE: MINTEL RESEARCH

FIGURE 6.1 New foods and beverages launched in the United States in 2006. (Courtesy of Mintel International Group Ltd., Chicago, Illinois.)

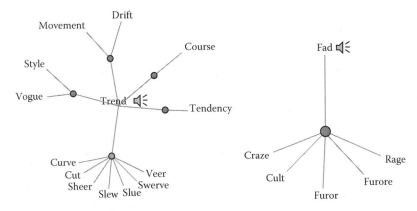

FIGURE 6.2   Comparison of trends and fads by the visual thesaurus (VisualThesaurus.com).

as "a passing fashion or craze." Visualthesaurus.com (Figure 6.2) demonstrates the nuances of the two terms, showing that, while they are very different, they are not mutually exclusive (vogue and style versus craze and rage). The problem, then, is not that the industry does not know the difference between the two words themselves, but that the term "trend" must be a word that has lost meaning and impact over time because it has been applied—mostly incorrectly—to so many phenomena, including those which are short lived and generally considered fads.

The other perspective that must be addressed is the apparent assumption that a trend is "good" and a fad is "bad." From a large manufacturer's point of view, of course, the ability to hold the consumer's interest for an extended period of time is desirable and economically necessary, because there is usually technology that must be developed or adapted to a new food product innovation. In addition, that technology or R&D development must be written off over an extended period of years to keep initial product prices reasonable. Olestra, originally developed and patented by Procter & Gamble, for example, took nearly 25 years to bring to the marketplace. Had P&G been concerned that interest in diet foods, which goes back several decades, was a passing fad, then it would not have invested so much time and capital into developing the technology and clearing it with the Food and Drug Administration (FDA). Note that this was not the case for the electronics industry, which charges premium prices to a small market of "early adopters" of products, and then lowers prices over time to attract a larger market as production costs go down.

In an increasing number of cases relating to new food products, a so-called fad (or short-lived interest in a product) can be a desirable and profitable business for companies that plan accordingly for such ephemeral opportunities. Take for example the proliferation of new food products (particularly those aimed at children) whose shapes, colors, flavors, cobranding, and introduction time are directly tied to the launch of a new blockbuster movie, such as the Spider-Man series in the early 2000s (Figure 6.3). These products were produced, priced, and marketed to move through stores over a short period of weeks or months, never to be seen again. It has also

FIGURE 6.3   Spider-Man Pop Tarts®, an example of a fad.

become common in the candy industry, among others, to see "limited edition" flavors that could not sustain themselves for years on store shelves, but that can be
produced and sold in a sufficient quantity to turn a profit for a company before being
discontinued. Therefore, whereas it is acknowledged that the so-called trends may
provide a long-term business opportunity, it is important to understand that "fads"
are not inherently bad or undesirable just because the window to capitalize on the
short-lived business prospect is a matter of months instead of years. The challenge,
however, and the subject of this chapter are to help companies identify which of the
emerging opportunities is likely to be long term.

Finally, we must define the categories of trends that are to be covered in this chapter. There are dozens of important consumer-related, demographically based trends
to consider in developing new food products, such as birthrates, generational cohorts,
income levels, household, ethnicity, education, population density, etc. By examining
these data points, one can see what kinds of factors may influence consumer-based
trends or fads. These data should be collected and considered to distinguish a trend
from a fad, but they do not represent the trends—or even the categories of trends.
Furthermore, there are also market dynamics that reflect changing consumer needs
and wants, such as the rise of new store formats, restaurant concepts, and even new
food product introductions. Seldom are these dynamics and their specific manifestations or occurrences the actual trend or fad themselves, but to emphasize, they are
simply the *manifestations* or evidence of the trends. And once again, certainly these
should be looked at to determine the strength and longevity of those consumer trends
that hold the most potential for profit.

The types of trends that this chapter presents are the so-called "needs-based" trends that consumers exhibit through their attitudes and, more importantly, in their buying behavior. The issue to address is how can the product developer sort out the consumer-based needs that are underdeveloped, fleeting, or mature, from those that are likely to be more substantial and long lived. Afterward, it is up to the company and its culture, size, predisposition for risk, available capital, etc. to determine whether or not the company has the resources and nimbleness to capitalize on passing fads, or whether it is more interested in investing in trends that are likely to be here for the long haul. In a way, following either of these two extremes is quite easy. It is the companies that try to catch a wave near the end of its peak that are likely to get consumed by the wave.

6.2 WHY TRENDS AND FADS MATTER IN PRODUCT DEVELOPMENT

Upon distinguishing between trends and fads, the natural follow-up question is "Why do food companies care about trends or fads at all?" It was not always this way with the food industry. The well-documented historical evolution of the consumer goods industry in any basic marketing textbook shows that the market began with a "manufacturing perspective," in which demand for products far outstripped supply, so manufacturers called the shots for many decades. Then, when the market began to catch up and there were several sources of, or an excess supply for, a desired product, it became necessary to move to a "selling perspective," in which price and features dominated the discussion and retailing became more important.

By the end of the twentieth century, however, the food products market had matured. This maturation combined with a highly educated and increasingly selective consumer base, moved the industry to a "consumer perspective," in which the consumer's point of view and needs dominate the discussion, not just for how to market a product, but indeed what new products and features to offer at all. In the first two perspectives—manufacturing and selling—manufacturers and retailers let their own perceptions and business equities drive their activities. They focused on how to teach the consumer what they needed and wanted.

In the newer consumer perspective, manufacturers watch, listen, interact, and ask questions of consumers about certain aspects of their lives, then take that information and translate it into products which fit the companies' brands and, hopefully, manufacturing capabilities. More advanced companies even maintain a feedback loop or establish a consumer panel or web-based community to keep the consumer-need and product-delivery gap as narrow as possible. Why all the trouble? Because food marketers these days (a preferred term to "manufacturer" or "supplier") have learned the hard way that if they do not listen to consumers then they will miss the mark, and their products will be ignored or rejected, and eventually fail. Compounded with this is the financial reality that new products cost millions of dollars to develop, commercialize, gain shelf placement for, and market to the consumer.

The challenge with tracking and addressing trends for the purpose of growing one's business is that since "true trends" stretch across years and decades, tracking them and looking for new changes to report and leverage is like watching oil-based paint dry. Linked to this is the question of knowing when to step into the market with new products that address emerging trends, neither at a time when it is too "niche" to represent a commercially viable venture, or just as bad, too "mature" as to have enough momentum and unmet need remaining with consumers to be profitable. Therefore, slow-to-develop trends are as frustrating for large food corporations to address as fast-burning fads can be (if not anticipated as such, like a limited-time-only product). And compounding all these questions is the good-news/bad-news reality that technology streamlines and lowers the entry barriers for all companies in completing the steps in product development, from information gathering to rollout. Therefore, the market's ability to identify and address opportunities quickly is accelerating; transforming what might have been considered a trend many decades ago into a fad that peaks quickly as a result of faster reaction times that are soon trumped by newer discoveries or technologies. Likewise, product life cycles are shrinking due to this accelerated information/technology/re-engineering pace, forcing companies to continually react and adapt.

6.3   ORIGINS AND EVOLUTION OF CONSUMER TREND ANALYSIS

It is a matter of speculation just how long food companies have been officially tracking and relying on consumer-driven behavior, needs and wants, etc. (and not just analyzing, population data, spending data, birthrates, etc.) to grow their business and influence decision making. The outside world is, of course, not privy to the internal activities of companies for the past five decades or so.

Historically, awareness of, and a perceived need to focus on, consumer-driven trends as a basis for developing products took shape in the 1960s, as recorded by books such as that of Yankelovich president J. Walker Smith's, *Generation Ageless: How Baby Boomers Are Changing the Way We Live Today.* However, it was probably the widespread acceptance of John Naisbitt's 1982 book, *MegaTrends* which truly "baked in" corporate America's interest in reading the tea leaves of consumer "need-based trends." Of course, in turn, this knowledge would allow a company to draw patterns and conclusions that could be useful to growing shareholder value.

The conversation and controversy over long-lasting trends versus short-lived fads in the food industry rose to the surface late in the 1980s, not long after Quaker Foods made its big push to promote research on the health benefits of oats/oat bran in the consumer diet. That effort was essentially the first time that a food ingredient available to anyone (versus a patented pharmaceutical or branded product formula) was shown by a substantial body of long-term, legitimate, medically based research, to link diet, and disease prevention. This health claim for oats, and specifically oat bran, later to be upheld by the FDA in its food product labeling laws, resonated with consumers, particularly Baby Boomers, who were beginning to feel the effects of middle age and wondering how to improve or affect their longevity. It sounds obvious 20 years later, but back then the concept was revolutionary and even controversial.

Quaker's success would change the food world. Within months, grocery shelves and fresh bakeries were packed with hundreds of products made with oats, including not just healthful cereals and breads, but also less virtuous cookies, fat- and calorie-laden muffins, and even oat bran beer. In 1989, more than 200 products bearing an oat bran callout were introduced, according to *Gorman's New Product News* (now Mintel). Quaker stock grew, oat bran sales grew, and consumers ate mountains of oat bran muffins (containing plenty of fat and sugar, by the way) with the expectation that not only would they avert heart disease, but possibly even lose weight and grow younger in the process. Then in 1990, a still-controversial study was released and widely promoted claiming that wheat bran was just as effective as oat bran in lowering cholesterol levels. In turn, the magic bubble that raised oat bran far above all other foods popped, and so did the industry's short-term gains. Nonetheless, now 20 years later, we see that interest in oats is still a strong and lasting product trend.

6.4 FIVE CORE MANIFESTATIONS OF CONSUMER NEEDS RELATING TO FOOD

Dr. Abraham Maslow laid out five core needs relating to human nature (physiological, safety, belonging, etc.) that have continued to be supported even after more than 50 years of scrutiny and application. Undeniably, these five core needs can be used (and have been used extensively by this author) to track trends and fads in the food marketplace.

However, there is a related set of "consumer food macro-needs" that the author has tracked and tested for more than 20 years. These macro-needs are more food-focused and are specifically tied to product and service-related attributes offered by the food industry. They have been tested on thousands of food professionals and dozens of new product consulting projects in this period of time. The needs represent the complete set of macro-need states under which all specific trends and fads can be subcategorized. Some of these can and have been treated mutually or exclusively in the past as a potential benefit for a specific food product. In recent decades, it has become more common to combine need states in order to enhance the appeal or relevance of a product once the overwhelming appeal of a single product benefit has subsided. For example, early introductions of microwaveable frozen dinners strongly focused on the convenience-driven need state, with less emphasis on taste and health attributes. Over time, however, manufacturers found it necessary to combine the microwave-focused convenience trend with other trends, such as health, taste, or value-related benefits, in order to distinguish their products from the sea of competitors' products.

6.4.1 TREND #1—CONVENIENCE

The convenience trend is probably the oldest and most tapped trend in the food industry. Centuries-old preservation methods, such as salting, freezing, and canning of foods, were invented long ago to extend the shelf life of perishable foodstuffs for consumption at a later time when those foods were not available in a fresh state. The twentieth century brought new technologies in manufacturing, as well as consumer appliances that opened the door to new levels of convenience. Manufacturers sold novelties like Jell-O mix, Oreos, and canned tomatoes that saved the housewife time in preparing meals made hitherto from perishable commodities.

Around the same time that the oat bran trend was developing, a different set of food marketers were focusing on the microwave oven appliance, whose presence in households had been growing steadily during the early 1980s. The new opportunity for consumers to heat their individual frozen dinners in a microwave oven in 5 min or less, instead of the traditional 30 min in a conventional oven, held a sufficient novelty for several years in the late 1980s to grow microwave-based new product introductions. The year 1986 saw 278 introductions, and 2 years later (in 1988) those had almost quadrupled to nearly 1000. Category sales grew almost as quickly. By 1992, sales started to level off well below the $4 billion mark that had been predicted years before. The apparent failure of consumers to stay devoted to purchasing and consuming frozen microwave dinners to meet the market's expectations could have been attributed to microwaveable foods being simply a passing novelty that consumers readily abandoned like that icon of fads, the Pet Rock.

However, further investigation into the category dynamics of frozen microwave, as well as examination of results from many consumer surveys, tells a different story. The following are the salient points from this trend:

1. Although early frozen dinner entrants like Campbell Soup Co. devoted a massive amount of R&D resources and science to creating dinners engineered to rethermalize in the microwave, other companies simply repackaged the existing product in microwaveable containers to ride the growing trend wave. Consumers, it turns out, were not pleased with the resulting sacrifice in taste quality.

2. A flood of thousands of new microwaveable foods in the course of a few years created competitive dynamics that led to heavy discounting in the category, further deflating sales expectations and forcing quality compromises. And just as manufacturers were learning by trial and error what products and technologies worked best together in the microwave, consumers were learning what foods they would accept and which ones they would devote more time and trouble to. Nowadays, it is virtually impossible to track microwaveable product sales because they have become a heating alternative to most new and existing frozen food products as opposed to the main feature/benefit.

3. Finally, and probably needless to say, the microwaveable "trend" is a significant component and driver of the $28 billion dollar frozen food category today, despite some bumps and bruises for companies in the early 1990s, and doomsday articles in the press denouncing the quality of microwave foods and the demise of the category.

6.4.2 Trend #2—Wellness

Wellness is perhaps one of the most debated and controversial trends for food marketers. At first, wellness was only an infant trend (vitamin fortification, digestive biscuits) until well after World War II. At the start of the 1950s, wellness began to become important, as American consumers amassed enough wealth and girth to cause them to move from worrying about having enough to eat to being concerned with which types of foods they should select to maximize their health. Furthermore, science moved slowly. Research studies substantiating the belief that the overconsumption or avoidance of certain foods could cause or avert life-threatening diseases did not emerge and reach consumer awareness until well into the 1980s. What makes the trend controversial today is not whether the industry or consumers now widely agree that food choices affect one's health, but rather what each group ought to do with this knowledge.

Manufacturers moved quickly to market on the tail of popular health news featured in the press in the 1980s and 1990s relating to oat bran, cholesterol, fat, sodium, sugar, and other "bad-for-you" ingredients. In turn, consumers flocked to these products with the hope and expectation that not only would these make them thin, but also taste good. In most cases, consumers were disappointed with the trade-off between taste and health. Foods with less fat, salt, and other hedonically appealing attributes did not taste as good. Furthermore, and just as important, the "results" of healthy eating did not show up quickly enough. Consumers failed to see "instant" results from these foods (eaten sporadically and as part of a diet still high in calories and low in exercise), so they abandoned them.

Probably the most significant rise and fallout in the health trend during this period of time was the "fat free" fad. The author happened to work for the company at that time that created the name and positioning for the big daddy of all fat-free food brands in the early 1990s: SnackWell's. Nabisco's SnackWell's was a line of reduced-fat and fat-free cookies and crackers that were analogs of popular, higher-fat, cookies, and crackers then on the market. One of the products in the line, a fat-free Devil's Food Cookie (see Figure 6.4), was produced by a copacker for Nabisco. Initial production was limited as the company tested the waters for interest in this product before expanding production. Consumers who initially tasted this product raved about its indulgent attributes despite its lack of fat. The available product quickly ran out, causing a

FIGURE 6.4   SnackWell's®, an example of a product for the wellness trend.

shortage. News quickly spread of consumers chasing SnackWell's trucks or camping out in stores and buying up product, creating a buzz that fed the frenzy of interest in fat-free foods. Within a couple of years, the SnackWell's brand grew from zero to $500 million dollars, and dozens of companies hitched a ride on SnackWell's rising star.

Once again, the same press that fed the fat-free trend finally stunted it. In the mid-1990s, studies began to emerge showing that consumption of fat-free foods was not making anyone thin or healthy. Just the contrary, in fact. It turned out that these foods had more sugar and calories in many cases and that, worse of all the permission that these "guilt-free," higher priced foods seemed to offer had caused many people to consume a higher quantity of these foods in one sitting. The unhappy result was that consumers risked actually gaining more weight than they had bargained for. Combined with the fact that many copycat, reduced-fat products also failed to live up to taste expectations, sales of SnackWell's and other reduced-fat foods began to decline quickly in the last half of the 1990s.

Ten years later, we see that the reduced-fat "trend" is not gone, but it has found its level and products that are still on the market deliver to consumers' more realistic taste and performance expectations. And since then, marketers have identified and latched onto a number of seeming ingredient-based product fads in the larger health trend, including functional ingredients like ginseng or caffeine, lower-carbohydrate foods, high-protein foods, 100 cal foods, high-fiber foods, low-sugar foods, genetically modified organism (GMO)-free foods, women's health foods, and vitamin-enhanced beverages, among others.

Even more in question than the consumer's potential fickleness about fat, carbohydrates, or other dietary considerations, is the deep, really unanswered question as to whether or not consumers are truly interested in purchasing products that have favorable nutritional composition. Said in the vernacular, many observers believe (and there is evidence to support the fact) that consumers "talk thin" but "eat fat." Though it is true that sales of low-fat milk, "diet" foods, and even fruits and vegetables have risen over the past two decades, it is also true that consumers are consuming more calories (approximately 50,000 on average by some estimates) in a year than they did in 1980, and that obesity levels are at their highest ever in the history of mankind. The growing awareness of this reality has, however, spurred industry and consumer alike to keep plugging away and to try and make at least incremental progress in changing the product mix as well as behavior.

6.4.3 TREND #3—QUALITY

Quality or "taste" related needs are perhaps the most elusive of the five core food trends, which is probably why they lack a seminal "killer fad" such as microwave foods or fat-free foods that the other two macro-trends have experienced. But this macro-trend of quality has had its share of fads and fleeting moments. The challenge with this group of usually qualitative and subjective product attributes is that virtually no company fails to claim that its ingredients are the highest quality, its production is the best, and that its taste is of course superb.

For purposes of trend-tracking, therefore, we exclude the persiflage coming from those vague and overused "highest quality" claims, which have lost all meaning and significance, or the equally meaningless terms such as "deluxe" and "new and improved." Rather, our quality-related trend and fad tracking are to be limited to those descriptive product claims that promise a specified improvement relating to better ingredients, cooking methods, or flavor callouts than do existing products. These quality cues, such as "imported" or "rising crust" or "made with real Hershey's syrup" usually come at a higher price, not only to offset the potentially higher cost of producing the product, but also to enhance the perceived value of the product. The one potential exception to this higher-price rule is the rise of private or store-label branded foods such as President's Choice from Canada, which cites lower marketing and development costs as a factor that enables stores to offer "better-than-national-brands" quality (with specific attributes called out, such as "33% more chocolate chips than the leading national brand") at an equal or lower price compared top-selling branded food products.

In recent years, marketers have found new superlatives to replace previous quality and deluxe callouts. These terms include "premium," "artisan," "gourmet," "restaurant quality," "real," and "select." They appear suddenly in one part of the grocery store and then filter through multiple categories from frozen foods, to baked goods and beverages or snacks before slowing disappearing or being replaced by new, more promising words. The frozen pizza category is a particularly interesting one to watch, which has adopted romance from restaurant menus to raise quality perceptions and price points, including "brick oven" baking techniques and gourmet ingredients such as portabella mushrooms, real Parmesan cheese, garlic, or other flavor-enhancing recipes.

Although the term "gourmet" has been well-worn since the Julia Child cooking shows of the 1960s, it remains a real and measurable opportunity to identify enhanced ingredient, flavor, or cooking techniques that can be used to alter the perception or better, the true, tasteable performance of a product versus its competitors. One category that has stood the test of time is the premium ice cream category. On the outside, premium ice cream packaging displays quality enhancements such as smaller size, gold printing, or gourmet ingredients like bittersweet chocolate, vanilla bean specks, or English walnuts. In turn, there are measurable, scientifically based attributes that can be measured to explain why a pint of premium ice cream tastes better for some consumers than half-gallon boxes of lesser-quality ice cream. These factors include fat levels from dairy or eggs, and product density (known as overrun in the frozen dairy industry).

6.4.4   TREND #4—EGO SATISFACTION NEEDS

In human history, there is a long precedent of foods serving as a status symbol. Up until the twentieth century, for instance, baked foods made with refined white flour were a delicacy reserved for the royalty and the rich. Now Wonder bread is available to the masses and white bread has lost its exclusive appeal in favor of "rustic" multigrain breads that ironically were the only baked goods available to the masses before the industrial revolution. Now these breads have become the new status symbol.

The mass market trend for food products that cater to the consumer's ego needs (versus their basic hunger needs or fears of an early death) is an even newer phenomenon than products with health-related benefits. As Maslow's hierarchy of needs points out, ego needs, also called status needs, are higher up the ladder than basic physiological or safety-related needs, and it has taken more time for a critical mass of American consumers to achieve sufficient spending power to make it profitable for food marketers to cater to them. But now, barring complete economic collapse, status needs present a long-term trend with great promise.

Within the arena of ego satisfaction, need states such as the need for expensive, branded, and/or luxury products to boosts one's ego include rare or scarce foods like wine, truffles, or even a limited-edition soda pop, as well as the need to be in control of one's food (such as made-to-order coffees, marble slab-mixed ice creams in custom flavors, and even custom-formulated cereals and vitamins ordered over the Internet).

A more recent subtrend within the ego-satisfaction world is the idea of cultural inclusiveness, manifested in products and marketing messages that cater to multicultural tastes, in deference to the rising spending power of America's Hispanic, African-American, and Asian populations, whose taste preferences have not been fully represented in the nation's food supply until recently. Given the expected accelerated growth of these populations in coming decades, it is likely this will prove out as a long-term trend versus a quick-fix fad.

Admittedly, there is a fine line between products that cater to taste/quality needs and ego-satisfaction needs. For example, if authentic, expensive Parmigiano-Reggiano, imported from Italy in limited quantities but notably superior in flavor to domestically produced Parmesan-style cheese, a product that strokes the ego or a product that delights the palate? The correct answer is that it satisfies different needs for different consumers seeking solutions to the priority need of the moment, so it has the potential to fit in both categories. The fact that it may satisfy both needs at once can indeed enhance its appeal and value to the consumer, just as bottled Evian water potentially satisfies a combination of convenience (portable beverage), health (safe source, no calories), taste (for those who can tell a difference), and ego ("imported" and expensive) needs all at once. One should keep this in mind in considering the evolution of quality benchmarks found in Figure 6.5. These benchmarks denote evolutions in the actual or at least perceived quality of a variety of popular foods, but the motivating factor that contributed to the success of these brands over the years is undoubtedly ego related.

6.4.5 TREND #5—VALUE NEEDS

The need to get a fair value from a food product in return for a portion of one's income is perhaps one of the most misunderstood—and underleveraged—trends in the core needs list. The most obvious and most relied-upon tactic in the food world to cater to this need has been to discount the price of a product or to offer a special, limited-time deal. Examples are buy one, get one; 99 cents; $1 off coupon, etc. The consistent ability of this tactic to generate a temporary increase in sales for a company would seem to secure its place as a guaranteed trend. However, the challenge is that it trains the consumer to focus on price and to buy on price appeal versus other types of value-rich propositions. Furthermore, a price-only focus cuts into profit margins

Benchmark "Trumping"

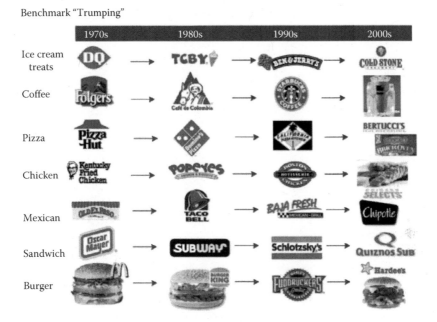

FIGURE 6.5 Benchmarks for products that satisfy the ego. (Courtesy of The Turover Straus Group, Inc., Springfield, Missouri.)

for a period of time, which runs contrary to the focus of this chapter to capitalize on "profit-building" trends. Therefore, we do not endorse it here.

There are other tried-and-true value propositions that offer good value in exchange for the consumer's dollar, without necessarily cutting into profit margins. Unfortunately, the CPG industry has not been as resourceful in tapping into these trends in the same way as have restaurant chains, car companies, and airlines. Airlines, of course, offer different classes of seating at widely disparate prices for essentially the same service, except for a few inexpensive bells and whistles like more legroom, warm nuts, a hot towel, and mixed drinks. Car companies have likewise invented classes of cars whose material costs are very similar, but whose design, branding, or other features provide a perceived value to status-conscious consumers for tens of thousands of dollars or more in sticker price levels. Finally, Hardee's is an example of a chain that redefined value for a hamburger by redefining its competitive set. Hardee's Thickburger was not positioned as the most expensive fast-food burger on the market at the time (though it was, in fact). Rather the Thickburger was the cheapest hamburger versus those sold at casual dining restaurants such as TGI Friday's. It was a quality claim that paid off for the chain and revived the brand.

Of course, the food industry is not completely devoid of value propositions that go beyond discounting. The continued success of Wal-Mart superstores, CostCo, and Sam's Warehouse Club Stores, along with a growing number of dollar stores, are a testament to the success that can come from appealing to a consumer's need to economize. Whereas overtly club stores appeal because they offer lower everyday prices in return for buying in larger quantities, in actuality these stores also have thrived because they constantly change the product mix, offer unique items not found elsewhere in food retail, and produce what has been called a "treasure hunt" atmosphere where consumers walk down aisle after aisle in search of a new discovery to appeal to their needs—at a great savings, too. Similarly, dollar stores offer a shock value of what can be had for "only" a dollar. However, a closer examination of products packaged in larger quantities found at regular supermarkets will often reveal that the consumer is paying more per ounce or pound of product, or buying off brands which, at the end of the day may not actually provide a superior value. But sales at these stores are growing nonetheless. Furthermore, these stores have begun to stock gourmet products at affordable prices, in order to move away from a strictly "cheap" focus.

Finally, within food retailing, there are rising stars like Trader Joe's from the West Coast, which offers mostly private label brands of gourmet and European-style products in a fun, offbeat seaside kind of attitude. Foodies appreciate the appealing prices of real maple syrup, quality cheeses, and unique bottled and canned foods not easily found on typical supermarket shelves. And bargain hunters love the idea of a decent $3 wine, affectionately referred to by insider shoppers as "Three-buck Chuck" (It originally was called "Two-buck Chuck"—but who's counting!!).

6.5　DISCIPLINE OF FOOD TREND-TRACKING— ITS METHODS AND APPLICATIONS

Before the invention and widespread adoption of the World Wide Web as a convenient source to connect people to mountains of research at the click of a button, trend analysis was a laborious and sporadic business that involved seeking out and exploiting unique data points or inventions that busy executives did not have time to collect and cipher themselves. Self-trained trendwatchers with an eye for seeing patterns and drawing connections over time perused obscure publications like the World Future Society's *The Futurist* magazine. These trendwatchers also traveled frequently while keeping their eyes open for new products, possibly tapping into an informal network of geographically dispersed people with like minds and interests who could connect some dots and propose potential trends that others could not easily perceive without time-consuming digging through a variety of data. The magic, however, was not so much the data itself, but the story that was woven to help guide marketers and give them inspiration for inventing new products to address untapped needs.

In a wired world, however, everyone has quick and easy access to the same, non-proprietary information, and lots of it to choose from. So now, the trick has become not "Where to find the information," but "How to sift through it." It is extremely difficult these days to be the first to discover something. In actuality, however, trend-tracking is less about discovering one new phenomenon, and more about discovering a lot of similar phenomena and how they point to a unique but consumer-relevant opportunity.

Thus, the first challenge in sorting trend from fad is to establish a reliable trend-sorting method (such as the five core food trends discussed above), and to use it for organizing and connecting dozens and hundreds of data points over an extended period of time from multiple sources. The second, and more critical challenge, is to determine who will collect and analyze these trends for your company, and to ensure that this pipeline does not deteriorate or get cut from a budget. As mentioned, the difficulty is not about having access or even about having any special skills over another businessperson. Rather the difficulty is maintaining the discipline or motivation to identify the data points at regular intervals and then to organize them into some kind of coherent document. The author's trend-tracking skills were developed across 15 years of full- and part-time journalism, as well as through periodic food industry speeches and planning sessions, where it was a constant necessity to produce and sell provocative trend perspectives—or be unemployed.

Given this caveat, what should one do to catch trends and understand them? There are databases that can be subscribed to for tens or hundreds of thousands of dollars that contain impressive volumes of clever perspectives on emerging trends in and outside of the food industry (nonfood information can be just as helpful in identifying trends as well). However, these databases do not search and retrieve the right information on their own. And even newsletters and special reports do not read and apply themselves. *Access* to the information is only the first step. Analysis and action are necessary follow-up requirements in trend sorting that must be a frequent part of the package. Not to do so puts a company in risk of misinterpreting a trend, its impact, and longevity. It is necessary to either create or outsource a dedicated resource, and hold that resource accountable for regular outputs of what is happening in the world. The information is available, so it is the discipline that is required.

Moving forward, let us assume that data points are being collected and sorted on a regular basis, but that the dedicated resource needs guidance and training on interpretation skills once some patterns have been identified. At this point, we are finally ready to engage in the sorting process of determining whether we have a trend or a fad. This is where we will utilize the five core foods needs discussed above (convenience, wellness, quality, ego satisfaction, and value) and some familiar example trends to use for practice. In some cases, we will have the benefit of hindsight on these trends, but it is more prudent to act as if we did not know the outcome.

For each so-called trend under evaluation, we should ask the following eight questions:

1. What core food need does this topic fulfill?
2. What specific subtrend does this relate to? What are the cultural, psychological underpinnings, or foundations of the trend (if any)?
3. What experience does the consumer have with the "trend" in the past?
4. Are there any countertrends fighting or contradicting this trend?
5. How "easy" is it for the consumer to adopt these products? Is there a big lifestyle change required, education, or compromises? How widespread is the interest?
6. What kinds of companies are currently manifesting the trend? What are the potential barriers to entry?
7. How long ago does the trend seem to have appeared? In the food business?
8. What demographic or other data points support or contradict this trend?

Example 1:   100 cal portions of foods and beverages: Fad or trend?

Analysis:

1. Core food need: Wellness.
2. Subtrend: Losing weight, portion control, and calorie watching has been of interest for many years, but there is no evidence that limiting snack consumption to 100 cal at a time contributes to weight loss or controls hunger.
3. Past experience: Some calorie control in entrees, but not widespread calorie focus.
4. Countertrends: Big portions, supersizing.
5. Ease: Makes portion control easier, but requires some sacrifice and a lot of self-control.
6. Companies: Led off by large food manufacturers like Nabisco, Frito, Coca-Cola. Few technology barriers, though some companies created new forms/formulations.
7. Trend maturity: First introduced in early 2006. Easily understood.
8. Trend drivers: Rising obesity concerns, failed diets, rise of between-meal snacking, skipped meals, meals on the run.

Prediction: The long-term trend is an interest in wellness, manifested as a desire to control intake and/or lose weight. The "100 cal bag" is the fad that is riding the underlying health-related needs. If consumers embrace this concept, then they will experiment on these products for the next year and most likely abandon them because they will not have lost any weight.

Action: Take the sustainable need states of wellness, portion control, and desire for weight loss to work on new products relating to one's brand and category that address these needs in an original and meaningful way.

Example 2:   Products with "sustainable," "carbon footprint," "ecofriendly" claims

Analysis:

1. Core food need: Ego satisfaction.
2. Subtrend: Stewardship. Also called "Going Green." There is a growing "pressure" by activists for all consumers to show responsibility for resources.
3. Past experience: Previous movements have focused on other methods, including recycling, reusing bags or refilling containers, and buying organic foods.
4. Countertrends: SUV driving, rise in disposable, and one-use products.
5. Ease: Outside of a small (and younger) core, more than half of Americans, according to a 2006 Landor study, do not consider themselves "green interested." Other studies show that they do not understand the terms or how to apply these in a meaningful way. Like the health trend, researchers say that there is evidence that even "interested" people talk green, but still act like energy hogs. Large corporations are legitimizing the trend, which makes it easy for consumers to passively comply.
6. Companies: Primarily grassroots companies at first, but now many large companies like PepsiCo, Wal-Mart, and McDonald's are making commitments.

7. Trend maturity: "Sustainable" has long been an agricultural buzzword, and stewardship goes back to the 1960s. The trend makes strides in strong economic times and fades during recessions. These terms began to hit the food product radar in 2006 and have continued.
8. Trend drivers: Concerns about global warming, humanitarianism, and waste of resources. More importantly, being green is becoming a status symbol for celebrities and businesses, which will fuel the trend.

Prediction: The stewardship movement is decades old, and has continued to build interest in recent economically turbulent times, meaning it seems finally to be hitting its stride. However, evidence of profit making has yet to come to light. Rather, companies are finding that it can be a big cost-savings move. Companies that are laying a strong foundation today will benefit in the long term, if only to avoid backlash of not converting.

Action: This is more than a product feature or claim-focused trend. Companies have to make a corporate-wide commitment. Take the trend seriously and determine how best to address it, avoiding sensational claims aimed at quick sales gains.

These examples are intended to illustrate the fact that sorting trend from fad is part science and part art. The method outlined above relies heavily on history to draw parallels, while addressing new dynamics in consumer attitudes, technology, and other environmental changes that must be considered as well. Once again, fads can be as profitable as long-term trends. What matters most in business is the ability to distinguish between the two.

BIBLIOGRAPHY

Microwave cooking faces dim future—decline in sales projections for microwave food products. *USA Today*, March 1995.

Capone, J., There's a word for it. *Media Magazine*, January 2008, p. 31.

Delroy, A., Manier, J., and Callahn, P., For every fad, another cookie: How science and diet crazes confuse consumers, reshape recipes and fail, ultimately, to reform eating habits. *Chicago Tribune*, August 23, 2005.

Dornblaser, L., A whole new wave–microwave foods. *Prepared Foods*, 1989 (annual).

Kavilanz, P.B., Gluten-free knocks low-carb fad off the shelf. *CNN Money*, February 2, 2007, CNNMoney.com.

Shapiro, E., New products clog food stores. *New York Times*, May 29, 1990.

7 Wide World of New Products: A Unique Perspective of New Products around the World That Have Hit Their Mark with the Consumer

Lynn Dornblaser

CONTENTS

| Success | = | Defining and meeting target consumer needs and expectations | × | The right food | × | Proper packaging and preparation | × | Positioned correctly at the shelf and in the media | × | Meet corporate logistics and financial imperatives |

7.1 INTRODUCTION

Creating and launching a successful new product is the Holy Grail all new product developers seek. As we all know, the percentage of new products that actually succeed is relatively small. And of all the thousands of products introduced into the marketplace every year, only a tiny percentage of those products are truly innovative.

Is that dire news? Not really, because consumers continue to demonstrate an interest in new products, variety, and new sensations. However, for a food developer or marketer, finding just the right concept to introduce can be a daunting task. One way to increase the odds of success is to look at emerging and established global trends and understanding how those trends can appeal to your consumers. One of the biggest reasons that it is essential to understand the global marketplace even if you are just focused on the U.S. market is that consumers increasingly are thinking and living globally.

Whereas looking at population life stage data is one very solid way to understand and define a target market, that measure may no longer be as valid as trying instead to understand some of the other issues and factors that influence consumers. Those influencers include the following:

1. Desire for variety and uniqueness, which is often seen on store shelves as more ethnic offerings, spicier foods, and unique flavor blends.
2. Desire to be the manager of one's own health, which translates into some products that offer specific, targeted health benefits and others that address overall well-being rather than specific health issues.
3. Desire for transparency, which often includes products revealing where they were processed, how the workers were treated, how the company behaves in a responsible manner, and the like.
4. Desire for community, which appears in marketing and promotion for products, often including games, contests, blogs, videos, and user-generated content.

Thus, if the broad influences on consumer behavior appear in most markets as they seem to do, then there is a benefit in looking at products and trends from all around the world to see what may fit in your company's portfolio. The smartest marketers work both globally and locally, adapting their global brands to local markets and also adapting local flavors or product forms and styles to a global audience. A clear success in this realm is PepsiCo with its lead in the world snack market. PepsiCo consistently has offered its global brands (e.g., Lays) in local markets with local flavors (e.g., Nori Seaweed in Indonesia), and expanded some of its local flavors to its global audience (e.g., Sweet Thai Chili in the United States).

7.2 NEW PRODUCT ACTIVITY CAN IDENTIFY TRENDS

Looking at new product introduction activity can help any product developer or marketer understand some of the key trends in the marketplace. New product introductions show where companies are spending their money on research and development, marketing, and product development. Therefore, the new product activity becomes a good indicator of what areas are hot and growing, and which ones are not, although it must be said that companies do sometimes make errors in judgment and back the wrong trends. Increases in products with mainstream health and wellness attributes indicate that companies are more focused than before on providing the basics to consumers. Some of the more esoteric or niche health and wellness claims may indicate that consumers are more knowledgeable about cutting-edge trends and may be willing to try them. There is also benefit in looking far beyond local borders to what is going on all around the world.

New product introductions in food and beverage globally have been increasing steadily year after year, as reported in the Global New Products Database (GNPD). With introduction data in 49 countries, the GNPD can offer a unique insight into the similarities and differences in activity and specific products around the world.

Looking at new product introductions for all 49 countries in our data base, shown in Figure 7.1 below, it is clear that the food and beverage industry is alive, vibrant, and growing, despite the impact of the recession. For GNPD, a new product introduction includes all flavors or varieties, and includes products sold in any country. Not included in the counts are products flagged as new and improved or package size changes. New product introductions saw increases in 2008 globally, as seen in Figure 7.1, despite the economic downturn affecting the industry in the last quarter of the year. Some of the most active categories were the ones to show some of the biggest increases, as seen in beverages, bakery, and confectionery.

	2006	2007	2008	Total	% Change, 2007–2008
Beverages	17,040	20,166	22,636	**59,842**	12.2
Bakery	13,120	14,961	16,591	**44,672**	10.9
Confectionery	10,720	12,982	14,938	**38,640**	15.1
Sauces and seasonings	9,669	11,829	12,700	**34,198**	7.4
Dairy	8,942	10,937	12,135	**32,014**	11.0
Snacks	9,203	10,887	11,099	**31,189**	1.9
Processed fish, meat, and egg products	7,468	8,532	8,739	**24,739**	2.4
Meals and meal centers	7,198	7,337	8,270	**22,805**	12.7
Desserts and ice cream	5,293	5,905	5,877	**17,075**	−0.5
Side dishes	3,294	3,953	4,460	**11,707**	12.8
Fruit and vegetables	3,269	3,841	3,658	**10,768**	−4.8
Spreads	2,519	2,927	3,515	**8,961**	20.1
Pet food	2,122	2,208	3,353	**7,683**	51.9
Breakfast cereals	1,882	2,086	2,398	**6,366**	15.0
Soup	1,747	2,073	2,246	**6,066**	8.3
Baby food	1,410	1,873	1,883	**5,166**	0.5
Sweeteners and sugar	417	610	669	**1,696**	9.7
Total	**105,313**	**123,107**	**135,167**	**363,587**	**9.8**

Note: When a new flavor of a beverage is introduced in five countries that counts as five new product introductions.

FIGURE 7.1   Food and beverage product introductions, global, 2006–2008. (Courtesy of Mintel GNPD, Chicago, U.S.)

A few brief comments are relevant about the most active new product categories. Beverages are normally the most active new product category, regardless of country or time period. This strong new product activity results from several factors:

1. Virtually any trend that can be identified (barring savory flavors) is a trend that can appear in the beverage category.
2. The types of products sold in beverages are ones that can be easily introduced by smaller companies (flavored waters, juices and juice drinks, etc.).
3. The beverage category appears to be the category where consumers look for novelty and variety. It is easy for them to try out something new in beverages—the cost is usually not prohibitive, and products usually are single serve. Therefore, there is no investment lost if the product does not perform. And because beverages are often single serve, they are a "personal" choice.

The other categories with the greatest number of product introductions are, for the most part, those in which we typically see a greater number of indulgent or "treat" products. Does that mean that consumers do not care about health and wellness anymore? Not likely. Rather, these are categories in which we see not only indulgent products but also those that may be positioned as better (not necessarily "good") for you. This is reflective of contradictory consumer behavior overall, wanting both treats and good-for-you offerings.

	2006	2007	2008	Total	% Change, 2007–2008
Beverages	2,983	2,981	3,664	**9,628**	22.91
Bakery	2,503	2,840	2,592	**7,935**	−8.73
Confectionery	2,348	2,578	2,531	**7,457**	−1.82
Sauces and Seasonings	2,302	2,201	1,933	**6,436**	−12.18
Snacks	2,109	2,198	2,199	**6,506**	0.05
Meals and meal centers	1,304	1,465	1,370	**4,139**	−6.48
Processed fish, meat, and egg products	1,198	1,233	1,198	**3,629**	−2.84
Dairy	1,082	1,188	1,223	**3,493**	2.95
Desserts and ice cream	1,115	976	807	**2,898**	−17.32
Pet food	671	667	851	**2,189**	27.59
Fruit and vegetables	559	705	672	**1,936**	−4.68
Spreads	651	692	536	**1,879**	−22.54
Side dishes	581	593	475	**1,649**	−19.90
Breakfast cereals	386	430	417	**1,233**	-3.02
Soup	319	352	315	**986**	−10.51
Baby food	134	183	187	**504**	2.19
Sweeteners and sugar	72	77	75	**224**	−2.60
Total	**20,317**	**21,359**	**21,045**	**62,721**	**−1.47**

FIGURE 7.2 Food and beverage product introductions, USA, 2006–2008. (Courtesy of Mintel GNPD, Chicago, U.S.)

It is worth noting the one category with the smallest percentage increases globally—meals and meal centers. This category is one that in many markets has become quite saturated and where shelf space is often most squeezed.

The world comprises different countries, and of course different patterns. A number of significant differences are worth noting between the United States and the rest of the world. Similar types of differences would appear when comparing countries or when comparing companies.

Looking at Figure 7.2, we see the following:

1. The US saw a slight decrease in new product introductions in 2008, driven by the economy and fewer introductions by smaller companies.
2. Beverages show greater increases in the US than the rest of the world in 2008, driven predominantly by increased functional and fortified drinks.
3. Bakery products saw a drop, as the US marketplace became saturated with wholegrain introductions and 100 calorie pack products.
4. Note also the increases in Pet Food, with products with more flavors and functionality, as consumers tend to indulge their four-legged family members.

7.3   WHAT IS HAPPENING WITH NEW PRODUCT TRENDS?

We continually look at new product trends, trying to understand which ones are most important, which ones are emerging, and continually attempt to predict where the future of new products is headed. Four of the most significant new product trends appear below.

7.3.1   SUSTAINABILITY

This trend encompasses a wide range of issues. Sustainability is probably the most important issue to face the food industry in the next 3–5 years. Increasingly, companies are engaging in various aspects of sustainability, including responsible formulations (organic, all natural, free range, pesticide free, and the like), responsible packaging (reduction, recyclability, reusability, and new package materials), and ethical manufacturing and marketing practices, such as fair trade. Each of these subtrends within sustainability should grow globally in the next few years, with many changes and modifications along the way. In the long term, however, watch for this broad area to equal and surpass the other trend areas.

7.3.2   WELL-BEING

Rather than "health and wellness" or "better for you," we are seeing claims and positionings related to "overall well-being" being promoted more in the last year or so. The conversion of Kraft's South Beach Diet line to South Beach Living, and Heinz's new Weight Watchers promotions calling "diet" a "four letter word" make it clear that the "diet" is out and general well-being is in. Many of the subtrends observed are those that focus on either fortification (vitamins and minerals, and sometimes herbal ingredients) or inherent goodness (e.g., the natural vitamin and mineral content of fruits and vegetables). Another key descriptor in this broad trend area is "satiety," with diverse products on the market now making that claim, from Kellogg's Special K$_2$O protein water to snack bars making the claim to help you feel full for longer.

7.3.3 Segmentation

Although companies increasingly look at a variety of behavioral and lifestyle attributes when segmenting the marketplace, we still see significant segmentation of products along the lines of age and gender. Very specifically, we are seeing significant increases in products positioned just to women, and changes in the types of products positioned to children. For women, the newest trend area deals with beauty, with more food and drink appearing on the market in Europe and North America, making claims of enhancing beauty from the inside out. Danone, for example, offers its Essensis yogurt drink in Italy with a claim that it helps "moisturize the skin." And most recently, Nestle USA is introducing Glowelle, also a beauty drink.

For children, the news is mostly in the U.K. marketplace, with a movement toward more "junk-free" foods for kids, meaning ones that are made without artificial colors, flavors, or preservatives, and often also formulated with less sodium and sugar. Whereas that specific language has not moved to the U.S. market, products that promote healthy attributes to moms while providing fun for kids continue to develop.

7.3.4 Flavor Variety

Flavor experimentation continues to develop around the world with an overall increase in more ethnic flavor profiles. Although the ethnicity of new flavors may vary from market to market, in general we see more intense flavors, more sweet and savory flavor blends, and overall more variety. Two emerging savory flavors in the United States include curry and Moroccan. Curry is relatively generic and can be applied to a range of ethnic cuisines. It is characterized sometimes by hotter flavor profiles, many vegetarian offerings, and blends of fruits and vegetables. Moroccan cuisine often seems familiar to U.S. consumers, yet different in subtle ways. It, too, often is spicier, and uses more "warm" spices, such as cinnamon and cardamom. It is more meat-focused than curry can be, but also includes components such as couscous and beans.

Two growing sweet flavors are both fruit-based and have shown significant activity in the West in the last year or more. Açaí is a berry from Brazil with an intense flavor and color. It is mostly blended with other flavors for a more palatable taste. Goji berry (sometimes called wolfberry) comes from Asia, also appears in a variety of flavor blends. Both fruits have appeal not only in terms of their flavor and exoticness, but also with their nutritional profile. Both fruits are high in antioxidants, which today have strong consumer appeal.

7.4 CASE STUDIES—THREE TREND EXAMPLES THAT ILLUSTRATE HOW TRENDS CHANGE AND MOVE OVER TIME AND SPACE

Trends often can be best understood by taking a look at a few specific examples from around the world. What is notable about these three examples is that they each illustrate how a trend changes and modifies itself over time and as it moves from one part of the world to another. Each trend adapts itself to the unique needs of each country or region, which is the best way for a product or a trend to be long lasting.

Example 1: Microwaving positioned as "steaming": Positioning remains consistent while functionality changes

The use of the microwave and of branded consumer products that are specifically intended for the microwave had its genesis in the United States, with the peak years of introductions of products between 1989 and 2001. During that time, virtually every major and minor consumer packaged goods company introduced microwaveable products into the market. In those earlier days, companies experimented with times, power levels, and complexity of directions. Many of those early products did not survive. As the market, ovens, package technology, and product formulation matured in the U.S. market, microwaveable foods became positioned almost solely as convenience foods. Most microwave products had a relatively short cooking time (3–7 min), usually with little or no participation by the consumer. Most of the U.S. product positioning was and continues to be solely on convenience.

Meanwhile, in other markets, use of the microwave oven lagged behind significantly. Ovens tended to be smaller, less powerful, more expensive, and less available. And manufacturers offered relatively few products suitable for microwave cooking. However, as the market continued to develop, especially in the United Kingdom, we began to see the positioning of microwaveable foods take on a new twist, one not seen in the U.S. market. For some chilled meals in the United Kingdom, the positioning was around the "best way to prepare the product" in order to offer the best taste and eating experience. The first company to position the microwave foods in this way in the United Kingdom was retailer Marks & Spencer with its Steam Cuisine line, introduced in early 2001. The products appeared in clear packaging, with a heavy film lid with a valve that enabled steam to escape. Consumers did not puncture the film, as is common in the United States, but rather allowed the valve to do the work. Consumers also were expected to open the package after it had been cooked and then plate the food, instead of eating the food directly out of the tray as U.S. consumers do.

The concept of steaming expanded to new markets, with Heinz introducing Steam Fresh frozen vegetables in Australia and Birds Eye introducing Steam & Serve frozen vegetables in the United States, both in November 2005. In the United States, most of the products with any sort of a steam positioning are frozen, which fits the U.S. marketplace better than do the United Kingdom's chilled meals. Today, ConAgra Foods offers its Café Steamers line (introduced early 2007), which take the concept of the original United Kingdom Steam Cuisine line to a different level. The product comes in a two-piece container, enabling the sauce to collect in the bottom portion of the container during cooking. The consumers pour the sauce over the top of the product once it is cooked. Whereas this is different from the U.K. expectation of separately plating the product, this level of consumer participation in a frozen meal is relatively unique in the U.S. marketplace.

The concept of steaming, then, has remained consistent in its positioning—"the best way to prepare the product"—but evolved in form and consumer participation as it moved from one region to another.

Example 2: Probiotic dairy drinks: Positioning changes as the product moves from one region to another

The concept of probiotic "shot" bottles (small bottles about 100 mL, sold in multipacks) traces its roots back to Japan in the 1950s with the first commercialized product, Yakult. At the time, Yakult was sold in 100 mL bottles, in a multipack of seven bottles. Consumers were urged to drink one per morning for the best health benefit.

Probiotics are beneficial bacteria that aid gut health. They are found in a number of products, mostly dairy, quite often yogurt. In the U.S. market until recently, probiotics were referred to as "active cultures," whereas in other countries, a more straightforward description has always been used.

Starting in the late 1990s, probiotics were beginning to be popular in the European market, with expanding flavors, positionings, and entry into the market by a variety of companies. As the market for probiotics matured in Europe, the concept of the 100 mL bottles expanded to include other types of functionality. One example is drinks with plant sterols or stanols to help reduce cholestrol. Another example is small bottles that helped consumers "top up" their consumption of fruits and vegetables. The newest extension of the "little bottle" format in Europe comes from Unilever with its Hunger Shot Mini Health Drink, a satiety drink positioned as a healthy alternative to a snack.

In the U.S. market, Dannon had experimented with the concept for several years, with the 2000 introduction of Actimel in Whole Foods Market stores; it was later rebranded DanActive in 2004. It was not until Dannon's introduction of Activia spoonable yogurt in early 2006 that consumers in the United States became more familiar with the benefits of probiotics. The Activia advertising strongly promotes the gut health benefits of probiotics. After the launch of Activia, the company expanded its distribution of DanActive into mainstream stores in late 2006. DanActive differentiates itself with an immunity claim, rather than with a gut health claim.

Interestingly, new advertising for Activia using actress Jamie Lee Curtis, focuses more strongly on irregularity, which is a bit of a breakthrough in the U.S. market on discussion of the benefits of probiotics. In Europe, probiotics are much more clearly positioned as aiding "intestinal transit," thus illustrating how the same product with the same benefits can advantageously modify its approach according to region.

The U.S. market is still very much in its infancy with probiotics and with the concept of 100 mL "shot" bottles. However, it appears that the market and the consumers are learning from Europe, as we now see Unilever introduce its Promise Activ Super Shots, 100 mL drinks formulated to help reduce cholesterol (the product concept has been sold in Europe under various brand names since about 2004).

Example 3: Gluten-free formulations: Better established outside the United States, but growing

Gluten-free as a claim is one that is far more prevalent in Europe, especially in the U.K. market, than it is elsewhere. New product introductions have appeared in Europe (mainly the United Kingdom) since 2001, increasing steadily year-on-year. This may be due in large part to the fact that consumers and health professionals in the United Kingdom are more aware of celiac disease.

Beginning in 2006, we saw the number of introductions globally of products making any sort of a gluten-free claim grow significantly, with the expansion not only into new parts of the world (Australia, New Zealand, Mexico, and Brazil), but also into new categories, including baby food, sauces and seasonings, and desserts. Before that point, gluten-free products had been clustered mainly in bakery, snacks, and side dishes, as expected.

One of the marks of a growing and often established trend is its appearance in less-than-obvious categories. In this case, the appearance of a gluten-free claim in categories such as confectionery and processed meat and poultry seems to indicate that the trend is well established.

The types of products we see on the market globally are quite similar; products in grain-based categories making gluten-free claims and selling at a price premium, combined with products in inherently gluten-free categories being flagged with the claim. The companies that typically engage in gluten-free products in the United States are smaller, niche companies, whereas in Europe, we see more major or mainstream companies with gluten-free lines (especially in the United Kingdom with activity from retailers in their private label lines).

Gluten-free's substantial growth is due only in part to the fact that more consumers are being diagnosed with celiac disease. The Mayo Clinic reports that about 1% of the U.S. population has celiac disease (3 million people), while other sources indicate that this number is quite low due to lack of diagnosis.

In addition to the diagnosis of celiac disease or gluten intolerance, increasingly some consumers have been switching to a gluten-free or reduced-gluten lifestyle believing that it is better for them. They cite reduction in intestinal upset and increased feelings of alertness and energy.

The number of new product introductions has dramatically risen globally. Introductions in Europe tripled between 2005 and 2007, while the U.S. introductions more than doubled. This is likely due to consumers making a lifestyle choice for gluten free, and also reflects caution on the part of manufacturers to ensure that consumers with allergies are able to make sound choices.

7.5 HOW TO IDENTIFY TRENDS AND HOW THEY MOVE AROUND THE WORLD

Whether a new product appears on the market at the beginning of a trend, in the middle, or toward the end will significantly impact how successful that product may be. Similarly, introducing products that tie in with long-lasting trends usually perform better than those that are positioned against a short-lived fad.

How can one know the difference between a trend and a fad? Which market movements signal trends, and which signal fads? Through several decades of thinking about, tracking, and predicting trends, we can offer at least eight points for consideration.

It is a fad when:

1. Restricted geography: The types of products that fulfill the fad appear in a very limited number of countries.
2. Few product categories: The types of products that fulfill the fad appear in a very limited number of categories (unless the fad/trend is only applicable to specific categories, e.g., gluten free).
3. Rate of introduction overly high: Product introductions skyrocket in a very short period of time, far in excess of normal new product introduction growth.
4. Sensationalist coverage: The consumer press tends to take a sensationalist view of the fad.

It is a trend when:

5. Wide geography: The products that fulfill the trend appear in a number of countries, often at or near the same time (although depending on the trends, they sometimes take quite a bit longer to move from Asia to the West, for example).
6. Many product categories: The products that fulfill the trend appear across a wide range of categories (unless the trend is applicable to only a few categories).
7. Slow but consistent growth: Product introduction growth is slower than a fad, although usually at a greater rate than overall new product introductions.
8. Coverage by respected professionals and the trade: Nutritionists, marketers, retailers, and the press take a more balanced view of the trend.

To put these guidelines to the test, consider these two trends/fads of recent years: low carb and wholegrain. On the face of it, these two areas may seem very similar. Both trends have their genesis in the U.S. marketplace. Low carb products were held to help weight loss, to help regulate insulin levels in the blood, and, in general, to promote good health. The same statements are also made related to wholegrain products. However, looking solely at the new product introduction figures, it is clear that whereas low carb was a fad, it could very well be that wholegrain is a trend that is here to stay.

Looking at overall introductions, low carb showed a steep rise and a steep fall. Figure 7.3 clearly shows the differences between low carb and wholegrain when looking at the spread across regions of the world.

The two charts (Figures 7.4 and 7.5) show that the low carb fad appeared almost exclusively in North America, specifically the United States. Europe caught the trend briefly in 2004 (almost exclusively in the United Kingdom), but it did not catch on. Wholegrain, on the other hand, has grown over the last 5 years in four of five regions

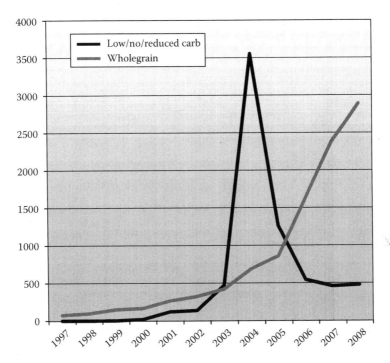

FIGURE 7.3 Low/no/reduced carb and wholegrain introductions, global, 1997–2008. (Courtesy of Mintel GNPD, Chicago, U.S.)

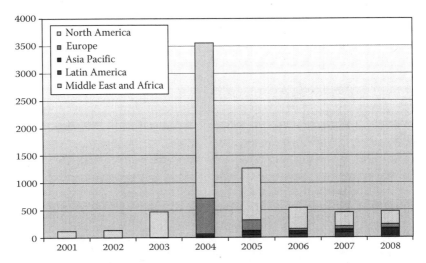

FIGURE 7.4 Low/no/reduced carb introductions, global by region, 2001–2008. (Courtesy of Mintel GNPD, Chicago, U.S.)

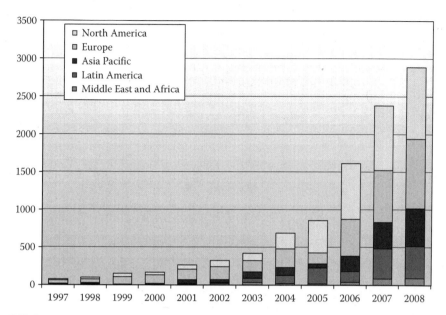

FIGURE 7.5 Wholegrain introductions, global by region, 2000–2008. (Courtesy of Mintel GNPD, Chicago, U.S.)

tracked by GNPD, with European introductions reaching similar levels as North America. In the case of wholegrain, Europe did not blindly follow the U.S. market as it seemed to do with low carb, but rather had a bit of a separate boost due to research showing that wholegrains also have a prebiotic effect. The dissemination of this research enhanced growth of the trend in Europe mainly in 2007. In addition, there are differences in how the trend is viewed in both markets: in the United States, wholegrain mainly means intrinsic goodness and heart health, whereas in Europe, the focus is also on the digestive health benefits.

7.6   HOW TRENDS CHANGE

Whereas the foregoing discussion of low carb versus wholegrain shows the differences between a fad and a trend, it also can be useful to look at how a trend changes and modifies as it moves from one part of the world to another. Many examples exist. One that shows this modification or better evolution in the trend from one part of the world to another very clearly is that of so-called beauty foods and drinks.

The concept of beauty foods and drinks (products that enhance beauty from the inside out) had its start in Asia. In Asia, the more common ingredients in these beauty food products are ceramide, collagen, and aloe vera. All three are ingredients that also are applied topically for beauty benefits. In Asia, they are positioned as offering the same or similar benefits when ingested. As the trend moved to Europe (and now to North America), the claims remain quite similar, but the ingredients changed. In Europe and North America, the claims still center on improving skin beauty from the inside out, but the claim statements also focus on moisturization and overall skin health. The ingredients also change. In Europe and North America, the ingredients include and focus on antioxidants and more familiar food ingredients rather than on ingredients found only in topical beauty products.

7.7 TO SUM UP

Clearly, then, understanding trends and predicting success are far more of an art than a science. For a food and beverage marketer to successfully introduce a product to market requires not only an in-depth knowledge of the consumer, the product category, and the market in general, but also understanding of the underlying consumer and market trends. And beyond that, quite often what is also required is a willingness to take a risk. Understanding as many of the less quantifiable market and consumer characteristics may make the difference in making a product a resounding success.

Part III

The Right Food

Success	=	Defining and meeting target consumer needs and expectations	×	The right food	×	Proper packaging and preparation	×	Positioned correctly at the shelf and in the media	×	Meet corporate logistics and financial imperatives

8 Consumer Packaged Goods Product Development Processes in the 21st Century: Product Lifecycle Management Emerges as a Key Innovation Driver

Chip Perry and Max Cochet

CONTENTS

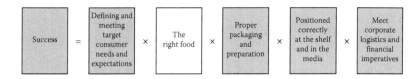

8.1   INTRODUCTION: THE CHALLENGES FACING CONSUMER PACKAGED GOODS COMPANIES

The increasingly competitive environment of the Consumer Packaged Goods (CPG) industry means more emphasis than ever will be placed on developing and sustaining new products. These must match both product functionality and consumer needs and desires in the most innovative way possible. Over the last decade, an accelerating number of challenges have influenced the way CPG companies approach new product development, pushing them to focus on innovation to maintain, or to gain competitive advantage.

At least four trends drive change and create new demands on product development:

Consumer requirements are becoming more elaborate. Consumers in today's developed countries are more educated than ever before about product quality and functionality. They require that the market offer them a greater choice from which to select and purchase. Today's demanding consumers no longer just want low-priced products, but also want them to be of higher quality and more enjoyable to consume or use. In today's innovation process, it is critical to identify key consumer insights at the front end of the product design cycle. It also requires reusing past insights and market research from existing, failed, or successful products. In a world where consumption is a primary activity, consumers, want to be able to choose the products that consistently meet their needs and desires.

Increased safety and traceability requirements. Government regulations are becoming increasingly stringent. An excellent example is the new Registration, Evaluation, and Authorization of Chemicals (REACH) regulation for managing chemicals in Europe. Both government agencies and public opinion now demand total traceability and responsibility in case a crisis arises with consumer products (e.g., lead paint, melamine in fish, *Escherichia coli*, etc.). Saying, "Sorry, our systems couldn't trace the issue" is an invitation to a lawsuit. Being able to not only contain, but also to avoid quality breaches, is going to be more important than ever. This is especially true because supply chains are becoming global. The media report will relish the latest product recall 24 × 7; all while many formerly loyal consumers punish brands that break their brand promises.

Fierce competition from private label brands and niche vendors. Private Label Brands (PLBs) or National Brand Equivalent (NBE) threaten consumer brands. Because of their flexibility and innovation capabilities, private label companies can innovate and adapt faster, with greater ease and flexibility, in response to changing consumer demographics and new trends. The result is simple—they can develop new products that are superior in meeting consumers' needs. These PLBs have abandoned the outworn strategy of competing just on price. Many of these retailer brands achieve hefty margins, particularly in the Food and Beverage sector, and have become more that just "equivalents" to national brands. Now they compete effectively against the higher-end national brands by delivering superior quality, while charging higher prices. In response, major manufacturers are starting to change their marketing strategies. For example, PepsiCo Inc. recently came up with a creative way to launch its new energy drink Fuelosophy in order to shake up traditional retailers and respond to niche vendors' ways of doing business. It initially launched the product through Whole Foods Market's retail channel instead of using a more traditional mass-market campaign and distributing the product through conventional retail channels.

Do more with less. In addition to the increasing competition of private labels and niche vendors, the recent rise of energy and commodity prices has shrunk top and bottom line growth potential in most segments of the Consumer Goods industry. This has forced consumer product companies to make budget cuts. In addition, their employees are required to increase their rate of innovation. For example, a leading CPG firm has an annual objective to increase innovation by 5% while holding their employee headcount flat. Due to the typical lack of organization of existing Research & Development (R&D) knowledge assets, which prevents any reuse of information, companies are constantly forced to reinvent the wheel. Most employees are limited to seeking product development information based on whom they know instead of searching their company's knowledge databases online. As a result, companies hoping to increase throughput and worker productivity face a daunting prospect.

8.2   THREE KEYS TO INNOVATION SUCCESS

All of these are common issues for CPG companies. Tomorrow's winning companies will be those that understand and accept these challenges, and find ways to address them through processes and solutions focused on new product innovation. New product success requires excellence in three areas: (1) reducing product development cycle time, (2) increasing product development innovation, and (3) reusing company knowledge assets. In the twenty-first century, the ability to innovate better and faster than the competition trumps all other areas of competitive advantage.

Cycle time or time-to-market determines how fast a company begins to recoup its investment in a new product. It also ensures competitive advantage by beating a competitor while seizing both market dominance and share. First-mover advantage often locks in significant profit margins for the life of the product.

Increased product development innovation determines the extent to which a product gains and holds its consumer base. It ensures that the company has the ability to build out the product franchise through successive product innovations, enhancements, and line extensions.

The ability to reuse knowledge assets drives product development costs down by reducing the costs of having to re-create knowledge that already exists. Reuse builds on existing knowledge, thus avoiding costly rework and repetitive mistakes. It is astonishing that how many companies continue to recreate insights and knowledge they have previously generated. One food company commissioned exactly the same consumer research project three times over 6 years as successive new brand managers took control of the brand. Each time the research resulted in the same insights!

To achieve success in these three areas, companies must look to the factors that drive innovation: people, knowledge, and systems. The types of employees working in innovation vary widely from visionary executive leaders, insightful problem solvers, and technical experts to product developers who ultimately design the product or formulate the detailed recipe. The knowledge that drives innovation comes from many sources—existing expertise in the company, individual experimentation, and discovery of what works and what does not. Additionally, there are ideas gained from external sources or knowledge that is documented and built upon, as a product is first developed and then managed during its life in the marketplace. Systems enable employees to efficiently leverage the company's expertise and knowledge, as well as effectively generate big ideas and profitable products.

In the twenty-first century, the innovation *status quo* is being disrupted as the marketplace requires ever-greater new product success rates and bigger, sustainable products. Yet, companies consistently delay or miss new product launch dates, over-promise business results, waste vast amounts of time and company resources, and undercapitalize on leveraging their existing product assets and knowledge. As leading companies come to terms with this, they are starting to make exponential changes to materially improve the business of innovation. As a result the practice of Product Lifecycle Management (PLM) is emerging as a key driver of innovation.

8.3   MOVING FROM PRODUCT CONCEPTION TO MANAGING PRODUCT LIFECYCLE

In order to understand how PLM has emerged as a key innovation driver, it is helpful to explore the history of product formulating and design to provide perspective. During the last 30 years, CPG corporations made huge productivity gains by utilizing powerful new business process software applications, starting in manufacturing. The leading example is enterprise resource planning (ERP) software (e.g., SAP, Oracle), which was developed to rationalize and better organize the core corporate functions of manufacturing, finance, and purchasing. Over time, HR, sales, and even marketing have benefited from the software that streamlined core functions, eliminated so-called siloed or individually isolated work processes, and built powerful databases that distribute standardized, accurate, and timely data and information to employees working in, and with, these related functions.

The last area of focus has typically been R&D—an area where the improvement potential of systems and processes supporting the product design or formulation is vast.

For over 30 years, designers of "discrete" products, i.e., products built from blueprints, have used powerful computer-aided design (CAD) software. The design and building of cars, planes, and electronics would be unimaginable today without the capabilities of CAD to rapidly design and share product specs across a wide group of collaborators located around the world.

In the CPG industry, product formulators have developed ideas through scientific experimentation. Many still use laboratory notebooks to record and document the experimental processes that led to a final formula, manually recording the endless trials to iterate a formula until the best combination of ingredients results in a winning product. Today, in addition to paper laboratory notebooks, companies still use spreadsheets as well as homegrown legacy systems for product formulation processes. Often these systems contain calculations or algorithms that have never been validated for accuracy. Yet, they continue to generate critical product data that the corporation believes to be accurate.

Importantly, data remaining in a paper form "dry dock" greatly hobbles the type of fast-paced, iterative, and collaborative bench-work necessary to remain competitive through innovation. The hard work of creating hundreds or even thousands of iterations of a product formula, then refining it until it is optimized, demands persistence, fortitude, and the adoption of processes that only today's software systems can make possible.

In the last 10 years, formula management systems, the process manufacturing equivalent to CAD systems, have emerged and matured. They offer sophisticated "management" capabilities to the product formulator as well as to others who interact with a formula, as it progresses from initial ideation to manufacturing. These systems, known as "authoring tools," are designed specifically to aid and enhance the design phase of product development. These software applications capture data as formula progresses from idea to final form, recording data that reflects the complete detail involved in each phase of work, as well as organizing the information so that all relevant data are linked to the final product.

The best authoring tools can eliminate recording in notebooks, enabling formulators to design, analyze, and iterate seamlessly until the right combination comes together to create or enhance a product that delights the intended consumer. Additionally, authoring tool applications open up the development process by encouraging cross-functional collaboration with specialists, which include, but are not limited to, raw material sourcing, regulatory, analytical testing, marketing, and many others.

Figure 8.1 shows a development idea progressing from concept to experimental to regulated states of product. Formulation designs mature as review and approval stages are archived. As the concept is refined and tested with consumers and customers during the experimental phase, samples, and prototypes are developed. These prototypes contribute to the increasingly complex data streams needed to track all changes made based on test results. Similarly, during the regulated phase, all formal tests are conducted and results studied and recorded to support regulatory approval of the final formula, which is then promoted to a production-ready state and handed off to manufacturing.

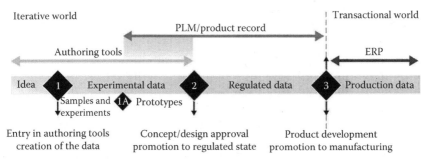

FIGURE 8.1 Formulated products and product lifecycle states.

Today, R&D organizational work processes focus on generating the final specification that is handed off downstream to manufacturing. This work generates the equivalent of 5% of the total information created. More importantly, 95% of the effort and knowledge created along the way, critically vital information, ends up being lost unless processes, systems, and databases are in place to capture these critical organizational assets.

Formula-based organizations serious about increasing innovation often begin by improving the primitive tools used by their formulators. The evolution beyond paper-based records and spreadsheets located on disparate, scattered laptops begins by improving the management of corporate knowledge and intellectual property assets. Effective authoring tools match the required underlying chemistry and formulation science expertise to the appropriate software vendors offering third-party formulation management tools. A number of authoring tools that possess industry-specific procedures have matured: they bring together industry standards and best practices, e.g., for the personal care and cosmetics, paints and coatings, and food and beverage industries, respectively.

Formulation software-specific functionality greatly increases productivity. For example, "where used" search capabilities scan across an enterprise to show the formulator who has worked with the ingredient of interest and what has been learned about it. The ability to compare a number of different formulas displayed in columns across the screen inspires the formulator to experiment, to learn, and to easily retrace steps, and determine exactly where the optimization occurs in a formula. Ultimately, sophisticated and powerful authoring tools deliver the right information to the right person at just the right time, regardless of time zone or location, enabling collaboration with peers, wherever they may work around the globe.

As an organization begins to advance its formulation and product design processes, the impact spreads beyond R&D. It becomes increasingly able to create connections and interactions across a broader array of business functions, such as packaging design, marketing, and manufacturing, in order to bring a new product to market. As a result, organizations begin to ask the questions that will lead, ultimately, to the PLM transformation such as

- How good are our processes?
- What are our strengths and where are processes underdeveloped?
- How does our company compare against the best in our industry?
- Which companies excel in organizing their product development resources and what benefits do they get as a result?

Asking these questions and benchmarking against other companies and industries have helped them to learn about and begin to value the discipline of PLM as a way to significantly improve their entire product development area.

8.4 EVOLUTION OF PRODUCT LIFECYCLE MANAGEMENT

Over the past decade, manufacturing companies in industries such as High Tech, Automotive, and Aerospace have aggressively adopted technology solutions to enhance the productivity and efficiency of their innovation and product development activities. They did so in order to respond to many innovation challenges caused by their competitive environments. Companies had to constantly reduce cycle times and time-to-market, increase throughput while improving product quality and safety, while ensuring that they were offering products that a customer would buy.

These technological solutions evolved over time from initial stand-alone product development authoring tools—such as CAD, file management, workflow, and collaboration organized around product data management (PDM). They developed into information technology platforms called PLM solutions. These solutions integrate complex networks of both point-solutions and collaboration work processes, supporting product development activities.

PLM solutions enable the management of product data from the early (and fuzzy) steps of ideation to the end-of-life of products on the market. The CPG PLM Platform comprises as many capabilities, applications, and work processes as are found in today's product development processes, functions, and work areas. A diagram of these different tools for the CPG industries appears in Figure 8.2.

Typically, the core areas represented in a general PLM model for CPG contain PPM (Project and Portfolio Management); Packaging, Artwork, and Labeling; Authoring tools (CAD/CAM tools for blueprint-designed products as used to design products like razors, blades and packaging); and Formula Management tools (tools for recipe/formula-developed products found in process industries). Other work capabilities such as regulatory approval, testing, claims, intellectual property, and patent management are included in more detailed depictions of PLM.

PLM data are much more complex and challenging to manage than the transactional data found in ERP systems. The assortment of activities is both structured (database structured in rows and columns, essentially the 1s and 0s of numerical depiction of data) and unstructured, that is, data and information created in tools such as Microsoft Office Word, Excel, and PowerPoint and other applications. This

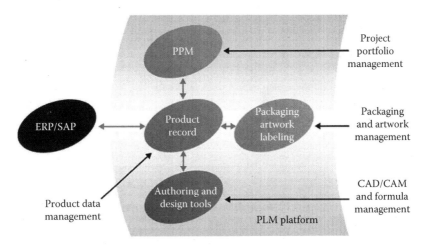

FIGURE 8.2 PLM defined for the CPG industry.

mix of data types always exists, so the PLM systems must be able to manage both types. In addition, the data generated within the PLM platform are also highly iterative, as later work builds on earlier versions until the final formula, package, or project plan is complete.

The key turning point in the evolution of PLM, and really the birth of PLM as a discipline, occurred when software vendors started building their solutions around a single concept. The concept was master data management for all data that relate to product development. Since all product development activities have at least one thing in common, namely they relate to a specific product, the data that they produce have been centrally organized by product in order to maximize access, reuse, and transfer of this information.

The heart of successful PLM is PDM, enabled by a data model called the Product Data Record (PDR). Whereas PDM does not have the scope and the capabilities of PLM, it contains the PDR, the building block which "pumps" essential information from and to the different applications composing the PLM platform. PLM is not possible without first creating a single version of the "truth" for all product data through the PDR.

8.5 AT A CROSSROADS: PLM AND THE CONSUMER GOODS INDUSTRY FINALLY READY FOR EACH OTHER

CPG companies lag discrete industries such as aerospace and automotive in terms of PLM adoption by about 10 years. But CPG companies have taken note of the vast improvements PLM has brought into discrete industries. CPG companies are beginning to understand how they can also benefit from such solutions to help them drive product innovation. PLM implementations can greatly enhance an enterprise's bottom line by delivering a potential 5%–10% revenue uplift* as well as considerable savings and productivity improvements in product development activities.

Typically PLM benefits are:

1. Condense product time-to-market and time-to-profit by at least 30%: CIMdata cites an example where the development time for a household product was reduced by 75%—from 18 to 4 months.[†] This equates to more revenues due to improved market penetration (e.g., in food and beverage, a 3 month difference in product launch can equate up to 30% in market share) but also frees up R&D capacity to work on other new product development projects and improve innovation throughput.
2. Reduce product development costs by 10%–40%: The building of prototypes can be reduced by 15%–30%. Pratt & Whitney Canada reports saving $500,000 per engine by eliminating maintenance prototypes due to PLM.[‡] Boeing affirms that it was easier to assemble the first 777 plane than it was

---

* CIMdata Report "Product Lifecycle Management: Empowering the Future of Business," 2002.
[†] CIMdata Report "Product Lifecycle Management: Empowering the Future of Business," 2002.
[‡] "V5 PLM in Aerospace: Enabling Innovative Products for a Better Future" by Alain Houard, Dassault Systèmes.

to assemble the 747 after 20 years of production, traceable in part to a 90% decrease in engineering changes.* The aerospace giant's new plane, the 787, will be designed completely virtually, using a PLM platform linking more than 300 suppliers.

3. Improve innovation workers' productivity by 20%–30% and augments collaboration: For instance, PLM systems can reduce the engineering review process by more than 80% or from 12 to 2 days.† Because of improved collaboration, projects spend less time in idle. Virtually no time is wasted doing the thankless process of reviewing and modifying out-of-date data.

4. Guarantee compliance and allow traceability back and forward from any lifecycle state of a product, ingredient, process, or idea from concept to commercially available product: Tyco Healthcare improved its new products compliance with the Food & Drug Administration's 21 CFR Part 11 regulations by 30%.‡ A recent Aberdeen report§ points out that companies that put in place PLM technologies can identify and meet compliance requirements early in the product design process and achieve significant results, such as a 27% product-recall reduction and a 31% improvement in the number of products in compliance.

5. Facilitate the reuse of existing knowledge assets (Knowledge Management): Managing all product data generated during development is a tremendous way to sustain innovation goals, while reducing development costs and risks. Furthermore, knowledge management helps a company stop "reinventing the wheel."

6. Rationalize IT infrastructure: Dell realized a 30% reduction of their global PDM infrastructure.¶ Industry research typically cites 15% reduction in IT support.

The six benefits identified above have been validated by multiple PLM implementations in the automotive, aerospace, and high-tech industries. In comparison, there are still very few large CPG companies that have "implemented" PLM to the same scale. Part of the reason is that the CPG industry is still a technologically reactive industry, as shown by its late adoption of ERP and other technology enabling solutions. But PLM solutions have improved so much, particularly in terms of virtualization (enabling modeling, testing, visualization, etc.) and data integration (making cross-functional data exchange and reuse, and traceability a reality), that significant possibilities are now achievable for CPG companies. Importantly, best-in-class PLM solutions have matured in their architecture, and now have a unique data model at their core, centrally defining and managing product data.

---

* "V5 PLM in Aerospace: Enabling Innovative Products for a Better Future" by Alain Houard, Dassault Systèmes.
† CIMdata Report "Product Lifecycle Management: Empowering the Future of Business," 2002.
‡ Oracle Agile Web site (http://www.agile.com/customers/index.asp).
§ Aberdeen Report "The Product Compliance Benchmark Report: Protecting the Environment, Protecting Profits," September 2006.
¶ Oracle Agile Web site (http://www.agile.com/customers/index.asp).

In the meantime, as discrete industries reach maturity in PLM, growth for software solutions sales is maturing and PLM vendors are looking for new industries to penetrate. They are just starting to understand the unique challenges to the CPG industry and are developing CPG-specific solutions as a result.

8.6 INNOVATION CHALLENGES MAINLY ARISE FROM THE DIFFICULTY IN ACCESSING AND REUSING EXISTING, PRODUCT-RELATED KNOWLEDGE

CPG companies characteristically lack internal product data organization. Whereas companies invest heavily in branding to ensure that the consumers can easily find their products in a crowded market, their internal product data are decentralized and scattered, making it impossible to find and reuse. This situation is exacerbated when data "creators" are transferred across brand assignments, functions, and geographic locations. As such, today's CPG search methods are ineffective, with employees seeking critical product data by looking for those who did the work, or possibly by a search of intranets and local file systems. When it becomes impossible to find critical knowledge assets quickly, employees default to recreating the data they seek, or alternatively, plunge forward based only on intuition. In fact, a recent study reveals that up to 77% of employees search for data in multiple places, then manually combine the results.*

The lack of information technology systems and formal processes to manage product information creates gaps between existing knowledge in the company and the amount of information needed to be created. Product development activities require specific expertise, resulting in the tendency for information to reside in silos. In the least technologically advanced CPG companies, this tendency toward silos means information is stored on people's hard drives, in handwritten notebooks, or even in their heads. For more advanced CPG companies, the multiplication of best-in-class IT systems has created new silos of information, and data reside in point-solutions. In such cases, integration between systems is only partial, even when it exists the infrastructure is inadequate for managing product development information.

Clearly, in all cases, employees cannot find information when they need it—either because they do not know it exists, they cannot find it, or they do not have access to it. With industry research indicating that engineers spend as much as 30% of their time searching for data,† the consequences of lost time and opportunities to reuse valuable information assets are significant. If they can address this problem, then CPG companies will be able to increase reuse of existing knowledge, increase traceability of information, reduce inaccuracies, and eliminate the manual rekeying of data. All of these benefits will lead to the reduced cycle-time, increased throughput, and information accuracy, and, therefore, help companies to better respond to the challenges they face.

---

*Kalypso Report "Advanced Information Access: The Application of Search Technology in Product Development," May 2008.
† Tenopir/King: Communication Patterns of Engineers, 2004.

8.7   PRODUCT DATA RECORD: THE HEART OF PLM

PLM solutions help tackle the challenges of managing product information, but only if they are guided by a vision and strategy leading to the adoption of a true technology platform. For CPG, developing a PLM strategy is the prerequisite for achieving substantial results comparable to those in other industries. The first step in defining a PLM strategy is to understand, define, and map the data needed and produced across the development lifecycle of a product. Some activities are done similarly across the CPG industry—e.g., testing or regulatory and safety analysis—but there is no one-size-fits-all solution. Companies have to map their processes and data individually. This begins with creating the PDR. The different constituencies who must be involved in the creation of this information map are shown in Figure 8.3.

A CPG company's most valuable innovation asset is its product development data. These data consist of not only information defining the product, but also all of the knowledge that was created during the development of the product, from ideation, to launch, through end-of-life of the product—even experimental iterations that either led to a final product or remained at an idea, concept, or semideveloped state.

The PDR is the single version of truth for a company's product data and associated product-related data. The PDR defines all the data elements and their relationships needed to fully describe a product lifecycle. As the single version of truth, the PDR is the heart of PLM, and successful product management is not possible without it.

In order to maximize the reuse of product data and take advantage of its value as a source of innovation, companies should manage product information as one data model. The PDR serves as a powerful information management tool that helps companies regain control of, and manage their product data. With the amount of information increasing as a project progresses through product development phases, it is crucial to manage the information lifecycle in order to keep track of all valuable product knowledge created (Figure 8.4). Additionally, there is a great value in keeping track of data from earlier stages, in order to go back and reuse these data for future development initiatives. Both successful and failed experiment results have to be captured and placed in searchable databases. Unused discoveries and conclusions for

Business Function	Business Function
Analytical	Microbiology
Artwork	Modeling/simulation
Brand	Nutrition
Claims	Packaging design
Consumer relations	Packaging development
Engineering	Process development
Formulation	Procurement/purchasing
Health, safety, and environment/sustainability	Quality assurance
Initiative and portfolio management	Raw materials management/specifications
Intellectual property/patent	Records management
Market research/consumer insights	Regulatory affairs/product stewardship

FIGURE 8.3   The different constituencies who will use the PDR.

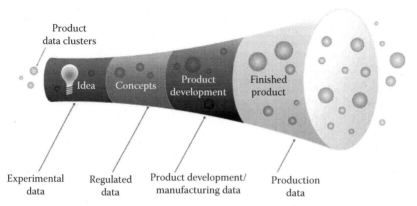

FIGURE 8.4   The nature of data that emerge during the development of a product.

one project can be extremely valuable jumping-off points for a new product concept development project, saving tremendous amounts of time and money.

The challenge for companies is to make sure that the single version of the truth—the most up-to-date, accurate information—is available at all times within R&D functions and beyond. The key to effectively organizing the product data is to organize it by product—a simple organizing principle that was overlooked for too long. Many companies, including leading innovators, have not found a successful way to organize product development information assets because they never considered this data management problem holistically. The PDR requires an enterprise-wide view of all innovation processes, which is a paradigm shift from organization by business units or functions to a true product-centric hierarchy.

The PDR is built on a conceptual data model, a map of information elements (*data entities* or *entity classes*) and their interrelationships (Figure 8.5). Specifically, the PDR describes data elements that are inputs or outputs of a work process, the characteristics that define them (*attributes*), and their relative interactions (*relationships*).

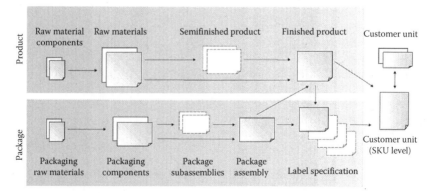

FIGURE 8.5   Product Data Record.

The Master Data model (or Level 2) shows all structured (information captured in a database) and unstructured (documents) data elements related to each Level 1 physical element.

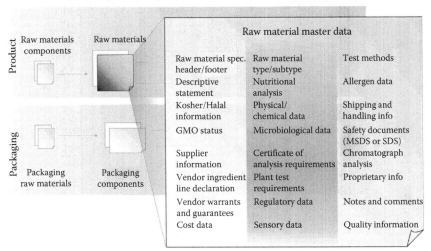

FIGURE 8.6   PDR Level 2: Master data elements.

It also documents the lexicon of an organization and its activities. This data model forms the basis of the central product database at the core of an implemented PLM system.

A PDR describes the high-level structural components of a finished product also called "Product Architecture" (e.g., raw materials, packaging subassemblies, formulation, and final packaging assembly) as we see in Figure 8.5. The PDR breaks down each of these physical Level 1 items into data entities. Figure 8.6 depicts an example of an exploded Level 1 and describes the data standards for defining raw materials information. The PDR characterizes the master data relating to each of the physical constituents of a product. At the lowest level of detail, the Data Field model, Level 3, references data fields associated with data elements and defines their standards as we see in Figure 8.7. Level 3 is where the PDR is translated into PLM software.

The first step in implementing a PLM system consists of defining the PDR, which is the foundation for a PLM strategy. The PDR forces companies to go back to the basics in terms of data management by defining (or redefining) data ownership or control of IT projects, how they fit within the overall strategy and how they answer true business requirements. This step offers an opportunity for IT and business managers to work together and understand each other's requirements, constraints, and strategies in terms of product development.

A list of what the PDR is and is not appears in Figure 8.8.

Once built, the PDR should be the documented "single version of the truth" for product data, acting as a reference for all current and future iterations of the product across the organization, work processes, and technology solutions. A best practice is to implement a PDR "sanity check" prior to going forward with any product data-related projects. Because the PDR and the PLM strategy must be synchronized, this

The data field model (or Level 3) references data fields associated with data elements and defines their standards. It helps the translation of the product data model into an application landscape

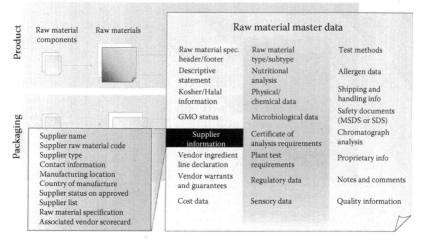

FIGURE 8.7   PDR Level 3: Data fields.

The PDR is...	The PDR is not...
The key to identifying information assets	A one-size-fits-all solution
The way to identify and understand the work that gets done between functions	A giant database
A framework for knowledge reuse	A system or application
A methodology to standardize types of information	A replacement for existing programs
The blueprint for PLM processes and systems	A process reengineering effort

FIGURE 8.8   What the PDR is and what the PDR is not.

verification keeps everyone in the company moving in the right direction, restating every initiative's tactical intentions as part of the higher-level objective.

PLM evolves into a framework that combines technology solutions with organizational and work process concepts and designs. The combination of diverse information integrates capabilities and further enables product development. This concept is sometimes hard to grasp for various functions dispersed across the company. Yet the PDR is a powerful tool that allows one to concretely document the PLM "blueprint" and map its scope, capabilities, and objectives. And very much like a building's plans, the PDR has different levels of details. Whereas the architects use a blueprint to build the house in detail, the owner just wants to make sure that the kitchen and the dining room are next to each other. Similarly, the PDR should be used as training materials for IT and business resources, because it draws a complete picture of how intellectual property assets are created and used in a company.

8.8   KEY FUNCTIONAL BENEFITS OF THE PRODUCT DATA RECORD

The information that the PDR provides is schematized in Figure 8.9.

Whereas there are many benefits to the PDR, here are a few from different business activities in the product development process:

Product design: The data contained in a formula or engineered product blueprint are the essential DNA sequence that describes what a product is and how to make it. Although preserving the final design is essential, it is of tremendous value to the organization to assess and learn from the iterative design work generated during the entire design process. This historical information often provides the raw materials and the toolbox for designing the next new product.

Packaging: The information on a consumer product's packaging comes from multiple locations. Best-in-class CPG companies have integrated their packaging activities downstream, from the package design to manufacturing. Most companies, however, still lack the upstream information from product development to package design. Although ingredient information comes directly from the finished product's formula, some information comes from suppliers. This information varies from allergen information for food products to controlled substances in personal care products. Product claims can also come from marketing or other dedicated departments. One of the major benefits of the PDR is integrating the two information branches of product development and packaging development to identify and characterize the business requirements, and as a result, document the best way to create those data linkages.

Product claims: The data supporting a claim, necessary to its defense, often reside in different places, and target different functions (e.g., project documents for Project and Portfolio Managers, recipes for formulators, design files or key lines for package

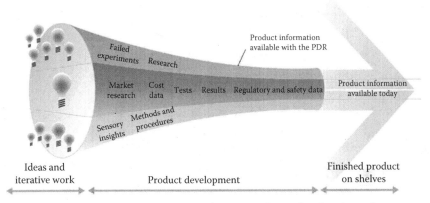

Each successful product launch takes the most superb innovators dozens of product ideas and an inconceivable number of iterations. But it does not mean that companies have to lose all this valuable knowledge and have to keep reinventing the wheel again and again. The PDR helps track this information back and re-use it for future product development work.

FIGURE 8.9   The types of information available in the PDR.

designers, etc.). The PDR creates the links between these functions, maps the information, and links it to the development of a product. The happy outcome is that information can flow between functions and changes can cascade quickly and effortlessly. Workflow can generate a robust claim justification and route the documentation for legal review.

Traceability: Traceability is an essential requirement for CPG companies. They must manage product data with the objective of consistently meeting or exceeding regulatory standards to avoid or quickly manage any crisis that could damage their brand, like *E. coli* outbreaks, or the recent Chinese pet food crisis. Because the PDR manages data across the full product lifecycle, from idea to manufacturing, it ensures compliance by enabling full traceability throughout the development phases and manufacturing, through to suppliers. This gives companies a process to react quickly during product recalls.

Figure 8.10 illustrates an example where a company receives a consumer complaint signaling the potential presence of an allergen in a candy bar, despite the label claim that the product is "allergen free." With the help of the PDR, the company can quickly identify which plant was involved in the manufacturing of this particular final product, as well as identify all the parties involved in the supply chain. The final and intermediate formulas link to ingredients, which trace back to suppliers. In a matter of minutes, the company can require all involved suppliers to confirm that none of their products contains the allergen.

Regulatory, environmental, and safety: A product Material Safety Data Sheet is an example of a document that contains information that is dispersed in different

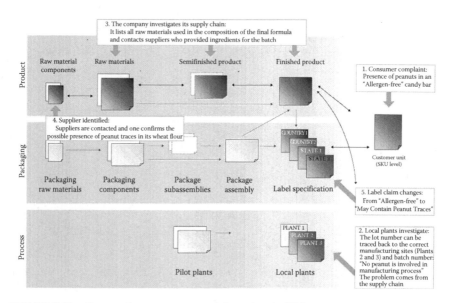

FIGURE 8.10 Course of a consumer complaint using the PDR.

locations and formats across the company and its customers. Data come from various sources, such as formulators, analysts, toxicologists, microbiologists, and procurement. Without central management, data exchange cannot be efficiently enabled, causing data duplication, version control, and data accuracy issues. The PDR drives standardization and classification of these data and builds a unique, evolving, and traceable data record.

Procurement: The PDR is the blueprint to understand what data can and should be shared across functions to improve the procurement process and maximize economies of scale and other savings. Examples of such savings include version control of purchasing forms, guaranteeing accuracy, and allowing reuse of existing purchasing knowledge. Most importantly, the PDR enables the delivery of real-time raw material costs to the formulators at the point when they select the supplier from which they will source the ingredient for this product formula. By selecting the lowest ingredient cost when the product is first formulated, formulators lock in the cost for the life of the product.

Project and portfolio management: Leading innovative companies in the CPG industry focus their efforts on innovation projects that are driven by market needs, with their actions driven by their product strategy. To attain this level of strategic planning, companies prioritize projects through sophisticated portfolio management. Companies anticipate needs and outcomes via technology roadmapping. The PDR allows companies to improve resource planning by seamlessly integrating information for product and packaging development, ingredient and materials selection, and production planning. The PDR also improves the vision of current and past product development projects, and links them to portfolio management. This linkage helps companies define clear development strategies that leverage product platforms and core competencies.

Knowledge management: All too often, employees cannot access information easily and on a timely basis. Many employees working on managing new product development projects consider it easier to start over than try to find past concept and project information. The PDR links existing knowledge with in-context product data. The PDR structure offers a way to track back to previous relevant initiatives by searching by-product characteristics, specifications, concepts, or just key words. This integration encourages greater reuse of information/design and increases general knowledge of what has been done in the past (i.e., company's knowledge assets) as well as ensures and maintains knowledge during typical team member transitions and personnel turnover.

8.9   SUMMARY

Building and managing the PDR is a prerequisite to PLM initiatives targeted at improving innovation. The PDR improves the innovation productivity and becomes the core of a PLM strategy for the following four reasons:

1. It contains all critical information necessary to design, produce, and modify the product, helping as a result to define a PLM strategy.
2. It defines the product hierarchy by using this concept as a key logical construct for the classification and management of product data. The PDR defines relationships between data elements as well as integrates structured and unstructured types of data.
3. It contains linkages to the authoring tools used for the conception and development of a product. The PDR allows traceability to previous iterations of the concept, and to previous lifecycle states of a product in development.
4. It identifies areas of opportunity to streamline, standardize, and integrate systems and processes to help accelerate speed-to-market, reduce costs, increase knowledge reuse, and guarantee data integrity.

The CPG industry has the opportunity to improve innovation productivity by adopting PLM solutions, following the proven example of other industries. By identifying the areas wherein they can make significant process improvements and by integrating their initiatives through the PDR, CPG companies can generate significant innovation gains and increase the likelihood of success in new product development. The most innovative CPG companies have already embarked on the PLM journey and will likely reap similar or greater productivity and cost savings benefits than companies in precursor industries because of three reasons:

1. They have access to proven PLM technologies (including virtualization, collaboration, and data management capabilities).
2. They will benefit from the lessons learned by discrete companies during their adoption of PLM, and concept similar to the PDR. The knowledge of "what works, and what doesn't" coupled with state-of-the-art models about data use will advance the design and potential of PLM implementations in process manufacturing.
3. PLM vendors are committing to tailoring their solutions to specifically meet the needs of CPG companies.

REFERENCES

Aberdeen Group Report, Profiting from PLM: Strategy and Delivery of the PLM Program. Boston, MA: Aberdeen Group, July 2007.

Aberdeen Group Report, The CPG Innovation Agenda: Business Value Research Series. Boston, MA: Aberdeen Group, December 2005.

Aberdeen Group Report, The Product Compliance Benchmark Report: Protecting the Environment, Protecting Profits. Boston, MA: Aberdeen Group, September 2006.

B. Asura and M. Deck, The new product lifecycle management systems: What are these PLM systems and how can they help your company do NPD better?, Visions Magazine (January 2007), http://www.pdma.org/visions/jan03/plm.html

CIMdata Report, Product Lifecycle Management: Empowering the Future of Business, 2002.

G. Young, Innovation in packaging. The Product Record: Agile Quarterly Publication 1(2), 2005.

M. Grieves, *Product Lifecycle Management: Driving the Next Generation of Lean Thinking.* New York: McGraw-Hill, 2006.

H. Chesbrough, *Open Innovation: The New Imperative for Creating and Profiting from Technology.* Cambridge, MA: Harvard Business School Press, 2003.

C. Perry, The product data record: The blueprint to guide product lifecycle management. *Consumer Goods Technology*, March 2008.

B. Hindo, At 3 M, a struggle between efficiency and creativity. How CEO George Buckley is managing the yin and yang of discipline and imagination. *Business Week*, June 2007.

A. Houard, *V5 PLM in Aerospace: Enabling Innovative Products for a Better Future.* Suresnes Cedex, France: Dassault Systèmes.

M. Kahn, The global approach, *Innovation: Making It Happen.* London: Caspian Publishing, 2002.

K. Sneader and E. Roth, Reinventing innovation at Consumer Goods Companies. *The McKinsey Quarterly*, November 2006.

M. Abramovici and O.C. Sieg, Status and development trends of product lifecycle management systems. Paper presented at the International Conference on Integrated Product and Process Development, Wroclaw, Poland, 2002.

M. Rao, *Knowledge Management Tools and Techniques, Practitioners and Experts Evaluate KM Solutions.* Burlington, MA: Elsevier Science & Technology Books, 2004.

R. Cooper, Perspective: The stage-gate idea-to-launch process—update, what's new, and NexGen systems. *The Journal of Product Innovation Management* 25(3), 213–232, 2008, (Published Online: 19 March 2008).

R. Cooper, *Winning at New Products: Accelerating the Process from Idea to Launch.* Reading, MA: Perseus Books, 2001.

S. Flanagan and C.-M. Lindahl, Driving growth in consumer goods. *The McKinsey Quarterly Web Exclusive*, October 2006.

S. Mc Kie, Collaborate to innovate. *Intelligent Enterprise*, March 2004.

J. Stark, *Product Lifecycle Management: 21st Century Paradigm for Product Realization.* New York: Springer, 2004.

D. Tapscott, *Competing with the Truth: Master Data Management in Discrete Manufacturing.* New Paradigm Learning Corporation, Toronto, 2006.

C. Tenopir and D.W. King, *Communication Patterns of Engineers.* New York: John Wiley & Sons, 2004.

9 Personalizing Foods

Heribert J. Watzke and J. Bruce German

CONTENTS

9.1   INTRODUCTION

The knowledge underlying the industrialization of modern foods is one of the most successful, yet least heralded achievements of science. Conversion of highly unstable, unpalatable, and frequently overtly toxic agricultural commodities into the diverse array of food products that comprise the core of modern urban diets requires a broad and deep understanding. The understanding must cover many fields, from the chemistry of biomolecules, the colloidal properties of biomaterials to the biology of plants, animals, and microorganisms. It is also necessary to understand the chemical, structural, and biological responses of molecules, biomaterial ensembles, and living tissues to the changes in a variety of external variables. The range of these external variables is large, often complex, including temperature, pressure, and shear forces.

Further, this knowledge of biomaterials must be understood in the context of foods. The molecular compositions of these biomaterials have been mapped onto the toxicological safety, nutritional composition, and multiple sensations that foods elicit in humans, when consumed. All this knowledge is used in the industrial process technologies that convert the unstable, unsafe, and sparingly nutritious agricultural commodities into safe, stable, and completely nourishing foods.

Till now the overall health of the consumers has not been an explicit target of food processing. The food products themselves, not the consumers, have been the overt targets and designs of the science of foods. This approach has been adequate to the task of providing a wide range of product choices in the food marketplace. This approach has also created an effective delivery system for essential nutrients through various fortification methods. The result is nutritionally adequate diets to populations. But this strategy is neither designed nor able to achieve optimal diets for individuals within populations.

In the past few years, the focus has changed, to a person, somewhat away from the food itself. Now, foods, diets, and health are undergoing a conceptual revolution. The variations among individual consumers, for instance in their existing health status and their aspirations for future health performance, are becoming the targets of all of the core knowledge of food science and nutrition. This trend in health science research is creating a new generation of technologies with commercial applications in human health assessment. Information technologies are being developed to interpret personal health data. The technologies are paving the way for a broad range of commercialized solution providers to enhance market opportunities from consumer medicine to food.

Growth in personalization of health will create many challenges to the food industry, but at the same time open the door to immense opportunities. By providing a greater diversity of customers, the technological capabilities of food material processing can expand in many respects. Food processing in today's sense means converting raw commodities to consumer products within very large and centralized factories, which then distribute identical products to entire regions of the world. As the diversity of consumption opportunities increases and individual health drives greater diversity into food processing, the processing of foods will move closer and closer to the actual consumer. Personal choices for products will be in part replaced by personal choice for processes. Point of purchase, point of decision, and point of consumption, each will become an operational term to describe a new range of options available to consumers.

All aspects of the agricultural enterprise will be influenced by the trends to personal diets. By valuing personal diversity, companies will create a diversity of foods. Agriculture, involving the use of plant and animal organisms, will take advantage of the diversity of life and the diverse biological materials as food inputs. For several decades, the quantity model of agriculture has encouraged commodity developers to seek the "optimal" commodities, including rice, soy, corn, and wheat. This model of pursuing a single "super" commodity with maximal agricultural yield in a single variety is, by its very nature, incompatible with the diversity of consumers and their varying health desires. Instead, the biological actions within varying agricultural materials will become a part of the basic value proposition of foods. Agriculture will reenergize the search for biological activities in the entire plant, animal, and microbial kingdoms. With different bioactivities related to health providing a core value of foods, processing methods that retain these biological properties will add a new dimension to the food processors tool set. Examples of biologically based or "soft technologies" maintain the biological activities of raw commodities, and yet ensure that the core value of safety will become an important target of the ongoing revolution in industrial food processing (Bertholet et al., 2005). Finally, biological activities themselves will become part of the processing stages of food materials (the so-called bioguided processing) as a means to further enhance the diverse health-promoting aspects of industrial foods.

This chapter deals with the needs for and the potential benefits to be gained by this new personalized approach to foods. The chapter examines the science and technologies that will need to be built and commercialized to bring the personalization of foods to practice.

9.2   CORE VALUES OF INDUSTRIALIZED FOODS

The basic sciences of food materials have developed quickly over the past century driven by the need to provide foods to the rapidly growing urbanized populations around the world. These sciences have continued striving to deliver on the basic values of safe, stable, affordable, convenient, fresh, delicious, and nutritious foods. Success in industrializing foods has enabled a fundamental change in agriculture and food. Significant advances in the chemistry of the biomolecules that make up food have provided the means to catalog the basic molecular composition of virtually all major agricultural commodities and food products (McCarthy et al., 2006).

Gradually, by gathering the extensive knowledge of the chemistries, physical properties and reactivities of those molecules, scientists have caused a shift in the industrial processing of basic commodities.

Agricultural commodities were traditionally processed with the intention to stabilize them for long-term storage as simple products purchased by consumers. Examples of these are grains, meats, milk, fruits, and vegetables. More recently, agricultural commodities began to be processed with the intention to disassemble them and separate the biological tissues into their biomaterial components, proteins, carbohydrates, oils, etc. These separated components, now as purified ingredients, can be used to formulate a vast array of food products most of which are unrecognizable as the original commodity (Tolstoguzov, 1986). Today, this approach of disassembling raw commodities has become a model for the industrial production of foodstuffs (Figure 9.1).

The pipeline of industrial production from commodities to foods is dominated by food safety. The goal of this pipeline is to formulate safe, generic food products according to the consumers' expectations for particular foods and food categories. For industrial foods, the failure of foods to be safe, even in rare, isolated events of food-borne illness, can be catastrophic to the core brand position of any major food company, product, or service. Thus, the highest priority of the entire process and every step therein is ruled by the need to select, process, and formulate in light of the overwhelming importance of safety and stability.

In this product-centric approach to food processing, various industrial suppliers in the highly competitive marketplace with comparable products on the same shelves, compete on the basis of lower prices. The primary economic driver of profitability for the industry becomes the cost of ingredients and processing. In this model, the food industry must constantly look for the means to reduce the cost of ingredients, processing, and distribution. The cost-driven model neither serves the agricultural producer being constantly driven to cheaper commodities nor the industry looking for cheaper ingredients, processes, and distribution channels. Finally, the products emerging from this enterprise are neither designed for nor consistent with the optimal health of consumers (see Figure 9.1).

It is in part because of this demand to be safe, that food processing first moved to a model in which commodities are disassembled into more stable components. These individual components can be separately processed in order to eliminate the risk of

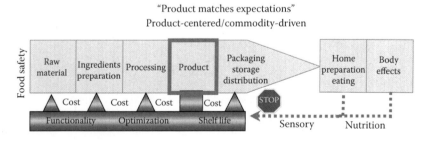

FIGURE 9.1 Current strategic pipeline for industrialized food processing from raw commodities to products, the marketplace, and the consumer.

microbial and chemical toxicities. Such a strategy also provides dramatically greater dexterity to the technologies of formulating foods that are ultimately directed to a larger set of core values. As a result, foods have become increasingly more stable, affordable, and convenient, freeing a larger segment of the population from the need to prepare foods for immediate consumption. Processed foods became the enabling products for the dynamic, independent lifestyles of the late twentieth century.

Freed from the constraints imposed by the complexities of raw commodities, the availability of isolated component ingredients provided food developers the means to pursue consumer preferences in the highly dynamic and competitive food marketplace. Foods became progressively more palatable. Eating is an innately physiologically pleasurable experience, and foods can provide considerable delight to daily life. As the science of describing food sensations matured, so did the means to formulate foods to exploit this new knowledge. Assigning molecules and structures to the complex and sophisticated sensory targets made it possible to formulate food ingredients to enhance the taste, flavor, texture, and even color of processed foods. As incomes rose, societies were able to afford more diversity in food choices. In some cases, this diversity was genuine. Access to rapid transportation and climate-controlled storage meant affluent urban centers had greater access to diverse commodities around the world throughout the year. However, some of the diversities were arguably false as apparently different foods could be assembled from the same ingredients and differences were solely associated with superficial appearance.

Industrialized foods provided the means to address nutrient deficiencies associated with the varying nutritional quality of raw commodities. An obvious advantage of formulating foods from component ingredients is the ability to explicitly add essential nutrients along with the overall formulation to achieve a complete adequate nutrient content. With the knowledge of, and ready access to, inexpensive vitamins and minerals, foods could be readily enriched with the essential nutrients. From the perspective on the adequacy of essential nutrients, foods had the means to become more nutritious. Unstated and perhaps even unappreciated by the consuming public, however, has been the inescapable fact that the increasing diversity of food choices placed the responsibility of obtaining a well-balanced, nourishing diet in the hands of each consumer. The enrichment of many foods with nutrients perpetuated a sense of confidence that the population was well nourished, and yet wider choices meant that subsets of the population could easily place themselves at risk, simply, of unbalanced diets. The population monitoring systems in the United States, like the NHANES population surveys documented that the population on average was adequately nourished, for certain nutrients, whereas subsets of the population were choosing diets that were not optimal (Bowman, 2002).

In the latter half of the twentieth century, a growing awareness of the ability of different diets and hence different food components to alter physiology, metabolism, and immunological functions has led to the development of "functional foods" in which these health values are explicitly added to foods targeted to subsets of consumers (e.g., athletes, infirmed, and elderly, Clydesdale, 1997).

The food industry has assembled a remarkable depth and breadth of knowledge of biomaterials making up foods as a result of a century of scientific research and development. Modern processed foods are possible only because of the detailed knowledge that this scientific activity has produced. The knowledge that makes the highly urbanized food supply possible consists in part of a relatively detailed compositional description of the agricultural commodities that underpin foods as well as the compositions of the assembled final products. The information base of commodities and foods includes the compositions of commodities; the amount and types of different proteins (from plant, animal, and microbial sources), carbohydrates (glucose, simple sugars, complex polymers of different sugar backbones and polysaccharides), lipids (triacylglycerides as fats and oils from plants and animals, complex membrane lipids, and sterols), nucleic and organic acids, alcohols and a wide array of plant secondary metabolites (carotenoids, flavonoids, anthocyanins, etc.).

The knowledge of the basic components of foods also includes a broad understanding of their macroscopic structures. The processing of basic commodities, separating various components, and then finally formulating these commodities and ingredients into food products change the relative composition of the basic molecules as well as their cellular and biological structures. Food processing is designed to modify and disassemble the commodities. Processing changes their basic compositions, stabilities, and safety, converting them into foods and food ingredients. Ingredients are reformulated, remodeled, and synthesized into new compositions and structures. Gels, fibers, emulsions, and foams are complex food structures that are explicitly formed during food processing and preparation. These structures contribute to the desired stability, shape, texture, and taste/flavor of the final foods. Considerable research has been directed toward understanding these basic structures, their formation, stability, and taste/flavor properties (Tolstoguzov, 1986).

The basic descriptive information base of the compositions of agricultural commodities is in the process of being broadened to include detailed genomic sequences of these same agricultural commodities (Catchpole et al., 2005; European Plant Science Organization, 2005; Womack, 2005; McCarthy et al., 2006). As scientific research builds the functional annotation of the genomes, these two basic knowledge sets are coming together to link the basic biology of the plants, animals, and microorganisms of agriculture with the compounds, structures, stabilities, and chemical interactions of these same commodities as the raw materials of foods (Tolstoguztov, 1986; Sagalowicz et al., 2006). Now, science is annotating the same genomes and compositions in terms of their nutritional, safety, and toxicological implications to the humans who finally consume them (German et al., 2006; McCarthy et al., 2006). This remarkably diverse and predictive scientific knowledge can be translated into practice by the myriad technologies that are used to process commodities into foods. Now, however, the knowledge is bringing the flexibility necessary to personalize formulations, ingredients, and structures to individual consumers.

The combination of biological knowledge and practical technologies will enable a fundamental shift in the way food and health are both produced and marketed. The society of the next decade has the unparalleled opportunity to build itself a personalized healthcare system in which the most logical input to health, diet, is guided throughout the entire agricultural system to the goal of improving individual health (Figure 9.2). Technologies that process foods range from careful control of heating and biological fermentations to highly complex engineering systems that

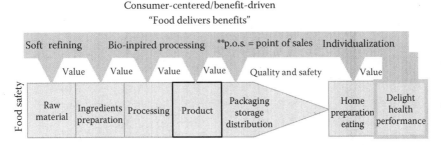

FIGURE 9.2 Proposed new approach to the pipeline of industrialized foods for the future, designed specifically for the consumer as the central focus of all processes.

simultaneously deliver the pressure, temperature, mechanical and biological inputs to high-throughput, multicomponent, sterile, process assembly lines (Bruin and Jongen, 2003). In the more personalized future, these technologies must effectively be brought closer to the individual consumer, to accommodate their unique health needs and their discrete performance occasions.

In the model schematized in Figure 9.2, the industry does not formulate products per se, but rather targets the consumer and the eating opportunities to provide a wide variety of benefits, using foods as the vehicles of those benefits. In this model, foods are designed to be more flexible in order to address the specific needs, desires, and aspirations of each consumer for their individual health and well-being. In this value-driven model, industries will compete to provide more value to consumers as will all entry points along the entire pipeline that can introduce values, from novel commodity ingredients, to innovative processing and customizable formulations for the consumers at their specific point of consumption.

9.3 DIVERSITY OF HUMAN HEALTH

Humans span an astonishing range of phenotypes. Healthy human adults vary in height, weight, activity, cognition, strength, endurance, flexibility, and in their preference for foods. We are most alike at birth, but as we progress through various life stages, we diversify into myriad lifestyles. A massive investment on research was made in the twentieth century to understand the basic biological processes of humans, detailing the complex interweaving pathways of biochemistry from genetics to physiology. Molecular biologists are still establishing the complex regulatory systems that maintain control of these pathways. Their goal is to understand biology at the level of the entire organism.

To date, scientific research has cataloged much of the basic principles of these biochemical processes in humans as a species. It has been only recently that scientists have begun to approach the research needed to catalog and understand the variations in the biochemistry of humans as individuals. As breakthroughs occur, it is necessary to bring the details of human variations into practice. To do this, industries will need to market diagnostics that can measure human health and empower individuals to recognize their differences and the implications of those differences

$$\text{Phenotype} = \text{Genotype} + \text{Environment} + \text{Genotype} \times \text{Environment}$$

FIGURE 9.3   Human phenotype equation.

to their health. How do humans differ, or more precisely how do their phenotypes differ? (see Figure 9.3)

Humans differ due to a wide range of basic biological variables. These variables are in some cases genetically based—chromosomal differences (e.g., male, female, or allelic variations in structural or regulatory regions of specific genes). In other aspects, human differences are due to the age and particular life stage of an individual (e.g., infancy, pregnancy, lactation, pre- and postmenopause, and puberty). In others, human variations are due to environmental influences, either exogenous and random (e.g., exposure to sunlight, toxins, allergens, bacterial inoculum, etc.), or endogenous and volitional, i.e., chosen lifestyle (e.g., exercise, athletic training, excess caloric intake and obesity, sedentary behavior, sleep cycle alteration, meal frequency, and temporal variation). When acting at a particular point in an individual's development or life stage, each of these variables may exert effects on various epigenetic or nongenetic elements. Such effects may then confer persistence of a particular phenotype through much of that individual's subsequent life and alter that individual's response to dietary components.

9.3.1 GENOTYPE

The field of nutrigenomics is rapidly discovering in molecular (sequence) detail those genetic differences across the human population that are related to differences in needs for or responses to various nutritional variables (Collins et al., 2003; Miller et al., 2004). The use of SNP technologies to measure an individual's genotype and provide nutritional recommendations based on known nutrigenomic variables is already a commercially viable practice. Although the total number of gene-nutrient factors remains small, it is increasing rapidly (Wei-Min et al., 2008). The great challenge for genotype as a means to provide sufficient personal information to guide health and dietary practices is the complexity of the human genome itself. Health is not simply a one-gene one-nutrient undertaking. The chances that a single polymorphism in a single gene will dictate dietary needs for an individual, even for a specific nutrient, are small. Nonetheless, since the genes that are responsible for human diversity are recognized, the capabilities of genotyping to provide individuals with actionable knowledge about their unique predispositions to diet, drugs, and lifestyle will increase as both a scientific and a commercial reality. Importantly, however, although this information will be valuable for some individuals, it will not be possible to build a complete health management system for the population, based on genotyping. Health depends on more than genotype.

9.3.2 ENVIRONMENT

In this chapter, the environment in which an individual lives is considered to include all aspects of exogenous inputs to phenotypes, including acute and chronic, random

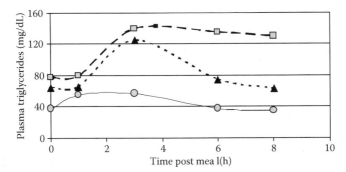

FIGURE 9.4 The levels of circulating triglycerides in postprandial plasma. Three subjects were measured at fasting (time 0) and at various time points (1, 3, 6, and 8 h) after consuming a standard liquid meal containing 20 g fat. (Data from Zivkovic et al., unpublished.)

and volitional, chemical and behavioral inputs. The nature of these inputs can be described or measured by their source (e.g., geographical distribution and solar UV irradiation) or their effects (e.g., vitamin D formation, McCarty, 2008). These inputs can be either random and unavoidable (urban pollution) or volitional (smoking), or therapeutic (Akamine et al., 2007). When considering the extent of exposure, some of the variable inputs can be generalized to a larger population (exposure to fluoridated drinking water) or unique to specific individuals (chronic consumption of sweetened beverages) (Malik et al., 2006).

One of the key aspects to the environment is, of course, diet. Quantifying and assigning to genetic or phenotypic causes the varying responses of humans to dietary components is emerging as a great success in the field of nutrigenomics. With the recognition that humans vary in response to diet, scientists are adding an obvious dimension to the measurement of phenotype: the individual response to dietary challenge. The varying responses among three normal, healthy individuals to a standard fat challenge test (Figure 9.4) illustrate this basic principle. Quite remarkable differences both in the quantity of triglycerides and the time of clearance can be seen in this very small group of individuals. The variation among these three individuals in both the maximum levels of triglycerides achieved and the overall time course of clearance of the additional triglycerides arises from the individual phenotypes despite consuming the same quantity of dietary fat. Given the importance of this variation for defining individual aspects of diet and health, it is likely that the response to dietary challenge will become an aspect of normal health assessment in future.

9.3.3 Life Stage

The nutritional requirements of humans vary according to each individual's stage in life. Early in life, the demands of growth and development change rapidly as do dietary needs and responses (Neu et al., 2007). Puberty and the transition to reproductive fertility cause a significant change in hormonal status and precipitate a diverse range of physiological and metabolic consequences, many of which alter the dietary needs and responses (Pietrobelli et al., 2008). In the reproductive cycle, estrus, pregnancy, lactation, and involution are all distinct physiological states for which varying nutritional implications are expanding beyond current guidelines (Bartley et al., 2005; Henriksen, 2006).

With the success of clinical medicine, vaccination, nutrient fortification, and sanitation engineering, life expectancy of humans has increased dramatically. One result of this success has been the emergence of elderly with a sustained and distinct life stage. With greater numbers of elderly humans, their unique physiological, metabolic, and even microbial states are recognized, including their unique nutritional needs (Labossiere and Bernard, 2008). Finally, various disease, injury, or pathological states intrude upon virtually every individual at some point in his or her life. Clinical nutrition is dedicated to understanding the unique metabolic demands of disease states. Furthermore, clinical nutrition seeks solutions to accelerate the rate of recovery and minimize the long-term consequences to individuals during these periods (Razack and Seidner, 2007).

9.3.4 LIFESTYLE

Humans are enjoying the fruits of affluence by embracing a remarkable diversity of lifestyles. For some, the freedom from the need for physical work and the proliferation of labor-saving devices has led to conspicuously sedentary behavior for most of their lives. For others, the opportunities to pursue vigorous professional or recreational activities have generated levels of human performance not seen in our history. Amid the myriad lifestyle factors that alter the nutritional status, one aspect affecting an individual's phenotype is the breadth of the choices and food alternatives within the modern food marketplace. This freedom of choice affords many individuals the opportunity to choose increasingly diverse diets throughout the year. However, for others, choice does not necessarily afford diversity for reasons varying from food preference to economics (Drewnowski and Darmon, 2005). Faced with a highly diverse food supply, individuals within societies are pursuing widely different dietary intakes of nutrients, caloric content, and macronutrient compositions. These are all in conspicuously diverse foods that vary in structure, complexity, and rheology. Furthermore, as lifestyles diversify, the eating habits of individuals are diversifying as well across wide temporal habits (from a single large meal per day to a dozen small, snack-like eating occasions) (Gilbert and Khokhar, 2008).

9.3.5 ENVIRONMENT × GENOTYPE = IMPRINTING

Environment at one stage of an individual's life can exert effects on his or her phenotype that persist later in life (imprinting, programming, memorization, or colonization; Bateson et al., 2004; Mathers, 2007). Scientists are beginning to detail the mechanisms by which the effects of environment can persist even after most cells and tissues in an individual have turned over repeatedly. The explicit covalent modifications of DNA that are passed onto subsequent replications of chromosomal DNA through multiple cell divisions are being detailed by the field of epigenetics (Burdge et al., 2007). The proliferation of specific cell and tissue types can be sustained through many stages of life. The development of preferences for sensory attributes of foods—principally olfactory preferences—persists through much of an individual's life, guiding his or her lifelong food choices. Finally, an unexpected environmental factor that exerts many effects on phenotype is the makeup and genetic diversity of the various microorganisms living on or in each individual.

Imprinting during fetal development has been shown to cause methylation differences across entire regions of an individual's genome and can result from various nutrient deficiencies in the mother (Burdge et al., 2007). As technologies emerged to measure epigenetic modifications in DNA, studies in animal models using various nutrient, environmental, and lifestyle factors have been shown to influence DNA methylation (Langley-Evans, 2006). Such studies imply that nutritional factors are also relevant to humans and would be predicted to result in similar persistent effects through epigenetic changes in an individual's genotype.

> The remodeling of cells and tissues as a form of imprinting has been documented, though not yet fully understood. Adipocyte hyperplasia early in development is proposed to be one of the factors that accounts for the high predisposition of adult obesity in children who are overweight (Ailhaud et al., 2008). These studies by Ailhaud's group have documented in animals that specific dietary factors early in life, of animals at least, produce or inhibit the proliferation of adipocytes that in and of itself could account for a persistent and altered response to diets later in life. Interestingly, the dietary factors ω6 and ω3 polyunsaturated fatty acids are apparently effective at altering cellular development within the range of normal human diets. Muscle mass is also responsive to the combination of conditioning and protein content of the diet (Haddad and Adams, 2006; Paddon-Jones et al., 2008). The persistence of this single tissue, muscle mass, its influence on whole body energy metabolism, its metabolic contributions post training, and a larger store of amino acids as muscle proteins would all be predicted to alter the response of an individual to various health states and diet (Petersen et al., 2007).

The imprinting of sensory preference is perhaps the least understood but most influential in the conditioning of modern humans to their habitual diets. Humans apparently do not rely on nutrient cues to guide their food choices. Instead, humans rely on a system of acquired food preferences to influence their dietary choices. The remarkable property of olfactory preference is the process by which positive and negative preferences for particular flavors are acquired in an individual as a series of complex, contextual memories early in life (German et al., 2007). This system of acquired flavor preferences underlies much of the cultural variation in foods and cuisines around the world. This also means that flavor preferences for foods with poor nutrient quality, if acquired by an individual early in life, will guide a lifelong habit of poor food choices, in spite of the fact that the sensations will continue to be perceived as highly positive.

Another means by which early diet can program an individual's long-term responses to foods is the capacity of early diet to influence an individual's microbiome (the ensemble of microorganisms coinhabiting each of us; Zoetendal et al., 2001). Early diet shapes the bacterial population by direct inoculation of particular microorganisms present in foods, or selectively influencing the competitive success of microorganisms by components that can only be fermented or utilized by certain bacterial populations (Bäckhed et al., 2005).

Until recently, the role of an individual's microflora was considered to be relatively minor to that person's overall health. Recent studies on the development and maturation of the immune system indicate that each individual's immune system is shaped in part by interacting with the endogenous microbial population in the intestine. Remarkably, the diversity and activity of the microflora appear to be important for the successful development and maintenance of the acquired immunity and for tolerance. These studies imply that foods will be used in future to guide an individual's immune system to improved surveillance and response to pathogens (i.e., vaccination response) and also to improved tolerance and prevention of allergies (Iweala and Nagler, 2006). Following the stunning revelations by Gordon's group on the influence of specific bacteria on energy metabolism and predisposition to obesity, a person's microflora is viewed as a pivotal factor in human metabolism, immunity, sensation, disease resistance, inflammation, and comfort (Turnbaugh et al., 2008).

9.4 TOOLS TO ASSESS HUMAN PHENOTYPIC DIVERSITY

The technologies addressing human health assessment are moving forward rapidly in many aspects, from detailed genetic sequencing and genotyping to precise and accurate measures of abundances of small molecules (metabolite) in biofluids to the spectacular advances in molecular-based imaging (Figure 9.5). Several books, articles, and chapters provide detailed information and analysis of the tools that are developed to assess human genotypes and phenotypes (Fay and German, 2008).

The equation of life: Assessing the diverse aspects of personal health			
Phenotype =	Genotype +	Environment +	G × E
Molecular imaging Metabolic profiling Immunologic assays Physiological cognitive performance testing	Genotyping SNP analysis Family history Sibling comparisons	Detailed dietary intake Activity monitoring Ambient data recording Toxic exposures	Maternal records Childhood records Cellular DNA methylation Tissue composition and cellularity Microflora composition Flavor preferences

FIGURE 9.5 The equation of biological life. Each individual's phenotype is the result of interactions between his or her genotype, environment, and the integrated interactions between genotype and environment accrued over a lifetime.

While measures of genotype (i.e., sequencing) are becoming commercially practical, the utility of sequence data in explaining existing phenotype and as a predictor of future phenotype in that individual will be limited. Discouragingly, measures of environmental influences on each human are not particularly useful at present and thus, most human health assessment will need to be based on measuring the phenotype itself. Metabolic, physiological, immunological, and performance assays are becoming available and commercially viable. Consequently, industrialization of human health assessment could become a reality within a few years.

9.4.1 NUTRIGENOMICS

Whole genome sequencing is scientifically possible; it is impractical for individuals today because of cost. The tools necessary to detect key aspects of the genetic variation in individual humans are being brought into commercial practice. As the genetic basis for important variations in human responses to diet is recognized by scientific research, companies respond by providing this information to consumers. This form of commercialization of personal information will likely grow. With each new discovery of a polymorphism that links aspects of human nutritional variation to specific genetic loci, these sites can be incorporated into technologies capable of analyzing tissue samples from consumers. For those individuals who contain those genetic variations that directly affect their responses to or needs for particular nutrients, this information will be valuable. However, for most individuals, the ability of genotyping alone to guide dietary choices is limited and will remain so for many decades. This limitation comes about because of the complexity of the human genome and its relation to health. There is a great deal of scientific discovery needed before our understanding of the genetic sequence basis of human variation in response to diet is in any sense either comprehensive or predictive.

9.4.2 PROTEOMICS

Proteomics is the science of measuring the proteins within a biological sample (biofluid, tissue). Proteins are intrinsic to the functions of biology at every level. Thus, if one could genuinely measure and interpret all the proteins within core tissues and biofluids (skin, blood, saliva, and tears), then the results might predict human health, both acutely and as an indication of the past and future trajectories. However, true proteomic analyses will be daunting to bring to commercial practice. The complexity of even a simple proteome is quite large. The human genome encodes approximately 30,000+ genes. Furthermore, these genes are thought to encode for several 100,000 proteins and an order of magnitude more of protein variants deriving from gene and protein splicing, posttranslational modifications, etc. The challenges are also analytical because of the large dynamic range, i.e., the wide concentration span between the most and the least abundant proteins. For these reasons each of which must be solved, proteomics will not be a central tool for health assessment in the immediate future.

Proteomics, however, as a technological platform is well suited to commercial applications in disease diagnostics. Those specific proteins presence and abundance of which are central to a disease process once discovered can be targeted for focused diagnostic platforms. For overall health assessment, however, the luxury of looking for single target is not possible. It is necessary to look at many proteins, and the overall dynamic range of individual protein levels spans 6 orders of magnitude in a cell and even 10 orders of magnitude in specific compartments of the human body, e.g., in plasma (Saha et al., 2008). More discouragingly, the most biologically active proteins, those that regulate higher metabolic, immunological, and physiological processes are the least abundant. The problems do not end there. Proteins function by their three-dimensional solution structure, and the same protein can adapt different structures, in some cases many functionally different structures. Protein structures are in principle predictable from their sequences, as Domingues et al. (2000) found that 66% of the proteins having a similar fold also exhibit a similar function. However, the function of a protein depends also on its environment, a circumstance that complicates these considerations substantially. Thus, single protein levels or even protein functionalities will not provide valid biomarkers of health for a given individual or even a subset of the population.

9.4.3 Metabolomics

The technologies developed by the field of analytical chemistry are changing the way biological systems are described and studied. In particular, some high-throughput methods can accurately measure a significant fraction of endogenous metabolites in fluids and tissues. Thus, metabolic profiling gives an instantaneous "snapshot" of the physiology of the cell or organism (Raamsdonk et al., 2001). Comprehensive metabolite profiling has created some confusion among readers by the use of two systems of nomenclature.

1. Metabonomics is defined by Nicholson and coworkers (Nicholson et al., 1999; Nicholson and Wilson, 2003) as the "quantitative measurement of time-related multiparametric responses of multicellular systems to pathophysiological stimuli or genetic modifications." Metabonomics is a diagnostic tool for metabolic classification of individuals, and uses primarily NMR spectroscopic detection. Metabonomics is used for metabolite profiling across a wide range of chemical classes combined with multivariate statistical methods.
2. On the other hand, the term metabolomics was indicated by Fiehn to be the "comprehensive and quantitative analysis of all metabolites" (Fiehn, 2001). From an analytical viewpoint, metabolomics is generally the quantitative analysis of intra- and extracellular metabolites of biological systems (including microbial, plant, and mammalian systems) using mass spectrometry.
3. Metabolomics and metabonomics both have been employed in preclinical and clinical research, for environmental, biomedical application, and in toxicology and nutritional research (German et al., 2005a).

9.4.4 Molecular Imaging

The sciences of molecule-specific detection and three-dimensional image reconstruction are combining into a wide variety of imaging methods for medicine. Exciting developments in chemistry and physics are combining to create innovative imaging systems and are moving rapidly from research laboratories to industrial development and clinical practice. For clinical medicine, these advances have created a revolution in routine practice with human images, both localized microscopic examinations and whole body scans, used in applications as diverse as disease diagnostics to surgical positioning to therapeutic solutions (Margolis et al., 2007).

For health assessment, the applications of molecular imaging have not followed as quickly, yet they are equally exciting in potential. Why are they not moving into health management and disease prevention? The cost of analysis, concerns over privacy, and the availability of qualified technical support are cited as limitations to the expansion of imaging into routine assessment strategies for preventative health care. These are not the most important limitations, however. The critical barriers are the abilities of imaging modalities currently used for disease diagnostics to discriminate the subtle variations in the health of healthy individuals. If imaging could genuinely identify those who, while seeming to be healthy, are in fact at risk of losing their quality of health, then it is likely that these technologies would quickly penetrate the luxury end of health care. Nonetheless, various scientific initiatives are moving into the most obvious areas of health for which imaging would be valuable to individual consumers, for example weight management, body composition, fat accumulation in inappropriate tissues, and muscle and fuel accumulation in athletes (Sanz and Fayad, 2008). As various forms of imaging develop with accuracies being able to distinguish individual differences in phenotype, our personal images will likely become the most "visible" aspect of routine health care.

9.4.5 PERFORMANCE ASSESSMENT

Human physiology and its ability to influence health and respond to various stressors and the ability of humans to engage in a wide variety of physical, sensory, and cognitive activities can be considered as central to human performance. As scientists have addressed the variation in human physiology and performance, they developed tools to measure them accurately. The body of knowledge that studies have established provides a substantial resource for bringing standardized performance tests to practice as assessors of health. Immunologic responsiveness, aerobic capacity, VO_2 max, strength tests, endurance tests, memory tests, visual, auditory, taste, and olfactory acuity tests, all represent increasingly accurate measures of the diversity of human phenotypes.

Scientists have explored the boundaries of human performance with the above-mentioned measures, along with the variables that influence them. Perhaps not surprisingly, diet has emerged as a critically important input. In the future, we will see explicit performance measurements used as the basis for evaluating diet and food functionality. As large datasets of human health begin to include performance measures and diet as dependent and independent variables, it will be possible to mine these datasets, build causal relationships between specific performance and specific dietary factors, and ultimately provide the means to predictably improve specific aspects of human performance, on a 1:1 personal basis. Furthermore, because diet and lifestyle interact to modify our physiology, metabolism, immunity, and tissue structures, the ability to combine diet and lifestyle guidance to literally sculpt an individual's health and enhance his or her quality of life should become an obvious outcome of greater personal assessment. It will be possible to provide food consumers with options not simply of improving their health as lowering the risk of disease, but improving health by enhancing desired performance.

9.4.6 PERSONALIZING FOOD VALUES

The challenges to personalizing food values are considerable. The benefits will be equally compelling. As previously discussed, today's industrialized foods integrate a number of values within a specific food product. The intensive competition in the marketplace means it is not merely sufficient that a food be just safe, convenient, delicious, or healthy. Foods must be simultaneously safe, nourishing, stable, convenient, affordable, and delightful. Furthermore, this requirement extends to any modification of existing food products. A subsequent enhancement in an existing food's value must be so without compromising the other value assets. These demands place a great deal of pressure on all aspects of the food development process. Improving stability and convenience cannot lead to deterioration of flavor. Nor can improvements in flavor compromise safety.

The first generation of functional foods failed because, although they contained added-health values, the ingredients and technologies necessary to add these values compromised the perceived values of the foods (taste, flavor, stability, and convenience). Hence, within the very competitive food marketplace, they simply failed to attract consumers away from their existing choices. The next stage of functional foods recognized this need and remedied the defects. The functional health values have been incorporated into foods that also deliver all of the values that consumers expect. As foods become increasingly personalized, this basic principle will continue. Food values will continue to be integrated, but now more personally for individual health and also personally delicious, safe, and convenient for the desired eating situation.

The other key principle in foods today is choice. The last 50 years has seen a continuous trend toward providing products to consumers that maximize their ability to make immediate choices about the foods they wish to eat (Moskowitz et al., 2005). This drive to provide immediate food choices has literally transformed the entire agriculture enterprise and the food marketplace. Not just convenience, but safety, stability, packaging, distribution channels, and now nutrition, have all been driven relentlessly forward by a marketplace that caters to individual consumers wishing to choose their personal eating opportunities. Consumers will increasingly demand health as well. The emerging demand for nutritious foods is now moving both away from foods that are perceived as unhealthy, and toward foods that can improve the personal health of consumers (Sloane, 2006). This desire for better health from foods is accelerated by the reality that poor choices in foods have been damaging the health of the average consumer.

The aging, sedentary population in the Western world is increasingly becoming overweight, hypertensive, and diabetic. Each of these conditions is propelled by poor diets (Alberti, 2001). In general, the diets that consumers choose are not overtly deficient in essential nutrients (vitamins, minerals), although some examples of poor vitamin and mineral status among specific subsets are particularly discouraging having known the way to prevent them. Rather than being deficient in essential nutrients, most diets today are instead unbalanced in terms of calories and macronutrients (Popkin, 2006). Research is only beginning to recognize that macronutrient imbalances lead to chronic disregulation of normal metabolism within susceptible individuals. Such metabolic disorders are eventually devastating to the health of the population promoting diseases, and are characterized as an epidemic of endogenous, noncommunicable diseases (Quam et al., 2006).

The foregoing nutrition "problems" are becoming evident to the consuming public. The reason is simple. Obesity, hypertension, and diabetes are readily sensed or measured. As a result, consumers are vividly aware of their own deteriorating health, although they do not know how to change their health trajectories (Petrovici and Ritson, 2006). Even though the formal policies and infrastructures of public health are still unable to discover population-wide solutions for these health issues, nonetheless, consumers increasingly demand that their food provide health protection. The major growth in the food industry is occurring in the sector devoted to health-promoting foods (Clydesdale, 2004). Consumers increasingly recognize the importance of diets to their health, and demand products and services to meet their desires.

9.4.7 FOOD VALUES MEETING CONSUMER DESIRES

Consumers themselves will be major drivers for personalization of foods. However, until the tools match the variation in consumers to variation in foods, health improvements will be slow and limited to the major (average) health issues of majority of the population. Consumers are already educated in these health issues such as cholesterol, hypertension, blood glucose, and allergies. Thus, the average consumer is increasingly aware of health needs and becoming aware of varying health opportunities. The key will be to build sufficient knowledge on the relations between personal health and appropriate diets to create a new marketplace reality. The opportunities to capture consumers and market share are massive. Growing scientific knowledge and technological advances will race to meet these needs and opportunities.

Disease prevention is considered to be a clinical, therapeutic industry, but in fact it is not. Food is the only possible engine for disease prevention. There is, however, an important economic argument underlying the economic reality of disease prevention that blocks the practice preventive medicine. How much will a consumer pay for a drug or therapeutic or bioactive substance to prevent a disease that, as a result, the consumer will never have? The answer is discouragingly simple, very little. Drugs are considered by consumers to be a valuable product category precisely because they are designed to cure existing disease. Consumers will gladly take a drug that effectively cures diseases that they already have. Prevention means that you are healthy and do not have a disease. Hence, the value of any single therapeutic preventive to each consumer is very low. In order to be successful, the commercialization of prevention will have to be very cost efficient. In fact, it is precisely because foods that are already routinely consumed, carry an existing value proposition and can easily deliver multiple biological activities that they are appropriate as the "engines of prevention."

Foods and the food marketplace will change dramatically, but how? It is not clear yet how the various forces acting on this rapidly changing situation will drive it. It is possible to anticipate the most logical and biologically understandable directions. However, who would have anticipated a decade ago, that the most toxic substance known to foods, Botulinum toxin would become the core of a cosmetic industry in 2008! Some examples of success make it possible to predict according to categories of innovations based on the three drivers of foods: (1) desired consumer values, (2) molecular scientific knowledge, and (3) efficient industrial technologies.

From the consumer perspective, the major trend is toward individualization. There has been a virtual revolution in the personalization of consumer products from all areas. Computing has moved from centralized mainframes to personal computers. Entertainment has moved from centralized theaters to home theaters to portable, personal music and movies players. Transportation has changed from mass transit to personal motoring. Apparel has moved from the standard gray suit to personal statements of power and freedom (German et al., 2005b). Personalization is changing the food distribution system. Groceries that consisted of aisles of shelves of prepackaged, family size products 25 years ago, have been replaced by multiple single serve items reaching for consumers at the point of eating occasions. Health will be increasingly important in this rush to personal values.

The detailed, molecular level understanding of human health is moving from laboratory to the population. Consumers are beginning to know more about their own health, the health risks that they face personally, and the potential of foods to alter them. However, as consumers begin to take charge of health, they will demand as much enhanced performance as reduced risk of disease. Performance is a broad term and unquestionably, consumers will define it more broadly in practice than scientists in the laboratory can today imagine. Nonetheless, foods are increasingly formulated and marketed to health as performance achievement, not as disease reduction. Examples are only the most obvious, but they will expand rapidly as the distribution systems to individual consumers expand. Now, we look for athletic performance products for rehydration, protein-rich products for muscle rebuilding, and carbohydrate products for refueling.

Personalization will also extend the concept of personal risk. A core value of processed foods is to ensure safety, i.e., foods are safe. As safety becomes a value of personalized foods, the basic principle will be extended to include the means to make consumers themselves safer. The first generation of personal safety will mean more specific control of food processing and specific labeling (no nuts, gluten free, etc.). Already research pipelines include components that make the consumer safer from pathogens, allergens, and toxins not even necessarily associated with the diet at all.

The growth of industrialized foods during the past 200 years has been possible because of the development of process engineering technologies. These technologies have been brought in to practice as large, centralized factories processing agricultural commodities into safe, packaged, shelf stable food products. The success of industrialization has inadvertently turned out to be an important hurdle to the practical personalization of foods. How can a single factory make all these different, personal products? This limitation is true if one assumes that industrialized foods can come to the consumer only as prepackaged products. However, combining the goods of foods with the services of personalization provides a wealth of opportunities for innovation. Innovations in the technologies that will enable a more personalization of eating will take advantage of not only the basic biology of the humans consuming foods and the unit operations of basic process engineering, but also benefiting from the basic physical, chemical, and biological properties of the organisms and biomaterials of which foods are composed of and from which they are formulated.

9.5   ASSETS FOR INNOVATION IN PERSONALIZING HEALTH AND FOODS

9.5.1   GENOMICS AND THE GENETIC BLUEPRINT OF EVOLUTION

Genomics and the sequencing of the entire genome of humans is the major growth dimension in life sciences. Genomics seeks to develop a biology-based understanding of the emergence, functions, and relatedness of life forms, of hundreds of viruses, bacteria, plants and animals, and the entire ecosystem (Miller et al., 2004; McCarthy et al., 2006; Morris, 2006). The goal of this research on nutrition is to gain a mechanistic understanding of diet, human health, and disease. Various aspects of biology are discovered at unprecedented rates—from the development of diseases in humans to the production of biofuels by plants and microorganisms (Collins et al., 2003, Pharkya et al., 2004).

One of the interesting outcomes of this knowledge explosion involves understanding the diversity within species. Research is revealing an increasing number of ways in which humans differ from each other. Certainly, one way of difference is the response to diet. The fields of nutrigenomics and nutrigenetics are rapidly compiling examples of genetic variation in the normal population that predicts variation in health outcomes. From dietary fat and lipoprotein atherogenicity (Krauss, 2001; Ordovas, 2003) to dietary carbohydrate and weight loss (Martinez et al., 2003), to dietary vitamins and risk of birth defects (Whetstine et al., 2002), genetic variation is emerging as a key component of the variation in human health that is ascribable to diet. For example,

> The most widely used index of metabolic health is the level of circulating lipids in blood, primarily cholesterol and triglycerides. Just by looking at the mechanisms by which lipids are cleared from blood, we see a vivid story of the genetic variation within the human population. Clearance of triglyceride-rich lipoproteins is affected by polymorphisms at several gene loci, including apoB (Rantala et al., 2000), apoAV (Lai et al., 2003), apoE (Ostos et al., 1998), and lipoprotein lipase (Mero et al., 1999). The consequences of these genotypic differences are that individuals can be assigned to varying responses to, for example, high- and low-fat diets. The size and atherogenicity of cholesterol-rich, low-density lipoproteins, and their response to diet are mitigated by heritable genes that can be demonstrated to persist in the offspring of phenotypic carriers (Krauss, 2001). Again, as a result of defining these genotypes, individuals who vary in response to diet will eventually be assigned/guided to diets with specific fat contents and compositions.

Genomics and the knowledge of biology made possible by this field will contribute in many ways to our understanding of human health with respect to diet and its applications to food development, although not just through human genomics. The progressive steps of evolution of many life forms are becoming clearer in terms of their genetic and molecular details. Since most inputs to the food supply are or were living organisms, studying the genomics of these organisms provides a remarkable window to the consequences of selective pressure on different life forms, and the way these organisms evolved within specific ecological niches. The scientific understanding that is emerging from the microbial, plant and animal physiology, metabolism, and protection is leading scientists to recognize the breadth of functions of different molecules and their possible actions within the broad genetic diversity of agricultural commodities that serve as inputs to the food chain (Delmer, 2005). The first application of evolutionary genomics will be a simple improvement in composition (Watkins et al., 2001). Understanding the biological actions of specific molecules will inform future selections and processing to capture, remove, or modify these molecules in food streams.

As our understanding of the full complexity of biological ecosystems improves, the principles of food production will be shaped by biological criteria. The most immediate and relevant model to guide this approach is mammalian milk. Milk has informed many of the compositional recommendations for foods and human diets through history. The mammary epithelium is the most remarkable bioreactor for animal food production, and its multiple biological processes will be a model for guiding the processing of food materials (German et al., 2006; Lemay et al., 2007b). Milk is not a simple fluid containing inert components. Many of the components of milk are active catalysts benefiting the infant through the catalytic activity that they provide within the intestine of the infant (e.g., bile salt stimulated lipase makes lipids more available for absorption). The goal of most processing today is to destroy or inactivate catalysts to improve the stability of the processed raw commodities (lipoxygenases, lipases, and proteases). In the future, processing approaches will be designed to maintain or augment these activities in commodity processing streams. The diversity of biological activities is remarkable. Yet in the near future, the benefits of these specific actions will be likely limited to a subset of consumers. The consequence is that such products need to be personalized to match the value chain from processing to product.

9.5.2 Computing and Scientific Knowledge Management

Technology in every aspect of the industrial world is now driven by computerization. Information technologies are moving through the food process industry in virtually every aspect. The power of computational technologies to control food-related processing from unit operations to biotechnologies will grow dramatically over the next few decades (Lange et al., 2007). Computational mathematics is bringing revolutionary understanding of basic biological processes (Grigorov, 2005). The next steps for such advances in computational modeling will be to make them directly applicable to the improvement of individual human health (Sajda, 2006). Computational models, however, need standardized datasets. To realize the possibility of modeling personal health, we will need to coordinate the database structures of all biological and medical fields so that advances in the larger fields of biology and medicine can be readily translated into food and nutrition (Lemay et al., 2007a). This very logical application of bioinformatics into human health will require some important conceptual changes by the scientists in food-related health research. Complete, publicly accessible datasets must become the outcomes of scientific research, not simple hypothesis solutions.

Fields in which database consolidation has occurred (communication and satellite imaging, for example) vividly illustrate the applications that can be brought in to commercial practice (mobile phone technologies) at astonishing cost efficiencies (mobile phone technologies have supplanted hard-line technologies in the developing world). The example of financial transactions (credit cards) now routinely executed around the world shows what is possible today in terms of personalized technologies, and especially how little value is necessary to make it commercially viable. The routine use of electronic communication systems that spread throughout the world, in order to carry out large financial transactions, is now used for on-the-spot payment for a single cup of coffee. "If such a trivial value is sufficient to drive globally accessible credit card technology into coffee shops, then it is only the imagination of the scientist that limits what would be possible for foods and health if we could move to a more standardized and computer-friendly information system" (German et al., 2005b).

9.5.3 Technologies for Biomaterial Processing into Foods

Industrial food processing of commodities is designed to remove toxic and antinutritive components, prolong stability, and enhance the concentration of desirable components. As food becomes more health oriented, process engineering is increasingly informed by additional knowledge streams including the diagnostic results of health of targeted consumers (e.g., cholesterol and glucose levels in blood, allergic susceptibility) and the bioactivities of particular input stream options (e.g., phytosterols, probiotics). As foods become simultaneously more personal, then information associated with the preferences, health needs, and convenience criteria of individual consumers will be input variables that must be matched to the technologies capable of formulating to such personal and immediate tasks, i.e., point of consumption. Further, the diversity of the biomaterials accessible in food materials can be broadened substantially as the diversity of consumers is recognized.

The genomics revolution has enhanced not only the knowledge of biology, but also has dramatically altered the way research in the life sciences is conducted. Genomics promises to change the way engineering of biomaterials is carried out in many disparate fields, ranging from pharmaceuticals to foods. The key to success in food processing innovations will be to consolidate breakthroughs in biological aspects of biomaterials into the basic principles of food processing (Figure 9.6).

Food processing can also be considered to be a form of predigestion of raw biological materials prior to consumption. Personal information about health will be critical new inputs to the processes of ingredient selection, product assembly, structure formation, and even packaging. This is not a conceptual leap for food in principle since sensory, stability, and safety concerns have already been consolidated into food processing. However, while in industrial processing to achieve economies of

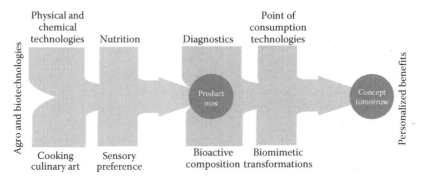

FIGURE 9.6 Knowledge streams in food processing.

scale, the goal is "bigger, faster, cheaper," for identical product qualities, the future of personalization will require "smaller, faster, cheaper" for "customized to consumer" quality attributes.

9.6 DIVERSITY OF FOODS: AXES ON WHICH PERSONALIZATION CAN DELIVER SOLUTIONS

9.6.1 COMPOSITION SPACE OF MULTIPHASE BIOMATERIALS

The basic biopolymers that make up foods interact in water to form complex structures. The composition and properties of these structures are critical to the quality, stability, and health properties of the foods as consumed. Hence, a great deal of research has been focused on understanding how to achieve particular product properties. Gels, fibers, emulsions, and foams are formed explicitly during food processing and preparation to provide products with distinct textural, mouthfeel, stability, and unique flavor-release properties.

Food ingredients assembled into complex, multiphase foods are not homogeneous, but instead spontaneously or with various energetic inputs form a wide variety of structures. This can be shown schematically as a generic space of food structures (Figure 9.7). In general, due to the immiscibility of most biomolecules and their solutions (lipids, proteins, and polysaccharides), foods are dispersions or suspensions in multicomponent and multiphase systems. Semisolid and solid foods are also best described as heterogeneous composite biomaterials.

The overall formulation space shown in Figure 9.7 forms a tetrahedron including all possible compositions of the three macronutrients (proteins, carbohydrates, and fats). Depending on the water content (the fourth apex on the tetrahedron) liquid foods (beverages) or semisolid and solid foods can also be represented within the solution space. This simplifying graph illustrates that changing a given food composition changes the expected structures in a predictable way, creating a set of new structures formed from a fusion of generic ones. These structures contribute to the stability, shape, texture, and sensory quality of the final foods, and are

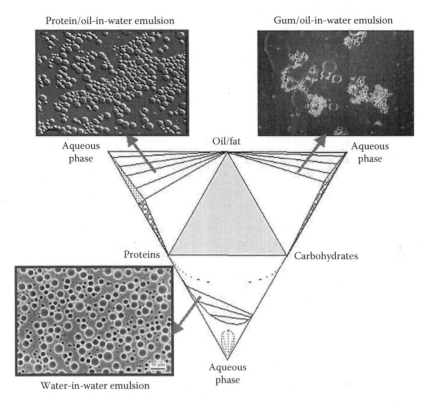

FIGURE 9.7 Food as a physical composite. This graph illustrates as a simplified phase diagram how changing a given food composition dictates a series of structures formed from a fusion of generic structures.

an indispensable part of the value of bread, dough, pasta, cheese, ice cream, etc. (German et al., 2005b).

The development of skills in basic food cuisine has led to considerable control of the structure of complex foods (either individual cooks or industrial process engineers). In simple terms, whether practiced in a kitchen or in a large industrial factory, the goal is to disconnect the specific biological functions of agricultural biomolecules (enzymes, muscle fibers, etc.) and gain the general material properties of their material classes (e.g., proteins and polysaccharides share with synthetic polymers their phase-separating behaviors in aqueous solutions, Tolstoguzov, 1986). The goal of food processing is to further extend these properties and transform the ingredient composition into distinct structural assemblies that control the attributes of the product (e.g., appearance, colors, flavors, and texture).

All possible combinations of food constituents (proteins, carbohydrates, lipids, and water) can be diagramed as a tetrahedron representing the overall generic space of food structures (see Figure 9.7). In most foods, that primarily contain water, the biopolymer structures are dispersions or suspensions mirroring the multicomponent and multiphase nature of the foods. In the foods lacking appreciable water, i.e., semisolids and solids, the structures are also describable as heterogeneous composite biomaterials. The final structures are a combination of the basic components of biomolecules, as well as the path to the final composition (temperature, pressure, mixing, etc.). With a considerable knowledge of how to combine different ingredients and control their assembly, it is possible to produce a wide variety of microscopic and macroscopic structures possessing desirable textures and specified rate of flavor release. This same knowledge can now be applied to vary the biomaterial composition and structures to explicitly meet the nutritional targets of individual consumers.

9.6.2 WHAT WE KNOW ABOUT STRUCTURE AND DIGESTION

Studies on fat digestion have pointed toward a strong influence of the dispersion structure both on the yield and dynamics of fat digestion and postprandial lipid state (Spelizini et al., 2005). Unfortunately to date, little research has attempted to elucidate the overall food structure "solution space," in terms of the effects that these multiphase mixtures exert on the digestion process. Furthermore, much less is known about how these effects influence overall health in individuals.

The best studied relationship between food biomaterial structure and intestinal functions is the digestion of carbohydrates. The widespread adoption of the term glycemic index in the marketing of complex carbohydrate rich foods attests to the recognition that structure relates to health. However, this simplifying approach ignores the fact that more complex structures within the intestine could influence digestion in unexpected ways. Combining thermodynamically incompatible protein solutions with polysaccharides would potential produce phase separating water-in-water emulsions.

In model studies, phase separated polymer solutions have been shown to induce partitioning of co-solutes (e.g. enzymes) between the separating phases. For example, the enzyme α-amylase partitions preferably into the dextran-rich phase in a polyethylen-glycol (PEG)/dextran aqueous two-phase system (Andersson and Hahn-Hägerdal, 1990). This now predictable behavior is used industrially to continuously convert dextran into simple reducing sugars which partition preferentially into the PEG phase. The amylase enzyme's favourable interaction with dextran enriches the dextran phase with the enzyme. Simultaneously, proteases including chymosin and pepsin show a spontaneous tendency to partition into the PEG-rich phase according to their strong PEG-protein interactions (Spelizini et al., 2005). Although these studies were performed in artificial systems (water soluble synthetic polymers) they provide a solution space description of the likely behaviour of complex food systems within the intestine, as well as the likely consequences to digestive reactions that can be expected to result from the physics of polymer phase separations.

9.6.3 STRUCTURE OF FOODS AS MULTIPHASE BIOMATERIALS

The importance of food structure for the overall metabolic response to foods has been examined in the experiments that analyze the metabolic effects of dietary carbohydrate. The varying metabolic responses to foods of constant overall composition but differing in carbohydrate structure have been so distinct that the variation in response to different carbohydrate-containing foods has been assigned a series of scientific and marketing terms including glycemic index and glycemic response.

These broad concepts notably the glycemic index (GI) provide a somewhat over-simplifying view to the overall complexity of the metabolic response to different carbohydrates. Nonetheless, it is clear that the composition and amount of carbohydrates alone do not correlate with postprandial glycemia (Dickinson and Brand-Miller, 2005). The fact that food products with the same composition but different structures generate different postprandial metabolism of carbohydrates, lipids, and amino acids and their various signaling hormones illustrates that the food structure itself is important to the overall metabolic regulation. Yet, nutritional sciences have not developed a detailed comprehensive understanding of how different aspects of food structure beyond that of gross composition affect the nutritional, metabolic, physiological, and immunological responses to foods.

The human digestive tract is evolved to survive on a variety of structures from liquid foods (milk) to solid particles (e.g., intact cereal grains). Under this selective pressure to obtain maximum fuel and nutrients from biomaterials in the environment, the intestine's responses are complex and varied. The success of humans throughout the world attests the remarkable abilities of our intestines to obtain sufficient fuel and nutrients from a wide range of plant and animal materials as foods. Studies taking a qualitative perspective to describe the digestive tract as a chemical reactor have been useful in providing a broad overview of the basic properties of this remarkable bioprocess engine (Logan et al., 2003; Wolesensky and Logan, 2006).

> To understand the links between food structure and intestinal properties and thus ultimately to understand metabolism, we should understand the multiple time- and structure-dependent processes of digestion and absorption. These models reveal that food structure affects the digestive breakdown of macromolecules and the absorption of the liberated monomers.
>
> Studies on the influence on digestion of the physical form of barley grains show clearly a strong influence of structure on the amount of absorbable glucose. Not surprisingly, intact cell walls slow the enzymatic degradation of starch (Livesey et al., 1995). Such results have been extended to various clinical subjects. For example, the amount and physical form of starch not digested in an ileostomy effluent, using Barley flakes, flour or starch, has been shown to depend on the presence of cell walls, both intact and in fragments (Botham et al., 1997). The starch contained in the effluent coming from barley flour is half of that coming from flakes. The digestion of barley flakes within the intestine leaves undigested the larger part of starch granules.
>
> Polysaccharide cell walls are not the only elements of food structure that influence subsequent digestive, absorptive and metabolic rates. Even small changes in food structure can influence subsequent metabolism. Armand et al. showed that a food material containing large-sized emulsion particles (low interfacial surface) produced a more rapid appearance of lipids in blood than the same quantity of dietary fat in smaller-sized emulsions (Armand et al., 1999).

9.6.4   WHERE WOULD WE START TO PERSONALIZE FOODS?

Foods are produced from biological raw materials. These biomolecules naturally make up the organized materials of cellular structures of agricultural commodities (plants, animals, and microorganisms), transformed and refined during food processing to achieve the safety, stability, and nutritional compositions of food products. The ingredients of foods, therefore, comprise both primary molecules of these commodities and their supramolecular structures. Processing can alter both the composition of the basic molecules, and importantly these cellular and biological superstructures.

One of the specific goals of traditional cuisines (either individual cooks or industrial process engineers) is to dissociate specific biological functions of agricultural biomolecules and gain the general material properties of their material classes (e.g., proteins and polysaccharides share with synthetic polymers their phase-separating behaviors in aqueous solutions; Tolstoguzov, 1986, 2004). Food processing transforms the ingredient composition in distinct structures which control all attributes of the product (e.g., appearance, colors, flavors, and texture).

One of the Darwinian pressures on the evolution of plants within the overall competitive biosphere is predation by animals. Avoiding this constant, aggressive threat to their genomes led to the elaboration of a vast array of secondary metabolites the presence of which was capable of discouraging the predation by a wide variety of biological strategies. Not the least of these strategies was the synthesis of antinutritious and overtly toxic compounds in potentially edible parts of the plants. Animals were thus placed in the equally constant and aggressive Darwinian pressure of dealing with these compounds. Digestive enzymes, toxin-binding proteins, bitter-taste receptors, and even commensal microorganisms emerged as competitive elements in animal genomes. The ongoing success and failure of their respective strategies have been responsible for the competitive balance of plants and animals throughout their long evolutionary history. Agricultural practices over the past several thousand years have joined this battle and the remarkable success of modern humans is the result. In the past few decades, human biotechnologies have literally began to turn the disparate elements in plant genomes against them and to take advantage of this vast array of secondary metabolites for the benefit of humans. A "bitter" irony indeed!

Early agriculture needed to target the overtly antinutritional components in food raw materials. These early efforts succeeded in part through empirical selective breeding. The least toxic, antinutritive, and palatable individual plants of each successive generation of crops were enriched and therefore, successful commodities emerged. However, the handling and processing of these commodities into foods was also critical in managing the toxicity and antinutritive components of plant materials. Separation of the most edible parts of plants, as well as heating and microbial fermentation were the toolsets by which processing of plant commodities produced more nutritious and safer food stuffs.

Throughout the long history of food process development, separation of components and their modification to safer, more stable chemical constituents became a driving force. Food processing was optimized to reduce the natural variability of the starting materials, providing constant quality and safety for the products. Among the losers in this strategy were biodiversity, complex structures, and the inherent biological activities of the plant components. Food structure was considered an important target for product quality. Yet, rather than using the basic structures of complex commodities, structure as texture (foams, gels, and emulsions) was reconstructed from basic biomaterial ingredients (proteins, lipids, and carbohydrates). The apparent goal was only to optimize palatability and stability of a product. The important role of food structure on the dynamics of digestion and its influence on the postprandial state and human health does not appear to have been expressly recognized or enhanced. Some interesting exceptions to this basic idea of the casual attitude of food structure in health can be seen in some culinary practices (al dente pasta, cheese) in which the benefits of food structure were arrived at empirically.

The human digestive tract as a chemical/biochemical reactor and the control of digestion has only recently been considered in basic nutrition research, and has not yet become a target for food development (Figure 9.8). Complex food materials pause and then are ejected from the stomach (1) as an initial physical and chemical converter (low pH, strong acid, active hydrolytic enzymes, and muscular turbulent flow). The upper portion of the small intestine (2) conducts an influx of homogenized emulsion phase, rapid neutralization of pH, and the addition of biological detergents (bile), catalysts (pancreatin), and muscular propulsions. Spontaneous disassembly of structures and the self-assembly of others are highly dynamic during the next period (3) and the intestine plays a very active role by selectively removing the majority of the absorbable nutrients capable of diffusing to the massive surface area of enterocytes. The metabolic, immune, endocrine, and muscular systems are actively sampling the luminal contents and modifying whole body functions, accordingly. Food

Food structure and composition and their influence
on gastrointestinal processes

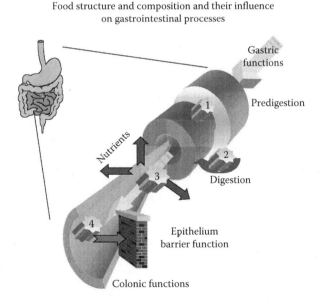

FIGURE 9.8   The gastrointestinal system as a model reactor.

composition and structure continue to play a key role in the colon (4) modifying the microflora, the populations of bacteria and their physiology, the intestinal crosstalk that results and the metabolic products of bacteria that fuel and alter the health of the epithelial lining.

The complexity of the gut and its myriad physiological, immunological, metabolic, and neurological processes makes it clear that the postprandial state is sensitive to more than the simple composition and to the energy density of the food. The structural dimensions of the food influence the amount of the nutrients and the time and location in which they are solubilized and adsorbed into the body. Further, the neuroendocrine signals that coordinate the subsequent responses to food intake are highly dependent on the dynamics by which the complex biomolecules pass along the gut and deliver small molecules to the intestinal surface for absorption and signaling (Mace et al., 2006).

> Historically, the structure of foods and the explicit role of food structure on health were not considered by nutrition research. One consequence is that it is not possible at present to take great health advantage by controlling food structure. Food structure alters digestive processes, intestinal physics and dynamics, the rates and extends of absorption. It is instructive to examine scientific evidence to gain insights into the possibilities.
>
> Taking an evolutionary approach, milk, the product of Darwinian selection for nourishment and health of mammals, suggests that the time dependency of nutrient delivery is apparently under remarkable control. All of the macronutrient classes in milk exhibit structure-specific dynamics different than would be expected from simple homogenous solution behaviour. The casein proteins in milk are random coil protein polymers. It would be anticipated from this molecular structure that proteolytic digestion of these proteins would proceed rapidly. It does. However, the supramolecular organization of caseins as colloidal casein micelles confers a distinct digestive property in vivo. Acidic conditions destabilize casein micelles inducing a large protein coagulum (curd) that delays clearance from the stomach.
>
> The Darwinian pressure through evolutionary led to an even more remarkable property of milk protein digestion. K-casein is a glycoprotein conjugate that forms part of the overall casein micelle. In turn, chymosin is a highly selective protease enzyme (rennin). When chymosin, produced only in the mammalian infant stomach, encounters k-casein, the glycoconjugate subfragment of k-casein is cleaved off liberating a large glycosylated peptide (glycomacropeptide) which diffuses away from the micelle. The loss of this simple peptide destabilizes the entire casein micelle structure and once again, it coagulates into a supramolecular gel (curd). The net effect of these evolutionary elaborations of protein structure and enzymatic activity is to slow digestion by structure (gel) formation, and by using structure, control the rate of delivery of the major milk proteins and their absorbable amino acids, peptides, calcium, phosphorus, etc.

9.6.5 TECHNOLOGIES FOR FOOD PRODUCTION

The technological advances of the first half of the twentieth century brought the science of chemistry into practice for a wide range of applications. The petroleum chemistry revolutionized fuel and transportation. Fertilizer chemistry drove the green revolution in agriculture. Chemistry also revolutionized the food process industry.

Chemistry helped to design processes that could synthesize and separate chemicals as ingredients and processing aids to improve the safety, quality, and stability of agricultural commodities and food products (acids, gases, emulsifiers, colors, flavors, etc.). Vitamins were chemically synthesized, bringing freedom from nutrient deficiency diseases.

Today's research in food engineering continues to discover new approaches to processing materials within the existing unit operations, with the same goals in mind. Novel technologies that could potentially lower the cost and enhance the effectiveness of existing unit operations in food processing are studied. For example, it is possible to render pathogenic bacteria inactive by high pressure. Thus high-pressure sterilization of processed foods is studied intensively (Gould, 2000).

The major technology-based innovation that is transforming food processing is information itself. Computational technologies have come to dominate all industries of the twentieth century, based on the acquisition, storage, and mathematical manipulation of information. Computerized information and control now dominates food production, processing, distribution, and marketing. The technologies of information management are ready to combine food processing with biological activities (see Figure 9.6).

9.6.6 BIOGUIDED FOOD PROCESSING

As previously discussed, operations in food processing fractionate food components according to their physical characteristics, density, polarity, solubility, and size. These components are in turn exposed to varying, controlled environmental conditions, temperature, pressure, hydration, ionic strength, pH, and redox potential, all designed to maximize the stability and safety of the final products. Under such intensive process conditions, the biopolymers (proteins, lipids, and carbohydrates) are treated as their common material properties. Each class of molecules is viewed as valuable only for its bulk material properties, independent of the life forms from which they are derived and the biological activities for which they evolved. The goal of such intensive processing is logical in historical context. It was important to eliminate the antinutritive properties of plant materials including the protease inhibitors of legumes, the hydrolytic activities of all materials, and the oxidative activities of legumes, grains, and roots. However, eliminating these deleterious bioactivities also destroyed the potentially valuable structures, complexes, and activities that could be recruited to the health of individual human consumers.

Bioguided processing refers to the various strategies of agricultural and food processing in which the properties of raw materials are targeted, retained, and complemented to directly deliver specific biological health benefits to consumers (Ward et al., 2004). As a more personal perspective develops in diet, and foods are targeted to individual consumers and their unique needs and desires, all of the diversities of biology, all of the input variables of agriculture, and all of the technologies of biomaterials handling can be recruited to and focused on the goal of connecting bioactivities and their unique properties to individual health, performance, and delightful values (see Figure 9.9).

Transformation	Biological process	Benefit
Separation	Dissolution of specific molecular classes of lipids by soft oil processing Elimination of undesirable components	Yield Purity stability Safety
Complexation	Stabilize reactive compounds Maintain labile structures	Enhanced stability and availability
Self-assembly	Complex surfaces as binding agents Complex structures as releasing systems	Pathogen protection Controlled release

FIGURE 9.9   Biological transformations achieved during processing.

Bioguided processing enhances the ability to retain or enhance various molecular and structural properties of foods. When achieved with bioguided approaches, the basic processing operations of separation, complexation, and assembly offer considerably greater selectivity for biological structures. These processes generate biological activities for the consumer, not possible when using only physical- and chemical-based process conditions.

How will the basic concepts of biodiversity for the advantage of health be actually brought to practice? Taking such a perspective into new commodities and new foods runs the risk of retaining deleterious factors that have been successfully eliminated by traditional food processing. Therefore, bioguided processing requires knowing and understanding the specific details of both desired and undesired structures. Throughout history, empirical trial and error, presumably at the cost of considerable human sacrifice repeated over generations, was the means to achieve success. Trial and error food safety is not viable. A detailed scientific knowledge must replace facts-filled empiricism. Advances in comprehensive biological knowledge of organisms, genomics, combined with the detailed comprehensive analytics of biomolecules (proteomics and metabolomics) will provide the means to investigate the composition and structure of raw materials, their responses to various technological inputs, and their consequences to diet and health. Newly acquired knowledge of biosynthetic pathways will serve as a starting point for material separation and processing. One of the obvious examples of how this science will proceed is provided once again by the biomaterial that emerged through evolution as a bioprocessed food: milk.

The mammary gland as a bioreactor is the tissue on which most of evolutionary pressure was directed for milk production. Perhaps first and foremost, milk must provide all essential nutrients to the infant from birth to weaning. This very obvious target puts a metabolic and nutritional stress at the same time on the mother for whom these nutrients are essential as well. The pressures on evolution are thus to provide not only the quantities of each nutrient that are necessary for a particular infant at each point in its development, but also provide the nutrients in as bioavailable form as possible. This latter pressure provides a most vivid example of bioguided processing in mammalian milk. When examined closely, the multiple complex forms of nutrients found in milk provide remarkable lessons in how the intestine can be assisted with the very difficult process of acquiring unstable, sparingly soluble and difficult to absorb, vitamins, minerals, amino acids, and fatty acids.

> Lactoferrin is a very well studied model of mineral binding in a nourishing format. The 2 moles of iron bound to lactoferrin are maintained within a coordinate ligand geometry within the protein that prevents the iron from exerting its redox cycling properties that are so destructive to other biomolecules. Recent research has revealed that the human infant intestine expresses a lactoferrin receptor that binds and internalizes the protein intact, bringing with it, its complexed minerals (Suzuki et al., 2005). Thus the mammary gland achieves the successful, targeted delivery of iron to the infant's tissues through this distinct bio-guided process of lactoferrin assembly. The scientific insights in nutrition provided by such examples from milk will continue to inform the first generation of strategies for bio-guided nutrition (Figure 9.10). As scientists build greater knowledge of how to bioguide food materials, plants, microorganisms, and animal tissues will become seen as biological assets for their specificity and bioactivity and not just their agricultural yield of simple micro and macronutrients.

The ability to specifically introduce biological structures and activities into foods through more gentle (soft) processing allows a unique set of components to be recognized for their specific activities. What activities could soft processing add?

Activity	Mechanism	Benefit
Catalytic enzymes	Release encrypted peptides, oligosaccharides, polynucleotides, lipids	Enhanced efficacy
	Breakdown deleterious components	Improved safety
Receptor ligands	Bind to targets activating signal transduction pathways	Metabolic, immunological, physiological regulation
	Bind and antagonize signals	
Colloidal structures	Complex surfaces as binding agents	Pathogen protection
	Complex structures as releasing systems	Controlled release
Binding	Complex formation between proteins and ligands milk Casein and Ca++	Nutrient absorption
	Toxin binding by specific proteins, oligosaccharides	Toxin elimination

FIGURE 9.10 Biological activities guided to personal targets of health.

For example, many components known to exist in mammalian milk are destroyed by traditional processing. Each of these benefits is known to be achieved for newborn infants and yet, would also be of value to subsets of food consumers according to their particular health situation.

9.6.7 SELF-ASSEMBLY PROCESSES IN FOOD BIOMATERIALS

The personalization of foods will mean a gradual migration of food processing. The migration will begin with today's large centralized factories producing large numbers of identical products to tomorrow's processing that is capable of assembling more customized formulations appropriate for more individualized foods. In general, this will require that foods be assembled in smaller unit operations. While it is currently assumed that this assembly will require factory-like operations, such is not necessarily true. The goals of current processing of ensuring microbial safety while maintaining basic nutrient composition will broaden to include the added value of retaining many of the inherent biological properties of the commodities from which the biomaterials originally derive (Ward et al., 2004). Furthermore, considerable external energy is input during processing to produce multiple phases (foams, emulsions) as the means to formulate, stabilize, and texturize food products. Many of these desired goals could be achieved without macroscopic, high-energy unit operations currently used. In fact, to the extent possible, food processing will take advantage of the properties of molecules within biomaterials to spontaneously self-assemble into desired forms (Leser et al., 2006).

Self-assembly of biomolecules is not a new concept invented for the food sciences. Indeed, life itself is based upon it. The most remarkable property of life is that every organism, from the smallest to the largest, is ostensibly self-assembled from simple biopolymers. As the sciences of genomics begin to understand the more structure-specific properties of molecules as gene products, the self-assembly properties of biomolecules, even complex colloids, will be understood and controlled (de Campo et al., 2004). Many examples illustrate the principle of guiding the natural properties of biopolymers to the net benefit of food products. Cheese making illustrates how to recruit the biological properties of casein micelles to aggregate into three-dimensional gel networks outside an infant's stomach.

9.7 CONSUMER HEALTH, BIOLOGICAL KNOWLEDGE, AND ADVANCED TECHNOLOGIES PRACTICE AS PERSONALIZED FOODS

9.7.1 INDIVIDUAL HEALTH

The personalization of health will require three elements: assessment technologies, health trajectory prediction, and intervention strategies. Assessment was discussed above. It is becoming clear from the rapid rise in personalized medicine that the technologies to assess health are emerging for therapeutic medicine and will be immediately translated into other aspects of health.

Managing an existing disease is, however, not the same as projecting the health trajectory of healthy individuals according to an assessment even of a large number of metabolites. Decades of research were needed to recognize that high levels of cholesterol in blood could predict future heart disease that altered hepatic cholesterol metabolism was responsible for high-cholesterol levels in blood and that intervening in this metabolism through diet or drugs led to improved health—i.e., lower risk of heart disease (Plosch et al., 2005). Nonetheless, this success provides confidence that it is possible to act on systemic health without having a full mathematical, systems biology solution for the entire physiology of the organism.

Looking forward to many aspects of health, solutions will be pursued largely empirically, both in populations and in individuals. This empirical approach will be successful when the following four conditions are met:

1. Assessment technologies are used to measure large numbers of individuals.
2. Databases chronicling the outcomes of assessment with health outcomes are constructed and annotated appropriately.
3. Individual assessment results can be compared with outcomes from the larger databases.
4. Assessment converted to projections and interventions as appropriate.

The advantages of a combined empirical–mechanistic approach to health assessment and improvement are that we can start now—the discovery process will "find" the most common health issues first, thus providing value to the largest number of individuals immediately. The marketplace for these values will be competitive and push health improvements and commercial developments continuously with time and as results emerge.

9.7.2   Personalizing Foods and Diets

The food processing industry is responsible for providing safe and affordable foods based on manufacturing large quantities of relatively standardized products in factory operations, with these factories engineered to deliver well-defined and validated unit operations. The economies of scale that are possible with this process model have substantially increased the yield and reduced the costs of producing foods. Cost reductions extend from agricultural commodity production to food product assembly to market distribution. The food industry has made dramatic improvements in many aspects of manufacturing, waste management, environmental protection, and worker safety.

Is the foregoing success by the food industry compatible with a personalized approach to food production? Is personalization of diets according to individual needs for health possible both technologically and within the economic framework of modern society? We cannot be certain how the development of personalizing foods will proceed. Models of existing success in personalizing consumer products based on science, technology, and consumer values can help. One example of what is not only possible but available today as a personalized consumer product is GPS-based golfing. Motorized carts that transport individual golfers around the course are equipped with a GPS-locating system. The technology necessary to this product is absolutely state of the art. Orbiting satellites equipped with atomic clocks, using Einsteinian mathematics to triangulate distance coordinates, accurately pinpoint the devices against the massive database of the Earth's geographical surface. What is most remarkable is the value proposition. The consumer value that all of this technology is assembled to provide is to indicate the distance from the cart to the hole. This "personalized golf location" system is already in place as an example. How then can one argue that a more personalized approach to dietary guidance for health is either scientifically or technologically impossible? The question is simply how will it be done?

9.7.3 FUTURE PERSPECTIVES ON PERSONALIZING FOOD

Considerable technical hurdles remain to convert industrial food processing into a more personalized food delivery system. And yet, in many ways food will return to a more personal past. Food preparation has been very personal as each individual within the family unit ate according to the expertise of the family representative (typically mother). In this traditional model, the mother assumed the role of a diet designer for each member of the family. It was the mother's responsibility to process raw commodities into foods for the family meals. Anecdotal and cultural history combined with close individual inspection as the basis of diet decisions.

The question that industrial food must face is thus not whether diets and foods can be personalized. Rather, the question becomes how can the modern, highly efficient industrial processing that gained its value largely by centralizing food processing and gaining economies of scale be converted to an industrial system of personal food production. The values—safety, affordability, convenience, and delight—that brought industrialized foods to such success will continue to operate. To succeed, food production will invariably need to maintain all of these values and yet, become more personal with respect to health.

For any personalization of foods to be successful in altering health, it must be flexible in product composition. No matter what may be the differences in individual human health problems and requirements, to a great degree they will be resolved in large part by consuming diets of different molecular compositions. How then can one imagine that different compositions of foods, now decided at the stage of factory formulation, can be delivered to different consumers at the point of their purchase, or even at the point of consumption?

One model solution would be to increase the number and diversity of foods manufactured by the existing factory processes. In such a model, customization would be performed at the point of distribution using a system of guided choices. A customer would employ a personal knowledge device (e.g., handheld computer) to make specific choices appropriate for that individual. Such a model might seem attractive in creating minimal disruption in existing corporate structures, but it fails in many respects to deliver many of the potential benefits that personalized foods could and should provide. From the business viewpoint, it is also unlikely that the added inventory control, dilution of market space, and factory flexibility necessary to such a system would be cost effective.

An alternative model would be to alter the processes of food formulation such that much of the final compositions would be assembled at the point of purchase, preparation, or consumption. In this model, intelligent systems would assemble foods at the point where, and at the time, they are consumed. There are examples where such a conceptual change has occurred. Nespresso™ is a product in which a substantial part of the processing of espresso coffee (highly controlled temperature and pressure extraction of ground beans) is now provided by a device in the kitchen or even at the table of the consumer. The coffee drinking experience is "personalized," by different flavors, with the processing in large part performed immediately before consumption. In fact, moving from a one-size-fits-all model to a more plastic and dynamic system of food formulation would not be limited to the compositions for health. Flavor would be as personalized as health, simultaneously.

The example of Nespresso provides a proof of the concept for both flavor and health. If the composition of a meal/food was customized for a single individual, then its composition for health (macronutrients, micronutrients, bioactive molecules) could be controlled, but so too would its composition for a more personal preference for flavor, texture, and convenience. Even safety could be customized ensuring that individuals with allergies or intolerances to particular food components were assured that these were avoided.

9.8 CONCLUSIONS

The food industry in the twenty-first century will continue to deliver safe, convenient, affordable, and delicious food products that provide the means to assemble diets that maintain and improve the health of consumers. Nutrition and food processing research is designed to solve the health problems of the twenty-first century consumer.

Parallels to the science and consumer of the twentieth century are vivid. At the end of the nineteenth century, diseases caused by nutrient deficiencies were epidemic. They were solved by the success of identifying the essential nutrients; recognizing their absolute requirements; and retaining, enriching, and fortifying the food supply to ensure that deficiency diseases could be prevented. The sheer elegance of this success becomes clear when we realize how completely solved these historical, now rare nutrition deficiency diseases are. Much of the developed world has never seen the phenotypes of diseases caused by classic deficiency diseases: iodine, goiter; vitamin C, scurvy; vitamin A, blindness. We do not fear these diseases; we only know them from historical accounts.

The nutrition diseases of today are different. During the past three decades, diseases caused by unbalanced diets have become epidemic across the developed world (Alberti, 2001). Diseases including atherosclerosis, obesity, diabetes, hypertension, and osteoporosis are attributed at least in part to food choices. Food and nutrition research must adapt to these challenges and consider strategies to solve them. A combination of research knowledge and food applications must provide consumers with diets continuing the value systems of safety, quality, stability, convenience, and cost. These values are not sufficient. Diets must also ensure that each consumer as an individual maintains optimal health within the lifestyle that they choose to pursue.

The growing scientific knowledge of the diversity of human genetics, lifestyles, and environments shows that people are different. The same diet will not be optimal for all, and some form of personal diets is necessary. This realization would appear to be devastating to the current system of industrial food production based on large factories producing homogeneous products to high tolerances of safety, quality, and cost.

Now, to what must be done. The steps necessary for personalizing diet and health are sizeable.

1. Personal human health assessment is necessary. As the first example, measuring cholesterol, proved to be the key to identifying individuals with high cholesterol and at risk of future heart disease. Now, it will be necessary to assess many other aspects of health and identify those individuals who are at risk of metabolic disorders (obesity, diabetes, hypertension, and osteoporosis) and other health implications, including, for example, immunological issues such as immune senescence, allergy, and inflammation.

2. Food solutions are necessary. Once an individual's health has been assessed and the scientific knowledge of what the assessment results implied in terms of dietary composition has been put in place, it will be necessary to provide that individual with foods customized to that composition. Such a customization of industrial food processing seems at first impractical. Nonetheless, the food industry is already highly plastic in many aspects. Dispensing technologies apportion sweetener and whiteners to coffees, and expanding such capabilities into personal dispensing systems could be done immediately. The rate of arrival and acceptance of these technologies will be determined by the ingenuity and creativity of scientists, technologists, and engineers throughout the food industry, and the values that they provide to consumers.

Industrialized foods have revolutionized the human condition, freeing people from the need to spend hours everyday in food preparation and creating spectacular advances in each individual's capabilities to succeed in life and to enjoy life's success. Nonetheless, freedom must be matched by responsibility. The growth of metabolic diseases in the population (obesity, diabetes, and atherosclerosis) has proven that diet cannot be treated casually. Consumers need to act responsibly, to educate themselves in the consequences of actions. It is more important than ever to invest in scientific research and technologies. Scientists must build the knowledge that can guide the decisions that impact human lives.

A future in which consumer desires, scientific research, and technological innovation are combined is very attractive. Foods will be personally more healthful and simultaneously delightful. Their safety, convenience, and even appearance will be similarly more personal. With such a personal approach to delivering health solutions, it will become possible to consider health more broadly. Health is not only the freedom from disease. Individuals will also define health as their ability to achieve their goals for quality of life, for physical and athletic performance, for cognition and learning, and for comfort and freedom from fears of diseases, allergies, and intolerances. Society is on the threshold of a new era of food. Our lives will not be limited by the foods we make. Instead, we can look forward to a future when we will select the foods that truly enable us to achieve the lives we would like to lead.

REFERENCES

Ailhaud, G., P. Guesnet, and S.C. Cunnane, Feb 28, 2008, Emerging risk factor for obesity: Does disequilibrium of polyunsaturated fatty acid metabolism contribute to excessive adipose tissue development? *British Journal of Nutrition.* 100, 461–470.

Akamine, D., M.K. Filho, and C.M. Peres, May 2007, Drug-nutrient interactions in elderly people, *Current Opinion in Clinical Nutrition and Metabolic Care.* 10(3), 304–310. Review.

Alberti, G., 2001, Noncommunicable diseases: Tomorrow's pandemics, *Bulletin of the World Health Organization,* 79, 907.

Andersson, E. and B. Hahn-Hägerdal, 1990, Bioconversions in aqueous two-phase systems, *Enzyme and Microbial Technology.* 12, 242–254.

Armand, M., B. Pasquier, M. André, P. Borel, M. Senft, J. Peyrot, J. Salducci, H. Portugal, V. Jaussan, and D. Lairon, 1999, Digestion and absorption of 2 fat emulsions with different droplet sizes in the human digestive tract, *American Journal of Clinical Nutrition.* 70, 1096–1106.

Bäckhed, F., R.E. Ley, J.L. Sonnenburg, D.A. Peterson, and J.I. Gordon, Mar 25, 2005, Host-bacterial mutualism in the human intestine, *Science.* 307, 1915–1920. Review.

Bartley, K.A., B.A. Underwood, and R.J. Deckelbaum, May 2005, A life cycle micronutrient perspective for women's health, *American Journal of Clinical Nutrition.* 81(5), 1188S–1193S. Review.

Bateson, P., D. Barker, T. Clutton-Brock, D. Deb, B.D'Udine, R.A, Foley, P. Gluckman, K. Godfrey, T. Kirkwood, M.M. Lahr, J. McNamara, N.B. Metcalfe, P. Monaghan, H.G. Spencer, and S.E. Sultan, 2004, Developmental plasticity and human health, *Nature.* 430(6998), 419–421.

Bertholet, R., Junkuan-Wang, H. Watzke and J.B. German, 2005, Process for production of an oil containing long chain polyunsaturated fatty acids derived from the biomass, and foods, dietary products, cosmetics and pharmaceutical products containing it. World Patent Number EP1396533A1.

Botham, R.L., P. Cairns, R.M. Faulks, G. Livesey, V.J. Morris, T.R. Noel, and S.G. Ring, 1997, Physicochemical characterization of barley carbohydrates resistant to digestion in a human ileostomate, *Cereal Chemistry.* 74(1), 29–33.

Bowman, S.A., Sept 2002, Beverage choices of young females: Changes and impact on nutrient intakes, *Journal of the American Dietetic Association.* 102(9), 1234–1239.

Bruin, S. and Th.R.G. Jongen, 2003, Food process engineering: The last 25 years and the challenges ahead, *Critical Reviews of Food Science and Food Safety.* 2, 42–81.

Burdge, G.C., M.A. Hanson, J.L. Slater-Jefferies, and K.A. Lillycrop, June 2007, Epigenetic regulation of transcription: A mechanism for inducing variations in phenotype (fetal programming) by differences in nutrition during early life? *British Journal of Nutrition.* 97(6), 1036–1046. Epub. 2007 Mar 7. Review.

Catchpole, G.S., M. Beckmann, D.P. Enot, M. Mondhe, B. Zywicki, J. Taylor, N. Hardy, A. Smith, R.D. King, D.B. Kell, O. Fiehn, and J. Draper, 2005, Hierarchical metabolomics demonstrates substantial compositional similarity between genetically modified and conventional potato crops, *Proceedings of the National Academy of Science USA.* 102, 14458–14462.

Clydesdale, F.M., 1997, A proposal for the establishment of scientific criteria for health claims for functional foods, *Nutrition Review.* 55, 413–422.

Clydesdale, F., 2004, IFT expert panel report on functional foods: Opportunities and challenges, *Institute of Food Technologists.* 58, 35–40.

Collins, F.S., E.D. Green, A.E. Guttmacher, and M.S. Guyer, April 24, 2003, A vision for the future of genomics research, *Nature.* 422(6934), 835–847.

de Campo, L., A. Yaghmur, L. Sagalowicz, M.E. Leser, H. Watzke, and O. Glatter, June 22, 2004, Reversible phase transitions in emulsified nanostructured lipid systems, *Langmuir.* 20(13), 5254–5261.

Delmer, D.P., 2005, Agriculture in the developing world: Connecting innovations in plant research to downstream applications, *Proceedings. National Academy of Science. USA.* 102, 15739–15746.

Dickinson, S. and J. Brand-Miller, 2005, Glycemic index, postprandial glycemia and cardiovascular disease. *Current Opinion in Lipidology.* 16, 69–75.

Domingues, F.S., W.A. Koppensteiner, and M.J. Sippl, 2000, The role of protein structure in genomics, *FEBS Letters.* 476, 98–102.

Drewnowski, A. and N. Darmon, July 2005, The economics of obesity: Dietary energy density and energy cost, *American Journal of Clinical Nutrition.* 82(1 Supplement), 265S–273S. Review.

European Plant Science Organization, 2005, European plant science: A field of opportunities, *Journal of Experimental Botany.* 56, 1699–1709.

Fay, L.-B. and German, J.B., Apr 2008, Personalizing foods: Is genotype necessary? *Current Opinion in Biotechnology.* 19(2), 121–128.

Fiehn,. O., 2001, Combining genomics, metabolome analysis, and biochemical modelling to understand metabolic networks, *Comparative and Functional Genomics.* 2, 155–168.

German, J.B., F.L. Schanbacher, B. Lönnerdal, J. Medrano, M.A. McGuire, J.L. McManaman, D.M. Rocke, T.P. Smith, M.C. Neville, P. Donnelly, M.C. Lange and R.E. Ward, 2006, International milk genomics consortium, *Trends in Food Science Technology.* 17, 656–661.

German, J.B., B.D. Hammock, and S.M. Watkins, 2005a, Metabolomics: Building on a century of biochemistry to guide human health, *Metabolomics.* 1, 3–8.

German, J.B., C. Yeretzian, and H.J. Watzke, 2005b, Personalizing foods for health and preference, *Food Technology.* 58, 26–31.

German, J.B., C. Yeritzian, and V.B. Tolstoguzov, 2007, Olfaction, where nutrition, memory and immunity intersect, in *Flavours and Fragrances: Chemistry, Bioprocessing and Sustainability,* R.G. Berger (ed.). Springer, Berlin, pp. 25–23.

Gilbert, P.A. and S. Khokhar, April 2008, Changing dietary habits of ethnic groups in Europe and implications for health. *Nutrition Reviews.* 66(4), 203–215. Review.

Gould, G.W., 2000, Preservation: Past, present and future, *British Medical Bulletin.* 56, 84–96.

Grigorov, M.G., 2005, Global properties of biological networks, *Drug Discovery Today.* 10, 365–372.

Haddad, F. and G.R. Adams, April 2000, Aging-sensitive cellular and molecular mechanisms associated with skeletal muscle hypertrophy, *Journal of Applied Physiology.* 100(4), 1188–1203.

Henriksen, T., May 2006, Nutrition and pregnancy outcome, *Nutrition Reviews.* 64(5 Pt 2), S19–S23; discussion S72–S91. Review.

Krauss, R.M., 2001, Dietary and genetic effects on low-density lipoprotein heterogeneity, *Annual Review of Nutrition.* 21, 283–295.

Iweala, O.I. and C.R. Nagler, Oct 2006, Immune privilege in the gut: The establishment and maintenance of non-responsiveness to dietary antigens and commensal flora. *Immunology Review.* 213, 82–100. Review.

Labossiere, R. and M.A. Bernard, Jan 2008, Nutritional considerations in institutionalized elders. *Current Opinions in Clinical Nutrition Metabolic Care.* 11(1), 1–6. Review.

Lai, C.Q., E.S. Tai, C.E. Tan, J. Cutter, S.K. Chew, Y.P. Zhu, X. Adiconis, and J.M. Ordovas, 2003, The apolipoprotein A5 locus is a strong determinant of plasma triglyceride concentrations across ethnic groups in Singapore, *Journal of Lipid Research*. 44, 2365–2373.

Lange, M.C., D.A. Lemay, and J.B. German, June 2007, Multi-ontology framework to guide agriculture and food toward diet and health, *Journal of Science. Food Agriculture*. 87(8), 1427–1434.

Langley-Evans, S.C., Feb 2006, Developmental programming of health and disease, *Proceedings of the Nutrition Society*. 65(1), 97–105.

Lemay, D.G., A.M. Zivkovic, and J.B. German, 2007a, Building the bridges to bioinformatics in nutrition research, *American Journal of Clinical Nutrition*. 86, 1261.

Lemay, D.G., M.C. Neville, M.C. Rudolph, K.S. Pollard, and J.B. German, 2007b, Gene regulatory networks in lactation: identification of global principles using bioinformatics, *BMC Systems Biology*. 1, 56. Epub. Nov 2007b.

Leser, M.E., L. Sagalowicz, M. Michel, and H.J. Watzke, Nov 16, 2006, Self-assembly of polar food lipids, *Advances in Colloid and Interface Science*. 123–126, 125–136.

Livesey, G., J.A. Wilkinson, M. Roe, R. Faulks, S. Clark, J.C. Brown, H. Kennedy and M. Elia, 1995, Influence of the physical form of barley grain on the digestion of its starch in the human small intestine and the implication for health, *American Journal of Clinical Nutrition*. 61, 75–81.

Logan, J.D., A. Joern, and W. Wolesensky, 2003, Chemical reactor models of optimal digestion efficiency with constant foraging costs, *Ecological Modelling*, 168, 25–38.

Mace, K., I. Corthésy-Theulaz, L.-B. Fay, H. Watzke, P. van Bladeren, and J.B. German, 2006, Effects of food on metabolic regulation and disorders, *Nature Insight Obesity and Diabetes*. 444(7121) 839–888.

Malik, V.S., M.B. Schulze, and F.B. Hu, Aug 2006, Intake of sugar-sweetened beverages and weight gain: A systematic review, *American Journal of Clinical Nutrition*. 84(2), 274–288. Review.

Margolis, D.J., J.M. Hoffman, R.J. Herfkens, R.B. Jeffrey, A. Quon, and S.S. Gambhir, Nov 2007, Molecular imaging techniques in body imaging, *Radiology*. 245(2), 333–356. Review.

Martinez, J.A., M.S. Corbalan, A. Sanchez-Villegas, L. Forga, A. Marti, and M.A. Martinez-Gonzalez, 2003, Obesity risk is associated with carbohydrate intake in women carrying the Gln27Glu beta2-adrenoceptor polymorphism, *Journal of Nutrition*. 133, 2549–2554.

Mathers, J.C., 2007, Early nutrition: Impact on epigenetics, 2007, *Forum of Nutrition*. 60, 42–48. Review.

McCarty, C.A., Apr 2008, Sunlight exposure assessment: Can we accurately assess vitamin D exposure from sunlight questionnaires? *American Journal of Clinical Nutrition*. 87(4), 1097S–1101S. Review.

McCarthy, F.M., N. Wang, G.B. Magee, B. Nanduri, M.L. Lawrence, E.B. Camon, D.G. Barrell, D.P. Hill, M.E. Dolan, W.P. Williams, D.S. Luthe, S.M. Bridges, and S.C. Burgess, 2006, AgBase: A functional genomics resource for agriculture, *BMC Genomics*. 7, 229.

Mero, N., L. Suurinkeroinen, M. Syvanne, P. Knudsen, H. Yki-Jarvinen, and M.R. Taskinen, 1999, Delayed clearance of postprandial large TG-rich particles in normolipidemic carriers of LPL Asn291Ser gene variant, *Journal of Lipid Research*. 40, 1663–1670.

Miller, W., K.D. Makova, A. Nekrutenko, and R.C. Hardison, 2004, Comparative genomics, *Annual Review of Genomics and Human Genetics*. 5, 15–56.

Morris, R.M., 2006, Environmental genomics: Exploring ecological sequence space, *Current Biology*. 6, R499–R501.

Moskowitz, H.R., J.B. German, and I.S. Saguy, 2005, Unveiling health attitudes and creating good-for-you foods: The genomics metaphor, consumer innovative web-based technologies, *Critical Review of Food Science Nutrition*. 45,165–191.

Neu, J., N. Hauser, and M. Douglas-Escobar, Feb 2007, Postnatal nutrition and adult health programming. *Seminars in Fetal and Neonatal Medicine.* 12(1), 78–86.

Nicholson, J.K., J.C. Lindon, and E. Holmes, 1999, 'Metabonomics': Understanding the metabolic responses of living systems to pathophysiological stimuli via multivariate statistical analysis of biological NMR spectroscopic data, *Xenobiotica.* 29, 1181–1189.

Nicholson, J.K. and I.D. Wilson, 2003, Opinion: Understanding 'global' systems biology: Metabonomics and the continuum of metabolism, *Nature Reviews Drug Discovery.* 2, 668–676.

Ordovas, J.M., 2003, Cardiovascular disease genetics: A long and winding road, *Current Opinions in Lipidology.* 14, 47–54.

Ostos, M.A., J. Lopez-Miranda, J.M. Ordovas, C. Marin, A. Blanco, P. Castro, F. Lopez-Segura, J. Jimenez-Pereperez, and F.J. Perez-Jimenez, 1998, Dietary fat clearance is modulated by genetic variation in apolipoprotein A-IV gene locus, *Journal of Lipid Research.* 39, 2493–2500.

Paddon-Jones, D., E. Westman, R.D. Mattes, R.R. Wolfe, A. Astrup, and M. Westerterp-Plantenga, 2008, Protein, weight management, and satiety, *American Journal of Clinical Nutrition*, 87 (suppl), 1558S–1561S.

Petersen, K.F., S. Dufour, D.B. Savage, S. Bilz, G. Solomon, S. Yonemitsu, G.W. Cline, D. Befroy, L. Zemany, B.B. Kahn, X. Papademetris, D.L. Rothman, and G.I. Shulman, July 31, 2007, The role of skeletal muscle insulin resistance in the pathogenesis of the metabolic syndrome, *Proceedings of the National Academy of Science USA.* 104(31), 12587–12594.

Petrovici, D.A. and C. Ritson, 2006, Factors influencing consumer dietary health preventative behaviours, *BMC Public Health.* 6, 222.

Pietrobelli, A., M. Malavolti, N.C. Battistini, and N. Fuiano, 2008, Metabolic syndrome: A child is not a small adult. *International Journal of Pediatric Obesity.* 3(1 Supplement), 67–71. Review.

Pharkya, P., A.P. Burgard, and C.D. Maranas, 2004, OptStrain: A computational framework for redesign of microbial production systems. *Genome Research.* 14(11), 2367–2376.

Plosch, T., A. Kosters, A.K. Groen, and F. Kuipers, 2005, The ABC of hepatic and intestinal cholesterol transport, *Handbook of Experimental Pharmacology.* 170, 465–482.

Popkin, B.M., 2006, Global nutrition dynamics: The world is shifting rapidly toward a diet linked with noncommunicable diseases, *American Journal of Clinical Nutrition.* 84, 289–298.

Quam, L., R. Smith, and D. Yach, 2006, Rising to the global challenge of the chronic disease epidemic, *Lancet.* 368, 1221–1223.

Raamsdonk, L.M., B. Teusink, D. Broadhurst, N. Zhang, A. Hayes, M.C. Walsh, J.A. Berden, K.M. Brindle, D.B. Kell, J.J. Rowland, H.V. Westerhoff, K. van Dam, and S.G. Oliver, 2001, A functional genomics strategy that uses metabolome data to reveal the phenotype of silent mutations, *Nature Biotechnology.* 19, 45–50.

Rantala, M., T.T. Rantala, M.J. Savolainen, Y. Friedlander, and Y.A. Kesaniemi, 2000, Apolipoprotein B gene polymorphisms and serum lipids: Meta-analysis of the role of genetic variation in responsiveness to diet, *American Journal of Clinical Nutrition.* 71, 713–724.

Razack, R. and D.L. Seidner, July 2007, Nutrition in inflammatory bowel disease, *Current Opinions in Gastroenterology.* 23(4), 400–405. Review.

Sagalowicz, L., M.E. Leser, H.J. Watzke, and M. Michel, 2006, Monoglyceride self-assembly structures as delivery vehicles, *Trends in Food Science Technology.* 17, 204–214.

Saha, S., S.H. Harrison, C. Shen, H. Tang, P. Radivojac, R.J. Arnold, X. Zhang, and J.Y. Chen, April 25, 2008, HIP2: An online database of human plasma proteins from healthy individuals. *BMC Medical Genomics.* 1, 12.

Sajda, P., 2006, Machine learning for detection and diagnosis of disease, *Annual Review of Biomedical Engineering.* 8, 537–565.

Sanz, J. and Z.A. Fayad, Feb 21, 2008, Imaging of atherosclerotic cardiovascular disease, *Nature*. 451(7181), 953–957. Review.

Sloane, E.A., 2006, Top ten functional food trends, *Food Technology*. 60, 33.

Spelizini, D., B. Farruggia, and G. Picó, 2005, Features of acid protease partition in aqueous two-phase systems of polyethylene glycol-phosphate: Chymosin and pepsin, *Journal of Chromatography*. B 821, 60–66.

Suzuki, Y.A., V. Lopez, and B. Lönnerdal, Nov 2005, Mammalian lactoferrin receptors: Structure and function, *Cellular and Molecular Life Science*. 62(22), 2560–2575. Review.

Tolstoguzov, V., 1986, Functional properties of protein-polysaccharide mixtures, in *Functional Properties of Food Marcomolecules*, J.R. Mitchell and D.A. Ledward (eds.). Elsevier Applied Science, London, U.K., pp. 385–415.

Tolstoguzov, V., 2004, Why were polysaccharides necessary? *Origins of Life and Evolution of the Biosphere*. 34, 571–597.

Turnbaugh, P.J., Bäckhed, F., Fulton, L., and Gordon, J.I., Apr 17, 2008, Diet-induced obesity is linked to marked but reversible alterations in the mouse distal gut microbiome, *Cell Host and Microbe*. 3(4), 213–223.

Ward, R.E, H.J. Watzke, R. Jiménez-Flores, and J.B. German, 2004, Bioguided processing: A paradigm change in food production, *Food Technology*. 58, 44–48.

Watkins, S.M., B.D. Hammock, J.W. Newman, and J.B. German, 2001, Individual metabolism should guide agriculture toward foods for improved health and nutrition, *American Journal of Clinical Nutrition*. 74(3), 283–286.

Whetstine, J.R., T.L. Witt, and L.H. Matherly, 2002, The human reduced folate carrier gene is regulated by the AP2 and sp1 transcription factor families and a functional 61-base pair polymorphism, *Journal of Biological Chemistry*. 277, 43873–43880.

Wei-Min C., M.R. Erdos, A.U. Jackson, R. Saxena, S. Sanna, K.D. Silver, N.J. Timpson, T. Hansen, M. Orrù, M.G. Piras, L.L. Bonnycastle, C.J. Willer, V. Lyssenko, H. Shen, J. Kuusisto, S. Ebrahim, N. Sestu, W.L. Duren, M.C. Spada, H.M. Stringham, L.J. Scott, N. Olla, A.J. Swift, S. Najjar, B.D. Mitchell, D.A. Lawlor, G.D. Smith, Y. Ben-Shlomo, G. Andersen, K. Borch-Johnsen, T. Jørgensen, J. Saramies, T.T. Valle, T.A. Buchanan, A.R. Shuldiner, E. Lakatta, R.N. Bergman, M. Uda, J. Tuomilehto, O. Pedersen, A. Cao, L. Groop, K.L. Mohlke, M. Laakso, D. Schlessinger, F.S. Collins, D. Altshuler, G.R. Abecasis, M. Boehnke, A. Scuteri and R.M. Watanabe, 2008, Variations in the G6PC2/ABCB11 genomic region are associated with fasting glucose levels, *Journal of Clinical Investigation*. 118(7), 2620–2628.

Wolesensky, W. and J.D. Logan, 2006, Chemical reactor models of digestion modulation, in *Focus on Ecology Research*, Burk, A.R. (ed.). Nova Science Publishers, Inc., New York, pp. 197–247.

Womack, J.E., 2005, Advances in livestock genomics: Opening the barn door. *Genome Research*. 15, 1699–1705.

Zoetendal, E.G., A.D.L. Akkermans, W.M. Akkermans-van Vliet, J.A.G.M. de Visser, and W.M. de Vos, 2001, The host genotype affects the bacterial community in the human gastrointestinal tract, *Microbial Ecology in Health and Disease*. 13(3), 129–134.

10 Creating Food Concepts to Guide Product Development and Marketing

Howard R. Moskowitz, Michele Reisner, and Andrea Maier

CONTENTS

Success = Defining and meeting target consumer needs and expectations × The right food × Proper packaging and preparation × Positioned correctly at the shelf and in the media × Meet corporate logistics and financial imperatives

10.1 WHAT IS A CONCEPT IN THE WORLD OF PRODUCT DESIGN, AND WHY IS IT IMPORTANT?

In most companies, foods and beverages do not typically appear by themselves. They come from the efforts of product developers, marketers, trend spotters, and researchers to identify consumer needs and wants. Of course, there are many small boutique companies; the one- and two-product businesses that grew out of an individual's own recipe. However, these individual companies are, in general, temporary and if successful, soon swallowed by bigger organizations.

Concepts are blueprints that tell the manufacturer what attributes the product should have. They force discipline on the development and marketing process by producing a set of specifications, not necessarily technical, laid out in ideas (Cooper, 1993, Bacon and Butler, 1998). Look at Figure 10.1 to get a sense of a product concept in development. This paragraph talks about the product in terms of what it looks like, tastes like, what it has, etc.

Look closely at the two concepts in Figure 10.1. The product concept comprises virtually only the specification of the product, e.g., only what is in the product.

Concept—Product Description

Introducing Hearth Delights®

Hearth Delights is a new type of bread. Made from the most nutritious grains, the bread is also filled with the all sorts of wonderful ingredients that people love—nuts, raisins, fruit pieces

This new bread is fortified with vitamins and minerals for good health.

More than that... Hearth Delights comes in three flavors...
Regular—great whole grains, and great tasting
Sweet—the sweet flavors that you love, from fruit pieces
Savory—the richness of European cheese

Hearth Delights also come in the new FlavFresh® stay-fresh bag

At $1.99 per loaf, you will find this to be an affordable, fun treat

Concept—Positioning + Product Description

Introducing Hearth Delights®

Who of us can resist fresh-tasting bread with delicious ingredients

Do you remember when you first tasted that sophisticated, intriguing bread with all sorts of good things inside?

Now there's Hearth Delights—a bread that captures that first wonderful feeling when you discovered a world of bread beyond the regular white, rye and whole wheat

Hearth Delights comes in three great tasting flavors...
Regular—great whole grains, and great tasting
Sweet—the sweet flavors that you love, from fruit pieces
Savory—the richness of European cheese

At $1.99 per loaf, you will find this to be an affordable, fun treat

FIGURE 10.1   Example of a product concept versus a product + positioning concept.

The positioning + product concept is a bit different. There are other things going on—ideas that do not seem to have anything to do with the product, but more with the consumer. These other parts, positioning elements, give "more" to the product than just its features. There are statements, which we can best classify as emotion, are the reasons to buy, together with pictures that set a tone for the concept. Usually, product developers feel most comfortable with the product concept, as it specifies the product, albeit in terms meaningful to consumers. Developers can also work with positioning concepts. However, the developer must interpret in formulation what the concept promises but does not specify.

The concept is not a dry set of specifications, residing in a laboratory or factory, but rather something to which the typical consumer can react. There is a large worldwide market research business devoted to "concept testing" and "optimization." Concept testing focuses on the question—"is this a good product idea for our company, what drives this idea, and how can we make it even better?"

As companies get bigger, the competitive frame becomes increasingly overcrowded. The search continues to differentiate one's firm from the corporate pack. As such, the product concept becomes increasingly more important. In a world where there are only one or two competitors, a company can pretty much throw out any product that is tasty and be assured that, for the right price, consumers will buy it. Not so today... . People want new products, not just new flavors of the same old products. Product concepts help here; they give the manufacturer a chance to test the "idea" of the product before investing in product development itself. The judicious manufacturer rarely goes to market anymore unless the product concept scores sufficiently well. This scoring, as we will see later on, comes from comparing how well the consumer panelists rating the concept would rate other concepts the historic performance of which in the marketplace is well known. Once the concept scores sufficiently well, corporate management usually directs the R&D group to create prototypes. It is still not clear whether the best strategy is to first create a concept with a prototype to follow, or create a well-liked prototype first, with a concept afterwards that does justice to the prototype (Moskowitz et al., 2005a).

10.2   WHERE DO YOU LEARN ABOUT CONCEPTS?

Food and beverage developers have many resources to devise new product ideas. There is basic scientific literature which tells these developers about the physical properties of products. There is technical, but popular, business literature from companies and trade magazines which provide recipes, ideas for new products, key benefits of using new ingredients, and the like. Finally, there are the annual science and trade conventions, such as the Institute of Food Technologists (the "big one"), but also discipline-specific and product-specific conferences, where technologists meet to discuss the general area (e.g., baking, juices) and specific issues (e.g., sourcing flavor ingredients, etc.).

But where does the developer go to learn about product concepts? Look at any food science library in a university and you are not likely to find more than one or two books devoted to the topic (e.g., Fuller, 1994; Moskowitz et al., 2005a). Go to a company's library and you will find the same thing—lots of information about the technical aspects of the food and maybe whole articles about the food, but nothing about how to write a succinct, powerful, consumer-oriented product concept. Go to the marketing department and you are likely to find some reports and a few marketing books with general chapters about how to test concepts. There is an appalling paucity of information about how to create concepts and how to measure them. There are no "off the shelf" databases that one can buy to see "what works and what doesn't," although recently Moskowitz and Gofman (2007) suggested creating a "on the shelf" database of the consumer mind where one could go expressly to the topic and find out "what works."

10.3   FOUNDATIONS OF CONCEPTS—"BENEFIT STATEMENTS"

Some companies derive concepts from an earlier step called benefit screening, promise testing, or the like. No matter what the title, the meaning is the same. Rather than testing complete concepts, practitioners in these companies prefer to begin with single elements or ideas. Respondents sort the ideas into important versus not important, or rate/rank them. Sometimes, the respondents evaluate interest and other attributes such as uniqueness and importance. The reason for the multiple responses is straightforward. It is one thing to produce a product concept the components of which are "interesting." The manufacturer wants the product to differ from the other products in the market. Respondents rate the concepts on uniqueness, giving the developer a sense of how these ideas differ from what already exists in the marketplace. We can get a sense of the difference between interest, uniqueness, and ability to solve certain problems from Figure 10.2, which shows the results from the so-called benefit screen.

Many companies use benefit screens in order to sort through different ideas, as they do their homework prior to the preparation for developing full-blown concepts. When the screening is done correctly, the outcome is a rank order of promising ideas that can later be incorporated into concepts. The benefit screen does not measure how well these elements or benefits will perform in the body of a concept where the benefit has to interact with other messages. Its one thing to rank order ideas alone, but quite another to assess their performance in a concept. For instance, brand names are thought to be very important. However, all too often the brand names perform relatively poorly (i.e., are fairly irrelevant) in the body of a concept.

10.4   FROM SIMPLE IDEAS TO CONCEPTS AND THE NOTION OF CONCEPT SCREENING

Almost all products are initially tested through "concept screening." Companies come up with dozens of concepts, from looking at marketplace trends, sometimes by seeing what chefs are doing in restaurants and, quite often, from innovation or ideation sessions where a group of participants develop many different ideas. Sometimes, the concepts come from observing lead users, those who take products and modify them to new uses in order to solve problems (Urban and von Hippel, 1988). The notion of screening becomes very important because it is really not clear which ideas will "stick" and which ideas are perhaps nice, but do not resonate with the consumers.

	Interest	Unique	Solves Problem
Bread that tastes fresh all the way to the end of the loaf	7.1	6.7	6.9
Bread that is demonstrably less fattening	7.1	7.2	7.8
Bread with vitamins, nuts, raisins that could be eaten as a snack but all natural	6.8	6.5	6.8
High fiber content in a soft moist good tasting bread	6.4	5.1	6.1
Good European-style breads, i.e., real Italian bread, rye breads, and pumpernickel	6.4	5.7	5.7
No artificial ingredients	6.1	5.6	5.9
True garlic bread—instead of adding garlic flavor on top and then baking, have the garlic baked in and simply toast and butter	5.3	6.3	6.1
Packaging that you can see through to make sure the bread is OK	5.3	6.1	6.9
Fresh bread with lots of iron in it but…it must taste like fresh baked bread	5.1	4.6	5.2
Bread with cheese baked in	5.1	6.7	5.8
A bread with lots of fiber, vitamins, and minerals	4.9	5.3	6.7
Bread that has a real cinnamon sugar taste	4.7	5.6	5.7
A wholesome full grained bread that satisfies hunger quickly	4.7	6.7	5.8
A fresher taste, thick, and still soft	4.7	5.8	5.9
Crustless bread	4.7	6.4	5.1

FIGURE 10.2 Partial results from a benefit screen for bread, showing ratings on a 9-point scale for interest, uniqueness, and "solves a problem I often encounter."

In a typical concept screen (also abbreviated as ConScreen by industry practitioners for obvious reasons), the researcher presents the respondent with a number of different concepts. These may be in a simple format (e.g., bullet points, without much text), in an elaborate paragraph format, or even in a small advertisement. There is no fixed number of concepts in a ConScreen. The judicious researcher balances efficiencies (more concepts tested at the same time increase research efficiency) and data quality (beyond a certain, often unknown, number of concepts, a single respondent gets tired and stops paying attention).

Researchers test concepts in a variety of ways; mail surveys were one of the earliest. The respondent would receive a book of concepts, page through them, and rate each concept in the order presented. A cadre of project staff was responsible for creating the books, with putting the concepts in a different order to prevent order bias. With a large number of concepts, a specific respondent may evaluate only a subset; the precise number of concepts does not matter as long as the respondent stays interested, and as long as the study secures a sufficiently large number of respondents per concept. Other methods for doing concept screening include prerecruiting respondents to participate in a focus group, prerecruiting a large group of respondents to evaluate the concepts simultaneously (Acupoll, 2005), or doing the study on the Internet.

Concept	Total	Use A	Use B	Use C	Use D	Age Young	Age Medium	Age Old
101	66	75	67	61	66	78	71	56
102	64	65	64	63	64	50	57	76
103	60	58	58	64	56	52	51	71
104	60	63	62	55	57	56	58	65
105	56	51	60	54	70	47	66	54
106	56	59	52	59	55	58	56	55
107	54	52	59	50	49	57	59	49
108	54	55	56	48	52	54	56	51
109	52	52	51	52	46	56	53	48
110	50	56	50	49	51	60	53	41
111	49	41	52	49	59	46	47	52
112	39	34	38	45	37	40	50	29
113	38	30	40	42	42	50	34	31
114	32	31	35	29	39	36	32	29
115	31	39	25	32	27	46	29	24

FIGURE 10.3 Conscreen results. Fifteen concepts were evaluated on a 5-point purchase intent scale (1 = definitely not purchase → definitely purchase). Data in table shows % of respondents in total panel and subgroups rating each concept as 4 or 5 on the 5-point scale.

ConScreen generates tabular results that can be inspected. These can help drive a business decision. The concepts can be arrayed as rows and attribute ratings as columns. Figure 10.3 presents an example of what the data might look like. Researchers working with different types of individuals in the test sample (e.g., males versus females, product users versus nonusers, etc.) often lay out the data so, that the results from the total panel fall into the first data column, followed by the results of each key subgroup.

Concept screening excites and frustrates the applied product developer. The excitement is easy to explain, as every developer wants to create product concepts that work, that excite the consumer, and that have something unique. This method throws the different concepts "against the wall" to discover which ones stick. One concept that wins is a cause for some happiness; several concepts that win in a single concept screening exercise are a cause for a great deal of celebration. On the practical end, good news means that the developer can move forward.

On the not-so-happy side, concept screening frustrates because, for the most part, it does not directly tell you, i.e., diagnose, what features of the concept drive the respondent's vote of acceptance or rejection. The respondent does not necessarily know why he likes or dislikes a concept, although despite that, a number of commercial researchers have tried to find out what "drives a concept" by having the respondent circle the phrases that are appealing. This circling can be done in focus groups or on the Web, and may give some the not-necessarily proven belief that the activity discovers key messages. Certainly, the effort to find what works in a concept can be strengthened beyond these methods, which rely on the respondent to "know" his own mind. Unfortunately, that happy state of affairs is not necessarily the case.

10.5 BEYOND CONCEPT SCREENING TO DESIGN OF EXPERIMENTS

Design of experiments (DOE), a branch of statistics, allows the researcher to identify the contribution of different parts of a product or process by systematically varying the combination of elements in a way that uncovers the contribution of each element. DOE lays out the different combinations, which the developer then creates, and the characteristics (e.g., output) of which are measured. Through regression analysis, the developer builds an equation showing the part-worth contribution of each element. The test involves the more world-realistic sets of combinations of elements, much as the elements might be found combined in actual products or processes. Various books have been written about the way to think about these experiments, especially when it comes to food products and concepts (e.g., Box et al., 1978). Statisticians have written copiously about different test "plans" or sets of combinations, and the appropriate statistics to use when analyzing the outputs of the experiments.

When it comes to food concepts, where the elements are "integer" (i.e., present or absent), DOE traces back about 40 years to a well-known area of mathematical psychology called "conjoint analysis." The objective of conjoint analysis is to measure the component stimuli of a mixture by measuring responses to the mixture. Originally conceived as a way to "fundamentally measure" stimuli in the world of experimental psychology (Luce and Tukey, 1964), conjoint analysis quickly migrated to the world of marketing, helped along by the vision of two pioneers in the marketing world, Professors Paul Green and Yoram (Jerry) Wind (Green and Srinivasan, 1990; Green and Krieger, 1991). Conjoint analysis has been widely used by researchers during the past 40 years to understand how the elements of concepts drive the responses to these concepts (Wittink and Cattin, 1989).

The basic premise of DOE for product concepts is quite straightforward. The researcher first creates the architectural structure of the concept and then populates the different parts of the architecture with "elements." We see an example of this in Figure 10.4. The top part of the panel shows the concept and the architecture. The bottom part of the panel shows the same concept, this time without the architecture superimposed, i.e., the unannotated concept. The respondent evaluates this unannotated concept.

With a proper statistical design underlying the set of test concepts, DOE prescribes a specific layout of test stimuli. The researcher creates the combinations according to the design, controlling the ever-present temptation to edit the combinations that just "don't seem right." At the end of the exercise, the data comes out in the form of a rectangular matrix. As shown in Figure 10.5, five test concepts out of the full set are displayed. The left part of the matrix comprises a set of 1s (to denote an element is present in a particular concept) or 0s (to denote that an element is missing in the concept). Of course, most concepts will have few elements. For any row, the majority of the numbers will be 0. The specific design in Figure 10.5 comes from 4 silos, each comprising 9 elements. The right side of the matrix (not shown) contains the rating assigned by the respondent to each of the concept.

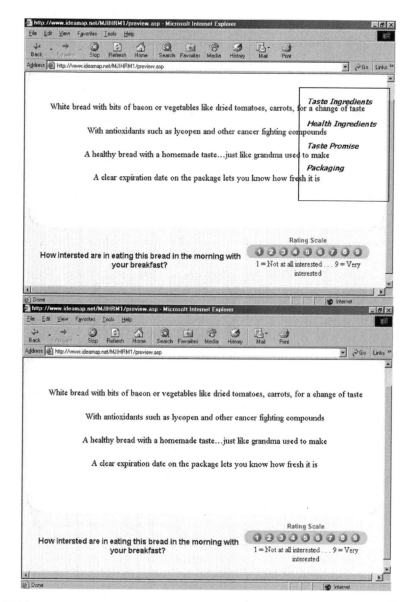

FIGURE 10.4 Example of test concept, first with the architecture and then as the respondent sees it.

10.6 AN EXAMPLE OF A DESIGNED EXPERIMENT FOR FOOD CONCEPTS—NEW BREAKFAST BREAD

The best way to understand the power and use of concept development tools is to review a case study. Our example describes the development of a nutritious bread, designed for eating in the morning, specifically breakfast but possibly as a mid-morning snack.

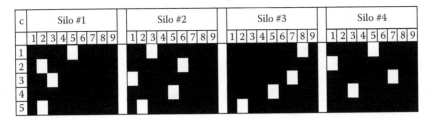

FIGURE 10.5   Example of the worksheet file, showing which concept elements are present and which are absent in a particular concept (C1–C5). Elements that are present are shown as white boxes.

The original study dealt with the features and benefits of this new product. Specifically, what should this bread be? How could a concept for this bread be crafted so that it appeals to the general public, is perceived to be reasonably new, and becomes the grist for a good product developer and marketer? Furthermore, within the area of bread products, could the concept research reveal the existence of new segments of individuals with different mind-sets? We know already from product work that people's sensory preferences vary widely, which is one of the key reasons for the proliferation of food products on the shelf within the same product category (e.g., orange juice, pasta sauce, pickles, and the like; Moskowitz, 1994). We also know that people have different preferences for concept statements; what appeals to one group may not appeal to another (Moskowitz, 1996).

We will use the case history approach, list the steps, give the rationale, and show some data.

Step 1: Define the problem, create the architecture, and populate the architecture with elements. In this first step the developer or researcher narrows the scope of the problem. The narrowing helps to focus attention on "what is appropriate." This project dealt with breakfast bread, generating four different silos. Each silo in turn comprised nine different options or elements. The actual number of options or elements within a silo can vary, but the research method used (IdeaMap.Net) works when the number of elements is the same for each silo (Moskowitz et al., 2001). The different elements in Figure 10.6 show two elements for each of the four silos. The full set of elements will appear later on.

Where these elements come from? The answer is "almost anywhere," ranging from competitive analysis (e.g., downloading web ads and news clippings to find out what the competitors are featuring), through focus groups and depth discussions (to find out hidden needs), as well as actual innovative sessions with the express goal of creating a set of these elements for later research and screening.

Step 2: Define the orientation page to the respondent, and the rating question. In concept testing without the product, it is important to define what the respondent must keep in mind while evaluating the product. Here the idea was to evaluate the product in the framework of a "breakfast bread." It is not really clear from the literature whether respondents can shift their focus from one end use to another, e.g., breakfast bread versus evening bread.

Taste/flavor ingredients

A new variety of sweet flavored bread like cinnamon, vanilla swirl, cinnamon raisin, almond raisin…mmm, the flavors will simply delight you

Real garlic bread…garlic and other herbs mixed into the dough and baked for a perfectly delicious bread

Health ingredients or features

Simply the best bread…high in fiber, low in carbohydrates tastes great and it is good for you

Enriched with lots of vitamins and nutrients as part of a balanced diet

Taste or use promise

Frozen bread that can be baked fresh at home

Great for sandwiches, lunches, dinners

Packaging

Comes in a Ziploc bag…no more twist ties

Has a simple, easy to understand label…so you know what you are getting

---

FIGURE 10.6   Two elements from each of the four concept silos.

Step 3: Acquire the data. The respondents are typically invited by cost-effective means. For today's high-level studies using experimental design, and with the increasing importance of visuals in concepts (not shown here), telephone interviewing is usually not a viable option. For this particular illustrative case history we used the Internet, because of four important features. First, the Internet-based research is quick, so that several hundred interviews may be collected in a few days. Speed in today's marketplace is exceptionally important. Second, the research is relatively inexpensive. Respondents need only to click on the link in the e-mail invitation to be guided to the study. There are no expensive mailing pieces or production pieces. Third, the data come back automatically tallied with the models created showing what elements "drive interest" for each person. Automatic and rapid data analyses are also a key benefit. Fourth, the specific research tool we used shows different test concepts to each respondent (see Figure 10.4, specifically the unannotated concept). Such ability to rotate and randomize elements into new concepts is best done by computer-based interview.

Different methods are used by researchers to acquire respondents for concept tests when they work in the Internet space. Some use pop-ups attached to Web sites so that every nth individual is invited to participate. Others use e-mail invitations to individuals who have agreed to participate. (This is what was done here.) Still others use existing panels of individuals, with each individual already agreeing to participate in studies, and to provide extensive information about himself or herself as well. For this study, 5000 invitations to participate were mailed out. Out of these, 448 respondents logged in and completed the study, for an almost 10% completion rate. The typical completion rate is between 5% and 10%, especially when the respondent is incentivized to participate by the chance to win a prize through a sweepstakes.

Step 4: Build a model (equation) at the individual respondent level showing how elements "drive" ratings. The experimental design creates a set of combinations that can be considered statistically "complete" for any individual. That is, the set of 60 concepts that a single respondent evaluated suffices to create an equation for that respondent showing how the presence/absence of each of the 36 elements drives the rating.

4.1 *First, begin with the experimental design.* As noted above, one gets a sense of the design by looking at Figure 10.3. For any concept, the vast majority of the elements are missing. This massive absence of elements from any concept is by statistical design. Furthermore, some concepts comprise one element from three, not four silos. A concept does not need to be complete—it can have one element from 2, 3, or all 4 silos present. A design mixes and matches all of the 36 elements so that each element appears exactly the same number of times against different backgrounds. In our case, each element appears three times. Furthermore, each respondent evaluates an almost totally unique set of combinations. That is, the structure of the experimental design is the same across all respondents, but each respondent sees a unique set of combinations obtained by permuting the basic experimental design (Moskowitz and Gofman, 2005).

4.2 *Second, recode the responses to go from the rating to the "interest value."* A rating for bread of 1–6 denotes relatively low interest in this particular combination health bread attributes. We recode that combination or test concept as the value 0. By contrast, a rating for the concept of 7–9 denotes high acceptance. We recode that combination as the value 9. Each row comprises a set of 36 elements (value 0 to denote absence; value 1 to denote presence in the concept). In addition, there are now two columns for the ratings. The first column is the original rating assigned by the respondent, so that the number in that column can be anywhere from a low of 1 to a high of 9. The second column is a binary rating, of either 0 or 100, with 0 denoting no or low interest, and 100 denoting high interest. We do most of the analysis with the data from the second column, which we define as the concept interest value.

4.3 *Third, apply the statistical method of regression analysis to the data of each person.* We create two models, one for the actual 9-point ratings as the dependent variable (not discussed here) and one for the interest value (the binary value of 0 or 100, respectively). Regression or curve fitting is a well-accepted statistical method found on all personal computers equipped with spread sheets, and, of course, those equipped with statistical models. Regression analysis in the very simplest of cases (two variables, X = dependent, Y = independent), fits a line of the form: $Y = k_0 + k_1 X$.

4.4 *Fourth, apply this same statistical thinking to the data we have collected.* Instead of one independent variable, X, we deal with 36 simultaneous independent variables, $X_1 \ldots X_{36}$. Each of these 36 elements corresponds to one of the 36 phrases about the bread. The independent variable could only appear at two levels—0 to denote that it was absent from a particular concept and 1 to denote that it was present in that concept. Statisticians call this a "dummy variable" which can only be absent or present. The data matrix will be analyzed by regression. The independent variables comprise 36 columns (one per variable). The dependent variables are either the actual 9-point ratings or the binary recoding (discussed here).

4.5 *Use dummy variable regression modeling.* The conventional analysis for concepts uses ordinary least squares regression to relate the 36 independent (so-called dummy) variables to the recoded binary variable that took on the value 0 or 100. In descriptive terms, we relate the presence/absence of our phrases describing bread to a person's saying "I'm not interested in this idea" or "I'm interested in this idea." The binary variable which is labeled interest versus noninterest, rather than the measure of degree of interest, is generally used as the dependent variable in these types of studies. The reason stems from the intellectual history of applied market research, in which concept testing is done as a specialty. Market researchers, with their heritage in sociology, focus on the number of people belonging to a group (interested, not interested in a concept). In general, market researchers are not particularly interested in the intensity of interest. They leave intensity of feeling to the psychologist.

Step 5: Summarize the results for total panel and key subgroups. We analyzed the data from each of our 448 respondents. For the working data, there is a large matrix of 448 rows, one per respondent, and 37 columns, one for each bread element, and one for the additive constant. The numbers are the utility values, technically defined as the parameter in the regression model. The additive constant is the basic conditional probability that a person will find the concept interesting without any elements. It is a baseline. The utility value is interpreted as the additive conditional probability that a person will find the concept interesting (rate it 7, 8, or 9) if the element is present in the concept. One may combine utilities and estimate their joint performance by adding their values together arithmetically.

5.1 *Look at some data and verbalize what is learned about our healthful bread.* The data for the total panel and key subgroups is shown in Figure 10.7. There are representative data from two ages (18–30 years old), two markets (Northeast and Northwest), three groups that say they look for different things (those who think preference/acceptance is important, those who think price is important, and those who think actual product quality is important). Finally, you will see three columns marked Segments A, B, and C, respectively. The segments represent individuals with different mind-sets, who were divided by the pattern of their utilities (Vigneau et al., 2001). Each column shows the average additive constant, and the average utility value for all participants who belong in the specific group. The equation was developed for each respondent separately. The data for all respondents in a group is combined to estimate the mean, and come up with the data in the table. The equation is expressed as the simple weighted sum of the elements:

$$\text{Interest (Recoded binary)} = k_0(=\text{Additive Constant}) + k_1(\text{Element 1}) + k_2(\text{Element 2})...k_{36}(\text{Element 36})$$

5.2 The first-order of business in analyzing designed experiments with concepts looks at the additive constant in the additive model. The constant, k_0, is the *conditional probability of an average person being interested in the health bread concept if no elements were present in the concept.* This additive constant must be estimated statistically, since every concept comprises at least two elements. The additive constant provides us with a good sense of how receptive consumer respondents are going to

	Total	Age 1 18–30	Age 2 51–60	Market 1 Northwest	Market 2 Northeast	Import Preference	Import Price	Import Quality	Segment A Interested but Conservative	Segment B Sweet Bread Seeker	Segment C Like Inclusions in Bread
Base size (Number of respondents)	448	78	72	44	125	69	201	136	145	189	114
Constant (Basic interest in the bread without any elements)	46	47	38	57	36	48	45	52	52	51	29
Performance of the 36 elements—ranked by total panel performance											
A1 A new variety of sweet flavored bread like cinnamon, vanilla swirl, cinnamon raisin, almond raisin… mmm, the flavors will simply delight you	19	14	24	20	23	20	19	21	1	21	37

FIGURE 10.7   How the different concept elements perform by total panel and by key subgroups. The elements are ranked by performance among the 448 respondents from the total panel. Strong performing elements with utilities >6 are shaded.

(continued)

		Total	Age 1 18–30	Age 2 51–60	Market 1 Northwest	Market 2 Northeast	Import Preference	Import Price	Import Quality	Segment A Interested but Conservative	Segment B Sweet Bread Seeker	Segment C Like Inclusions in Bread
A5	A new line of bread with added fruit, like cranberry, blueberry, dates, or raisins	5	-2	8	4	6	3	7	7	-10	5	23
C9	A healthy bread with a homemade taste...just like grandma used to make	4	1	7	8	5	5	4	3	2	4	8
B4	Low in calories, high in vitamins... provides 100% of the daily value of 10 essential vitamins and minerals	3	-4	6	2	6	2	2	1	3	3	3
D2	Has a simple, easy to understand label...so you know what you are getting	3	0	1	1	6	3	3	4	4	4	0
B1	Simply the best bread...high in fiber, low in carbohydrates tastes great and it is good for you	3	0	7	-3	4	1	1	0	2	3	4

A8	Good European style bread… French, Italian, focaccia and sourdough	2	7	2	1	4	0	4	0	2	-6	16
B3	Guaranteed freshness without preservatives	2	1	4	-2	4	3	2	2	1	3	2
C1	Frozen bread that can be baked fresh at home	2	1	0	4	6	2	3	2	2	2	3
D8	Additional nutritional information lets you know the benefits of the ingredients such as oatmeal to help lower cholesterol	2	5	3	0	3	4	1	4	1	2	3
D1	Comes in a Ziploc bag…no more twist ties	2	-1	0	2	2	7	0	3	4	0	2
B5	With added minerals such as calcium, magnesium, zinc for strong bones and muscle tissue	2	1	2	0	4	3	1	1	2	3	-1

FIGURE 10.7 (continued)

(continued)

	Total	Age 1 18–30	Age 2 51–60	Market 1 Northwest	Market 2 Northeast	Import Preference	Import Price	Import Quality	Segment A Interested but Conservative	Segment B Sweet Bread Seeker	Segment C Like Inclusions in Bread
B9	Made with all natural ingredients…no artificial flavors, preservatives or any other additives	2	–1	3	–1	5	3	–1	3	1	3	0
D3	A clear expiration date on the package lets you know how fresh it is	2	–2	2	–1	3	0	0	–1	2	2	0
C7	A variety of small loaves in one bag…white, whole grain, rye, or oat	1	–1	0	–6	2	0	0	0	–1	1	4
A7	A large variety of whole grain bread like wheat, rye, or oat with added poppy, sunflower, or sesame seeds… for those who want to eat healthy and have some extra zing	1	–7	10	9	4	2	–1	2	–4	–5	17

Code	Description											
B6	With added soluble fiber to reduce the risk of coronary heart disease and lower your cholesterol	1	4	3	-2	4	4	-1	-2	1	2	-1
C2	Great for sandwiches, lunches, dinners	1	0	0	-1	2	3	2	-1	0	0	3
C6	Comes in half loaves for smaller families or single people	1	2	0	-1	5	-2	0	0	2	-1	2
C8	An essential source of nutrients in every slice…satisfies your hunger	1	-2	2	-1	3	0	1	1	1	0	1
B8	Low cholesterol, low fat, low sodium…but still tastes great	0	-1	2	0	3	-1	1	-1	0	2	-1
B2	Enriched with lots of vitamins and nutrients as part of a balanced diet	0	0	0	-2	2	2	-1	0	1	2	-3

FIGURE 10.7 (continued)

(continued)

		Age 1	Age 2	Market 1	Market 2	Import	Import	Import	Segment A	Segment B	Segment C
	Total	18–30	51–60	Northwest	Northeast	Preference	Price	Quality	Interested but Conservative	Sweet Bread Seeker	Like Inclusions in Bread
D6 Packaging is organized by flavor or health claims so it is easy to find what you like	0	−2	2	2	−1	0	0	1	2	−1	−1
A2 Real garlic bread…garlic and other herbs mixed into the dough and baked for a perfectly delicious bread	0	3	3	16	0	−7	5	−5	5	−25	35
D4 A see-through package and label…see how fresh it really is	0	0	1	−2	1	3	0	1	0	3	−6
D5 Comes in a clear carton-like package, so it will not get crushed with all your other groceries	0	1	2	0	4	0	−1	2	−2	0	2

Code	Concept											
A4	A large variety of whole grain bread with fruits and nuts	0	-10	9	3	2	2	-1	-2	-12	0	14
D7	Each package has a diet exchange table given by the American Heart Association	-1	1	2	-5	1	-3	-2	-1	1	0	-3
B7	With antioxidants such as lycopene and other cancer fighting compounds	-1	-4	4	-3	2	-1	-1	-1	-4	1	-1
C5	100% organic and GMO (genetically modified) free	-2	-2	-1	-5	0	-6	-4	-3	-2	-1	-5
A3	A variety of cheese bread… three cheeses bread, cheese and onions, the possibilities are endless	-2	-4	-6	6	-1	-4	3	-4	-1	-25	33
C3	Great for dipping and stuffing	-3	-3	-7	0	-3	-7	-3	-3	-6	-3	0

FIGURE 10.7 (continued)

(continued)

	Total	Age 1 18–30	Age 2 51–60	Market 1 Northwest	Market 2 Northeast	Import Preference	Import Price	Import Quality	Segment A Interested but Conservative	Segment B Sweet Bread Seeker	Segment C Like Inclusions in Bread
C4 Yeast free, lactose free, egg free	−5	−7	−7	−8	−2	−5	−6	−5	−6	−4	−4
D9 Vegan bread with no added ingredients such as eggs, lactose, animal fat, etc.	−6	−6	−7	−5	−5	−7	−7	−10	−7	−3	−9
A6 The flavor of rich and dark chocolate in bread...chunks of chocolate and nuts, try something new, get out of the ordinary	−10	3	−12	−18	−5	−15	−6	−12	−40	−3	14
A9 White bread with bits of bacon or vegetables like dried tomatoes, carrots, for a change of taste	−20	−24	−15	−21	−15	−21	−19	−23	−25	−34	9

FIGURE 10.7 (continued)

be when faced with the concept. Before looking at the numbers, keep in mind some landmark values. Credit cards, by now a commodity, have additive constants around 20–30. Many foods have constants around 40–60. Our total panel shows an average additive constant of 46, right about where many foods fall. Almost half the respondents are interested in a health bread, without even knowing what the ingredients of the bread will be.

5.3 Let us proceed to the 36 elements. How do these elements perform? Which ones do well? How about subgroups? With 36 elements in concept tests of this sort, there are many opportunities for ideas to break through. The respondents may not be able to articulate what is important to them, but the experimental design can discern which elements drive up the ratings. The individual numbers in the body of Table 4 are the part-worth utilities, *the conditional probability or percent of people who would change their vote from "not interested in the concept" to "interested" if the element appears in the concept.* Of course, the developer's goal is to identify those individual elements that do very well, i.e., that are highly positive, say seven or more. If there are three or four of these high scoring elements in a concept, then it can attract a large number of respondents, even if the basic interest in the concept is low to begin with, i.e., shown by a low additive constant.

5.4 People vary in what elements drive their interest. The additive constant differs across groups. A high additive constant means that the respondent accepts the basic idea of the healthful morning bread, even without elements, whereas a low additive constant means that the basic idea is not acceptable although the right collection of elements certainly can bring up the concept score. The younger respondent (ages 18–30) is certainly more interested in the basic idea rather than the older respondent appears to be (ages 51–60). All the ages were not examined, just these two ages. Continuing along the same track, the respondents in the northwest are really far more ready to accept the basic idea of a health bread than the respondents in the northeast (47 versus 36). On the other hand, attitudes about what is important, product acceptability (preference), price, and quality do not drive basic interest in the bread. Finally, we divided the respondents by the pattern of what elements are important (concept response segments). Segment C, the group that responds strongly and positively to inclusions in the health bread, with a constant of 29, certainly does not like the basic, unadorned idea of the healthful bread with an additive constant of 29. However, there are a number of elements that really drive the acceptance by Segment C, what can be done to enhance the basic bread.

5.5 Generally, strong performing elements as having a utility of +7 or higher. In other studies, utilities around +7 are both statistically significant from the analyses of variance and reflect a degree of relevance in other types of studies such as focus groups. To make the search easy, the 36 elements are sorted from high to low for the total panel. This sorting is perfectly legitimate, meaning the utility values of elements can be compared across the different silos, and in fact, the results of different studies can be compared with different people over different years. The utility values of two or more elements can be compared even when they come from different silos, because the utility values have true meaning in an absolute sense. The silos are only a way of making sure that contradictory elements from the same silo do not appear together.

For the total panel, only one element does well: *A new variety of sweet flavored bread like cinnamon, vanilla swirl, cinnamon raisin, almond raisin...mmm, the flavors will simply delight you.* Most of the remaining elements do modestly at best, some acceptable, some rejected and therefore decreasing the number of interested respondents. Two elements in particular do very poorly, both listing either the savory or sweet ingredients. *The flavor of rich and dark chocolate in bread...chunks of chocolate and nuts, try something new, get out of the ordinary, and White bread with bits of bacon or vegetables like dried tomatoes, carrots, for a change of taste.* From the poor scores of these two elements, the developer immediately realizes that the typical respondent is conservative, not interest in radical departures from the typical product. Also, it is important to know that the best and worst elements talk about the product itself. Even though the experimental balances the appearance of each element and puts together the combinations in different ways, the key elements that push through are the ones that deal with the product characteristics. The others simply do not have that power to make their way through as strong or weak performers.

5.6 Looking at the data from subgroups gives a different perspective of what is important. Market researchers break out data into meaningful groups, such as age, income, market, as well as groups holding specific self-stated attitudes, always with the hope that there is a group of individuals who will strongly respond to a concept. No concept generally scores well with all respondents, although there are many concepts that score poorly with all respondents. One of the key findings in this concept research is that dividing the respondents by the conventional ways used in market research simply do not point to mind-sets. There is no coherent pattern underlying the concept elements that perform well for a specific subgroup.

5.7 A more coherent pattern emerges when we divide the respondents by the patterns of their 36 individual utilities. Three segments or clusters of respondent are extracted through methods of the so-called cluster analysis that are elaborated elsewhere (Moskowitz et al., 2005a). The segmentation breaks apart the group of respondents in a surprisingly simple, and what will come to be, very reasonable way. Segment A responds to simple bread. Indeed, the winning element for the total panel (*A new variety of sweet flavored bread like cinnamon, vanilla swirl, cinnamon raisin, almond raisin...mmm, the flavors will simply delight you*) simply does not perform well for these individuals. Segment A responds most strongly to classic, simple bread. So, really, nothing that the developer does will really attract Segment A. By contrast, Segment B strongly responds to sweet inclusions, whereas Segment C strongly responds to most of the inclusions. The actual method to discover what the segment wants sorts the segment utilities, from highest to lowest, and then labels the segment based on the elements that perform most strongly (i.e., highest positive). Segmentation generally produces these simpler, more focused groups of respondents. The results "make intuitive sense."

10.7 CREATING CONCEPTS THAT DO WELL, USING THE OUTPUT OF EXPERIMENTAL DESIGN

The foregoing exercise of experimental design produces insight into the mind of the consumer. If we were to stop here, then we would know a great deal about what types of ideas that motivate the consumer, what mind-sets exist in the population, and the generally promising, as well as not so promising, areas. Experimental design is knowledge building. Indeed, an entire science of the human mind, which we call Mind Genomics®, can be and has been built from studies of this type (Moskowitz et al., 2006). For this chapter, we first concentrate on the creation of winning concepts, using the information from the experimental design of ideas.

How shall we construct our product concept for the healthful bread? Figure 10.8 presents a schematic of one approach. We see different rectangles, meant to show the construction of concepts. At the base of the concept, we find the additive constant. The higher that base, the greater the utility value. Above that base are the elements, also shown as rectangles with the height of the rectangle proportional to the utility value. Winning concepts comprise mutually compatible and strong performing concept elements. It always helps to have a large additive constant or base from which the concept is created, because this constant means a high basic acceptance of the idea, even without elements.

What type of concept should one build? The left-most concept is the least attractive. It begins with a low basic interest and combines three elements, each of which scores poorly. In a sense, this left-most concept is weak because it has nothing that attracts the customer. The basic idea is not particularly interesting and the individual elements are not very interesting either. Of course, one might build a "seemingly but not really strong concept," by adding on element after element, none of which is particularly strong. However, the misguided marketer or developer might think that with enough weak elements the concept's total score would be high enough to pass the hurdles set by the company. In actuality, the concept will fail due to the joint problems of denseness (too much information) and irrelevance (the

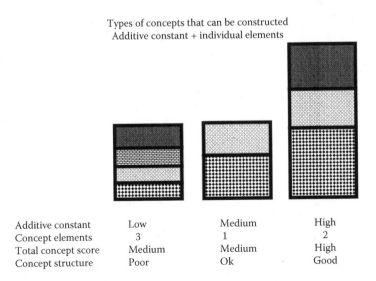

Types of concepts that can be constructed
Additive constant + individual elements

Additive constant	Low	Medium	High
Concept elements	3	1	2
Total concept score	Medium	Medium	High
Concept structure	Poor	Ok	Good

FIGURE 10.8 Construction of concept shown schematically as a set of blocks, whose height equals the sum of the utilities, beginning with the additive constant.

information does not excite the consumer). However, the food development business reality is that this type of concept is more prevalent in the industry than one might like. Many concept tests essentially put in block after block of text, hoping that the array of paragraphs will get consumers excited. It is a bit-like the old college strategy of test taking—if you do not know the answer, write everything you know about the topic in hopes that the professor will somehow happen upon the answer in this mass of poorly organized efforts.

The equally well-scoring middle concept has only one element (or perhaps two). Certainly, the concept is a lot more parsimonious—only one element does the work. Yet, the basic interest in the concept is higher, and the single element does almost as well as two of the elements of the concept we saw before. This concept is preferable—the elements are present, and there is little or no clutter to hinder performance.

The ideal, of course, is the right-most concept. Here, the basic interest is high and the developer has chosen two elements, both of which are high scoring and by judgment "fit together harmoniously." The happy outcome is that the concept is expected to score well because all the components working together to push toward a breakthrough. Although most marketers and product developers intuitively recognize this type of concept as the most desirable, they do not necessarily go about getting to this concept in a formalized manner. Those who do not do their homework with experimental design of ideas are forced to guess and hope that they happen upon this type of concept, rather than laying the groundwork for it through systematics.

Given this warning about making good concepts, let us create three concepts and see what is right versus what is wrong with each. To create the product concept, we choose the best or the most appropriate concept element from each of the four silos. We do not have to use each silo, and we will see that by allowing the developer a little freedom in rejecting some silos for a concept, the result may be a better performing concept. Furthermore, since we know how each element scores for each key group, we can choose any group(s) to be the target for which we create the concept, and at once estimate how the concept will perform among all groups. Figure 10.9 shows the results. From the practitioner's point of view, Figure 10.9 presents almost all that is needed. The table shows exactly how each element contributes to general interest.

Concept 1: A bread for the total panel. This bread scores a total of 76, primarily from two contributions, the additive constant (46) and a very strong product description from taste-oriented ingredients (19). The remaining silos about health ingredients, taste promise, and packaging hardly make a difference. Looking across the table at how this concept will perform among subgroups, we see that it will do quite well for the respondents from the Northwest (score of 87), and among respondents in Segment B (labeled the "sweet bread seeker"). Finally, we can look even more deeply into the performance of individual elements, to find that most of the elements are either positive or neutral. Only in one case, among the younger respondents (18–30 years old), do we discover a negatively performing element, which means putting that element into the concept actually drives people way (*Low in calories, high in vitamins... provides 100% of the daily value of 10 essential vitamins and minerals*, utility = −4 among the younger respondents, meaning 4% of the respondents will actually change their rating from interested to not interested if that element is incorporated

Morning Bread		Total	Age 1	Age 2	Market 1	Market 2	Import	Import	Import	Segment A	Segment B	Segment C
			18–30	51–60	Northwest	Northeast	Preference	Price	Quality	Interested but Conservative	Sweet Bread Seeker	Like Inclusions in Bread
Bread for total panel												
	Constant	46	47	38	57	36	48	45	52	52	51	29
A1	A new variety of sweet flavored bread like cinnamon, vanilla swirl, cinnamon raisin, almond raisin… mmm, the flavors will simply delight you	19	14	24	20	23	20	19	21	1	21	37
C9	A healthy bread with a homemade taste…just like grandma used to make	4	1	7	8	5	5	4	3	2	4	8

Expected Concept Performance

FIGURE 10.9 Construction of three concepts and the "diagnostics" that show how the additive constant and the selected elements "drive" concept acceptance.

(continued)

| | | Expected Concept Performance | | | | | | | | |
Morning Bread	Total	Age 1 18–30	Age 2 51–60	Market 1 Northwest	Market 2 Northeast	Import Preference	Import Price	Import Quality	Segment A Interested but Conservative	Segment B Sweet Bread Seeker	Segment C Like Inclusions in Bread
B4 Low in calories, high in vitamins… provides 100% of the daily value of 10 essential vitamins and minerals	3	4	6	2	6	2	2	1	3	3	3
D2 Has a simple, easy to understand label…so you know what you are getting	3	0	1	1	6	3	3	4	4	4	0
Sum: Bread for total panel	75	58	75	87	76	77	74	81	62	84	76
Bread for Segment A											
Constant	46	47	38	57	36	48	45	52	52	51	29

A2	Real garlic bread…garlic and other herbs mixed into the dough and baked for a perfectly delicious bread	35	25	5	5	5	7	0	16	3	3	0
D2	Has a simple, easy to understand label…so you know what you are getting	0	4	4	4	3	3	6	1	1	0	3
B4	Low in calories, high in vitamins… provides 100% of the daily value of 10 essential vitamins and minerals	3	3	3	1	2	2	6	2	6	4	3
C6	Comes in half loaves for smaller families or single people	2	1	2	0	0	2	5	1	0	2	1

(continued)

FIGURE 10.9 (continued)

Expected Concept Performance

Morning Bread	Total	Age 1	Age 2	Market 1	Market 2	Import	Import	Import	Segment A	Segment B	Segment C
		18–30	51–60	Northwest	Northeast	Preference	Price	Quality	Interested but Conservative	Sweet Bread Seeker	Like Inclusions in Bread
Sum: Bread for Segment A	53	48	47	75	52	43	56	53	66	33	68
Bread that strongly emphasizes health and quality benefits											
Constant	46	47	38	57	36	48	45	52	52	51	29
C9 A healthy bread with a homemade taste…just like grandma used to make	4	1	7	8	5	5	4	3	2	4	8
B4 Low in calories, high in vitamins…provides 100% of the daily value of 10 essential vitamins and minerals	3	4	6	2	6	2	2	1	3	3	3

A8	Good European style bread… French, Italian, focaccia and sourdough	2	7	2	1	4	0	0	4	0	2	6	16
D8	Additional nutritional information lets you know the benefits of the ingredients such as oatmeal to help lower cholesterol	2	5	3	0	3	4	4	1	4	1	2	3
	Sum: Bread that strongly emphasizes health	58	56	54	68	54	57	58	60	61	55	59

FIGURE 10.9 (continued)

into the concept). Allowing the developer to eliminate silos in a concept and making the concept a bit incomplete means to attract this group, and one might wish simply to skip that element, and perhaps not even to use an element from that silo.

Concept 2: A bread for Segment A (labeled the conservative segment). Segment A can be characterized as basically interested in the concept of a healthful morning bread with an additive constant of 52. There are no elements that really appeal to this segment. As a consequence, one might use almost any of the positive scoring elements, although it is probably a good idea to make a savory bread rather than a sweet bread. The description—*Real garlic bread...garlic and other herbs mixed into the dough and baked for a perfectly delicious bread*—is the strongest performing element.

Concept 3: A bread that communicates health and quality benefits. To create this concept, first decide the type of elements to select and then choose one or several groups whose utility values will be used as the criterion for accepting/rejecting the element. Since there is no clear segment that focuses on nutrition and wellness, we use the total panel. When we opt to use elements with a specific tonality, we create a more coherent concept, but we may lose some acceptance. This loss of acceptance is clear when we compare the total utility of concept 1 (best for total panel, utility sum = 75), with the concept for a bread that emphasis wellness and quality (utility = 48 for the total panel, a loss of 23% in concept acceptance).

10.8   CONCEPT TESTING AND PRODUCT INNOVATION IN THE FOOD AND BEVERAGE INDUSTRIES

Till now we have dealt with the evaluation of single ideas and either single concepts or experimentally designed concepts, generally for one product category. These concepts may represent entirely new products, but more often they represent line extensions, such as a new flavor, a new product format for the product, a new package, etc. What about concept development and testing as tools to enhance innovation, not just measure the output? Can the methods discussed above, ranging from benefit screening through concept screening to experimental design, be used on a proactive and practical basis to drive a program of product innovation?

If we focus on benefit screening, concept screening, or designed studies for a single product, then innovation occurs around the edges where the developer looks at new benefits, new product features, new concepts, or new elements in an existing product. There may be a better way for concept research to drive innovation. Consider the notion of experimental design of ideas, as we have just seen the illustration for health bread for breakfast. The product features come primarily from ingredients and packages that one might normally associate with bread, perhaps with a little "stretch" of the boundaries. What would happen, however, if we were to create a concept architecture comprising general features (e.g., appearance, flavor, special ingredients, packaging, merchandising, etc.) and then select elements appropriate to two or three different food products? Respondents participating in this type of study would be presented with concepts for novel foods that have features selected from several different foods. In a single concept, the elements might be appropriate to bread, cake, even candy, or a beverage. As long as the elements appear to make sense in the context of bread, and more or less "go together" even as a stretch of the imagination, this mixing/matching of elements from diverse products into a new product can become the source of innovation. The recombination of elements from different domains is called "mashing" (Moskowitz et al., 2005b).

The approach parallels the process we have just seen. The only difference is that the elements tested come from two or three different products, not from one product, so that the concept "mashes together" these disparate ideas into one new-to-the-world combination. Look at the elements in Figure 10.10, which show product features of

Low-Carb Bread	Cheese Cake	Yogurt	Selected for Mashing
		Appearance/texture	
Light and loaded with whole grains	Chunks of whole fruits baked in	With a yogurt center... smooth and thick... goes down easy	Light and loaded with whole grains
Dark, rich, multigrain bread	Moist and delicious... melts in your mouth	With a yogurt swirl... Naturally smooth... incredibly creamy	Dark, rich, multigrain bread
A low-carb white bread...what everybody's been looking for	Colorful swirls of chocolate cheesecake through and through	With a yogurt center... soft and light...a delicious part of a healthy lifestyle	Colorful swirls of chocolate cheesecake through and through
A beautiful, golden colored crust	Colorful swirls of strawberry and raspberry cheesecake... beautiful	Silky and inviting... melts in your mouth	Colorful swirls of strawberry and raspberry cheesecake... beautiful
A dark, whole grain bread	Thick and mouthwatering	Made with thick and creamy yogurt... appeals to all your senses	Made with thick and creamy yogurt... appeals to all your senses
Thinly presliced bread	Creamy and dreamy texture	Yogurt pockets... glossy and smooth... and so soothing	Yogurt pockets... glossy and smooth... and so soothing
		Ingredients/primary	
Coarse whole grain flour with caraway-tastes like rye	Made with 100% organic cream cheese...keeping all the goodness in	Made with 100% pure fruit and yogurt... enhances your daily fruit and dairy intake	Coarse whole grain flour with caraway-tastes like rye

FIGURE 10.10 Concept elements for three products, and the elements selected for mashing together to create a "new-to-the-world" product, comprising in part "low carb" bread, cheesecake, and yogurt, respectively.

(continued)

Low-Carb Bread	Cheese Cake	Yogurt	Selected for Mashing
Coarse whole grain flour with nuts and seeds	Made with only the fresh ingredients... eggs, pure creamery butter and fresh vanilla	Contains a delightful combination of yogurt and fresh fruit	Coarse whole grain flour with nuts and seeds
Made with nonhydrogenated oil	Sweetened with honey...a naturally sweet taste	Yogurt on top, fruit in the middle, nuts on the bottom...mix it all up for a satisfying snack	Sweetened with honey...a naturally sweet taste
Made with 100% whole wheat flour	Made with soy protein...dessert never tasted so healthy	Lowfat yogurt and fresh picked fruit... nutritiously delicious	Made with soy protein...bread never tasted so healthy
Made with pure, organic ingredients	Made with soymilk... feel good about what you feed your family	With fat-free yogurt... for those watching their calorie intake	With pockets of fat-free yogurt...for those watching their calorie intake
Made with soy flour and soy protein	Made with only egg whites...always a good source of protein	With 100% organic yogurt...natural and nutritious	With pockets of 100% organic yogurt... natural and nutritious

Ingredients/special

Low-Carb Bread	Cheese Cake	Yogurt	Selected for Mashing
Extra calcium and Vitamin D baked right in...to help build strong bones	GMO free	With vitamins A, C, and D...feel good knowing you are living healthy	Extra calcium and Vitamin D baked right in...to help build strong bones
Trace minerals and isoflavins added to protect again cancer, heart disease and diabetes	Rich in naturally occurring antioxidants...keeps you going	Added calcium... to give your bones a boost	Trace minerals and isoflavins added to protect again cancer, heart disease and diabetes
Soluble and insoluble fibers added for a healthy colon	Low carb. Fits into your healthy lifestyle	Organic fruits...so good for you	Low carb. Fits into your healthy lifestyle
Antioxidant vitamin E added to support a healthy immune system	Gluten free...for those with dietary restrictions	Acidophilus and ginseng added...put your digestive system on the right track	Gluten free...for those with dietary restrictions
Vitamins B6 and B12 added to maintain a healthy cardiovascular system	Rich in soy isoflavones...a natural source of sustainable energy	5% of whey protein... maintain the muscle but not the fat	Vitamins B6 and B12 added to maintain a healthy cardiovascular system

FIGURE 10.10 (continued)

Low-Carb Bread	Cheese Cake	Yogurt	Selected for Mashing
Chondroitin and glucosamine added to help your joints stay flexible	Sugar free…now that is cheesecake	10 times more cultures than ordinary yogurt…proven to help strengthen the body's defense system	Chondroitin and glucosamine added to help your joints stay flexible
		Flavors/tastes	
Chewy texture, rich in dates and nuts…great toasted	Mocha chocolate…an old fashioned favorite	The taste of peaches and cream…with a slightly tangy aftertaste	Chewy texture, rich in dates and nuts…great toasted
Strong, hearty country flavor	Pina colada…taste of summertime anytime	Chocolate cheesecake…Is it a snack or dessert?	Strong, hearty country flavor
Crunchy delicious bread filled with flax and sunflower seeds	Strawberry cheesecake… Gramma's favorite	Guava and mango… exotically profound	Crunchy delicious bread filled with flax and sunflower seeds
A nice, soft texture and a chewy crust, just as good as "real" bread	White chocolate…a decadent taste	Indian spice…excite your tastebuds	Indian spice…excite your tastebuds
Carb style bread— delicious whole wheat flavor	Kiwi lime…a taste of the tropic	Cucumber, rash and a touch of onion…enter a new dimension	Cucumber, rash and a touch of onion…enter a new dimension
Low-carb bread—the multigrain flavor you always look for	Plain Cheesecake… something to be said for tradition	White chocolate raspberry…a heavenly combination	Low-Carb bread—the multigrain flavor you always look for
		Packaging	
Comes in a clear, plastic box for protection	Comes in a resealable plastic box to preserve freshness	Packaged in plastic… designed to preserve freshness and guarantee purity	Packaged in plastic… designed to preserve freshness and guarantee purity
Comes double wrapped for freshness	Comes in a foil lined box to maintain freshness	In a clear rectangular container…specially designed to lock in all the flavor and guarantee freshness	Comes in a foil lined box to maintain freshness
Comes in a presliced loaf with paper between every two slices…so easy to separate	Has a resealable plastic bag to keep what you do not eat fresh	Packaged in foil containers…stays fresh longer	Has a resealable plastic bag to keep what you do not eat fresh
Comes in a heavy duty recyclable bag	Packaged as bite sized nuggets, eat what you want…freeze the rest	In a two-sided container…one side has the yogurt the other side has fresh fruit	Packaged as bite sized nuggets, eat what you want…freeze the rest

FIGURE 10.10 (continued)

(continued)

Low-Carb Bread	Cheese Cake	Yogurt	Selected for Mashing
Comes in a thick waxed bag to keep it fresh and soft in your refrigerator	Vacuum sealed for freshness	Round plastic container with the lid attached to it	Round plastic container with the lid attached to it
Comes in a special breathable wrapper... lasts for days without refrigeration	Comes in a recyclable box...keeping the environment safe	In a decorative plastic container...that you can keep and reuse	In a decorative plastic container...that you can keep and reuse

Size, store location

Low-Carb Bread	Cheese Cake	Yogurt	Selected for Mashing
Find it in the refrigerator case at your supermarket	Sold in bulk size... enough for a holiday get together	Available in the dairy section of your local supermarket	Find it in the refrigerator case at your supermarket
Find it right in the bread aisle of your supermarket	Sold in individual units...found in the freezer section	In the refrigerated section of your local convenience store	In the refrigerated section of your local convenience store
Find it flash-frozen for freshness in your freezer	Comes 10 to a box... great for an anytime snack	Can be found at most discount club stores nationwide	Comes 10 to a box... great for an anytime snack
Find it conveniently at your local deli	Available in a variety pack...two small sizes of your favorite flavors in one box	Available in an 8 oz. size...a full serving	Available in a variety pack...two small sizes of your favorite flavors in one box
Comes in a half size loaf...just the amount you need	Mini cheesecakes to grab for when you are on the run	Available in a 6 oz. size	Comes in a half size loaf...just the amount you need
Order on line... delivered right to your door	Found on the shelf in your local supermarket...needs no refrigeration	Available in vending machines	Available in vending machines

FIGURE 10.10 (continued)

"low carb" bread, cheese cake, and yogurt, respectively. The original elements are shown on the first three columns of the table. The elements that will be selected to mash together into new combinations appear on the fourth column. Each element in that fourth column can be traced to one of the original products. The respondents do not know the source of the elements in a particular concept, but that lack of knowledge does not matter. The respondents simply integrate their impressions and evaluate the entire combination as interesting, or not interesting, to them. The orientation page does not "give the game away" either, but simply says that this product will be a new snack or a new breakfast treat. The outcome comprises a set of winning elements, selected from different sources. These elements "win" or gain respondent acceptance in the context of a new-to-the-world set of combinations. From these elements, the innovating developer can select various combinations that are promising to the total panel, key subgroup, or to a marketing directive.

10.9 BEYOND CONCEPT TESTING TO THE NOTION OF "MIND GENOMICS" FOR THE FOOD AND BEVERAGE INDUSTRY

How can concept testing move beyond evaluating and optimizing for one product or product innovation? In the future and on an ongoing basis, practitioners and scientists in the industry can use databases for tracking consumers over time, across products, for both understanding and for more general, new product opportunities. Concept research in innovation can mash together ideas from multiple products to create new-to-the-world ideas.

These concept studies are really "one-off" tests, run by the developer or marketer at a certain time, to answer a specific development question. The concept studies presuppose no extant database of concept data, meaning that the developer must always run a study anew to answer certain problems. Indeed, there is no archival database of concept ideas, how well they perform for the total panel and key subgroups. By contrast, food science literature contains articles and books on how to create products, studies of the physical characteristics of different food ingredients, and a general body of "know-how," which if not particularly scientific and rigorous, nonetheless helps the developer who knows how to work within this world of ingredient information.

Recently, the author and colleagues (Beckley and Moskowitz, 2002) developed a system by which a database of studies is created for the food category and for the beverage category. These are the so-called It! Studies, with the food (and some beverage) category known as Crave It! and the beverage category known as Drink It! (Moskowitz et al., 2005c). A separate database was created for the category of good-for-you foods, known as Healthy You!

The actual implementation of the database followed these steps:

1. Create a set of studies with similar structure, so that as far as possible the elements would be "parallel" to each other, but of course tailored to a specific product. While keeping in mind the structure of the bread study above (four silos, each with nine elements), the researchers constructed 30 different studies for each of the three databases. All of the studies had the same structure as shown in Figure 10.11, so that across all studies in a database a specific element had approximately the same rationale for most, if not all, the elements. This rationale of a common structure created a database that could be analyzed across products to show how similar ideas perform in different products and what ideas "work" for a specific product.
2. Create a wall to list the available product categories, and invite a respondent to select the study of interest. The wall for the Drink It!® database appears in Figure 10.12. By giving the respondent the opportunity to choose the study topic that most interests them, the researcher increases the chances that the data for any study in the set of 30 will be populated by those actually interested in the particular product.
3. Look across products to find commonalities—what performs well, what does not. The It! Databases provide a rich source of information about what ideas "work" with different products. Segmentation allows the developer

	Element Rationale	Fruit-Based Smoothies	Energy Drinks	Flavored Water	White Wine
Silo #1: Flavor and product description					
E1	Basic beverage	Smoothies…a delicious slushy with fruit flavors	An energy drink, something other than caffeine…to get you going	Highly purified water with delicious flavors	Wine with a pale yellow color and a fruity flavor
E2	Basic beverage	Fruit smoothies… made with yogurt	An energy drink complete with a unique combination of ingredients for an energy kick	Flavored water made from pure, fresh spring water… directly from the source	A white wine blended with iced tea for a cool refreshing drink
E3	Slightly more elaborated beverage	Sweet, creamy smoothies with lots of flavor…so easy to drink	A light taste with complete balanced nutrition to help you stay healthy, active, and energetic	Lightly sweetened with natural fruit flavors	A lightly carbonated pale white wine with a hint of raspberries… so soft and sweet
E4	Slightly more elaborated beverage	Healthy, natural smoothies made with exotic juices	Engineered to make you more alert and maximize your memory… without the jitters	Refreshing flavors in a rainbow of colors	A light bodied white wine with a fresh, fruity flavor
E5	More elaborated beverage	Low-fat smoothies… high in vitamins and minerals	An energy drink with all the vitamins, minerals, and energizing ingredients you need	Lightly flavored and sweetened plus a little caffeine for a revitalizing taste	Wine with juice to make the perfect sangria

FIGURE 10.11 The 36 elements for 4 of the 30 product studies in the Drink It!® database, along with the rationale for the element.

	Element Rationale	Fruit-Based Smoothies	Energy Drinks	Flavored Water	White Wine
E6	More elaborated beverage	Wellness smoothies with vitamins, minerals and botanicals	An energy drink full of vitamins, minerals, and herbal energy to replace the essential body elements that are depleted during daily activities	Flavored spring water… with the antioxidants your body needs	Golden-colored wine with a sweet complex fruity flavor and delicate flowery fruit aroma
E7	More elaborated beverage	Smoothies blended with fruit, ice, and just a little froth	An energy drink with key amino acids	With a total of nine essential vitamins and minerals	A light bodied white wine blended with fruit juice
E8	More elaborated beverage	Smoothies with fruit and real brewed coffee	An energy drink blended with fruit flavors and juice for that feeling of the exotic	Flavored water with a splash of fizz	Icy cold frozen wine refresher
E9	More elaborated beverage	Already prepared— just shake and drink	With twice the jolt from caffeine… gives you just the added energy you need	Enhanced water that contains ingredients to energize you. …	Wine made with exotic flavors

Silo #2: Emotional promise from drinking, health, quality, mouthfullness, natural

	Element Rationale	Fruit-Based Smoothies	Energy Drinks	Flavored Water	White Wine
E10	Tempting and inviting	Drinking smoothies are cool and inviting	Drinking an energy drink is cool and inviting	Drinking enhanced water is cool and inviting	Drinking red wine is so relaxing
E11	Health stuff or quality	Made for the lactose intolerant	High in protein; low in carbohydrates	With calcium for strong bones and teeth	Made in the tradition of the greatest wine producers all over the world

FIGURE 10.11 (continued)

(continued)

	Element Rationale	Fruit-Based Smoothies	Energy Drinks	Flavored Water	White Wine
E12	Calories or caffeine	Low in calories	Low in calories	Low in calories	Lower in alcohol to manage your calories
E13	Premium quality	Premium quality	Premium quality	Premium quality	Premium quality
E14	With stuff or caffeine or calorie	Not too thin or too thick…just the right amount of creaminess	An energy drink that actually tastes good	Wellness water with natural botanicals	White wine with added antioxidants
E15	Natural	100% natural	100% natural	100% natural	100% natural
E16	Good for you ingredient	With all the goodness of soy	Energy drink designed especially for active women with isoflavones, and a separate one for men	Flavored water designed especially for active women with isoflavones, and a separate one for men	Fortified with vitamins and minerals
E17	Organic	Made with organic ingredients	Made with organic ingredients	Made with organic flavors	Made with organic grapes
E18	Refreshing, relaxing, drink more	So refreshing…you have to drink some more	So refreshing …you have to drink some more	So refreshing… you have to drink another	So refreshing… you have to drink some more

Silo #3: Drinking situation

	Element Rationale	Fruit-Based Smoothies	Energy Drinks	Flavored Water	White Wine
E19	Quick and fun…ready to drink	Quick and fun… ready to drink	Quick and fun…drinking does not have to be ordinary	Quick and fun…water is no longer boring!	Quick and fun…ready to drink, no bartender required
E20	Miscellaneous	Real fruit juice blended with the wholesomeness of tea	Cools you down	Cools you down	The aroma of white wine invites you to drink more
E21	The best	Simply the best	Simply the best	Simply the best	Simply the best
E22	Relaxes	Relaxes you after a busy day	Picks you up when you feel run down, tired, or stressed	Relaxes you after a busy day	Relaxes you after a busy day
E23	Take a break	Take a break from your busy day	Take a break from your busy day	Take a break from your busy day	Perfect ending to your busy day

FIGURE 10.11 (continued)

	Element Rationale	Fruit-Based Smoothies	Energy Drinks	Flavored Water	White Wine
E24	General promise	Looks great, smells great, tastes delicious	Looks great, smells great, tastes delicious	Looks great, smells great, tastes delicious	Looks great, smells great, tastes delicious
E25	Meal occasion/ starting	Helps you get your day started	Helps you get your day started	Perfect complement to your meal	Perfect complement to your meal
E26	Satisfaction	Pure satisfaction	Pure satisfaction	Pure satisfaction	Pure satisfaction
E27	Keeps you going	Thick, rich and smooth…keeps you going until mealtime	Keeps you going throughout the day	Keeps you going throughout the day	Great with friends and family
Silo #4: Brand, packaging, safety					
E28	Brand	From Tropicana	From Red Bull	From Glaceau	From Northern California
E29	Brand	From Snapple	From Capri Sun	From Aquafina	From Gallo
E30	Brand	From Capri Sun	From Gatorade	From Gatorade	From Kendal Jackson
E31	Brand	From Yoplait	From Starbucks	Keeps your body hydrated	Imported from France
E32	Brand	From Ocean Spray	From SoBe	From Wal-Mart	From Bella Sera
E33	Multiserve containers	Multiserve containers…so you always have enough!	Multiserve containers… so you always have enough!	Multiserve containers… so you always have enough!	Multiserve containers… so you always have enough!
E34	Miscellaneous	From Wendy's	Backed by the power of science	A great tasting alternative to sport drinks	For the sophisticated wine connoisseur
E35	Resealable	Resealable single serve container…to take with you on the go	Resealable single serve container…to take with you on the go	Resealable single serve container… to take with you on the go	Resealable single serve container
E36	Safety and trust	With the safety, care and quality that makes you trust it all the more	With the safety, care and quality that makes you trust it all the more	With the safety, care and quality that makes you trust it all the more	With the safety, care and quality that makes you trust it all the more

FIGURE 10.11 (continued)

Welcome to the DRINK IT!! survey.

We are interested in learning WHAT YOU DRINK
Please select the survey that you would like to participate in by clicking on one of the yellow buttons.
You can participate in as many surveys as you wish.
(You can participate in each survey only once)
Please send this link to all your family and friends.
Note: If you are under age 21, please do not complete this survey. The invite was sent to you in error.

Iced Tea	Yogurt Beverage	Soy Beverages
Soup	Carbonated Spritzers	Meal Replacement Beverages
Red Wine	Coolers	Hot Tea
Smoothies	Juice	Sports Drinks
Milk	Milk Smoothies	Energy Drinks
Flavored Low Alcohol Drinks	Flavored Cider	Cola
Coffee	Flavored Coffee	Kids Beverages
Flavored Beer	Flavored Tequila	Ready to Drink Flavored Coffee
Hot Chocolate	Shakes	Fiber Beverages
White Wine	Lemon Lime Soda	Enchanced Water II

2001 i-Novation Inc. All rights reserved. 1025 Westchester Ave., Suite 444, White Plains, NY 10604 Tel: 914-421-7444 Fax:914-428-8364 email:info@i-Novation.com.

FIGURE 10.12 Example of the "wall" for the Drink It! Database.

and marketer to uncover segments relevant to a particular product, as well as segments that transcend the product to define a full product category.

4. Assess the role of emotion. Emotion in product concepts is becoming of increasing interest. How do the brands do? How does reassurance work? Do promises of a great experience with the product drive acceptance of the concept?

5. Search for meaningful trends over time. What elements are beginning to "pop" up? Are brands increasing in their ability to drive the concept? What about the elements about warmth and reassurance, or about product quality and safety? Do these elements look like they are becoming increasing relevant and turning into stronger drivers? What does the trend mean in terms of the consumer mind, with respect to the product?

10.10 SUMMING UP—THE ROLE AND PROMISE OF CONCEPT TESTING

In this chapter, we saw examples of the different approaches, ranging from benefit testing to full-blown experimental design of ideas, and onto to concept research as a strategic method for innovation and for databases of the consumer mind. In the world of product development, concept testing has been given short shrift, despite its importance as providing a roadmap for the developer. Certainly, it plays a critical role in the decision process about a product. Company after company prides itself on efficient methods to understand the mind of the consumer, and by so doing crafts concepts that are presumed to drive consumer interest and, thus, promises success.

For the most part, concept research is done by corporations for specific purposes, usually new product development or product repositioning (taking an old product and giving it a new lease of life by changing its image). The fact that the data for concept research is primarily proprietary, in the repositories of companies, means that there is a lack of good scientific research about concepts, ranging from how to create their basics to how to use them for innovation. The authors hope that this chapter has added to the public archival literature, so that those who move forward with concept research need not "invent the wheel" yet another time.

ACKNOWLEDGMENT

Dr. Moskowitz wishes to thank his editorial assistant, Linda Lieberman, for finding and correcting errors and generally keeping him on production schedule.

REFERENCES

Acupoll (2005) Online: http://www.acupoll.com/homenglish/index.html

Bacon, Jr., F.R. and Butler, Jr., T.W. (1998) *Achieving Planned Innovation*. New York: The Free Press.

Beckley, J. and Moskowitz, H.R. (2002) Databasing the consumer mind: The Crave It!, Drink It!, Buy It! & Healthy You! Databases. Anaheim, CA: Institute of Food Technologists.

Box, G.E.P., Hunter, J., and Hunter, S. (1978) *Statistics For Experimenters*. New York: John Wiley.

Cooper, R.G. (1993) *Winning at New Products: Accelerating the Process from Idea to Launch* (2nd ed.). Reading, MA: Addison Wesley.

Fuller, G.W. (1994) *New Food Product Development from Concept to Marketplace*. Boca Raton, FL: CRC Press.

Green, P.E. and Krieger, A.M. (1991) Segmenting markets with conjoint analysis, *Journal of Marketing*, 55, 20–31.

Green, P.E. and Srinivasan, S. (1990) Conjoint analysis in consumer research: Issues and outlook, *Journal of Consumer Research*, 5, 103–123.

Luce, R.D. and Tukey, J.W. (1964) Conjoint analysis: A new form of fundamental measurement, *Journal of Mathematical Psychology*, 1, 1–36.

Moskowitz, H.R. (1994) *Food Concepts and Products: Just in Time Development*. Trumbull, CT: Food and Nutrition Press.

Moskowitz, H.R. (1996) Segmenting consumers world-wide: An application of multiple media conjoint methods, presented at the *49th ESOMAR Congress*. Istanbul, Turkey.

Moskowitz, H.R., German, B., and Saguy, I.S. (2005c) Unveiling health attitudes and creating good-for-you foods: The genomics metaphor and consumer innovative web-based technologies, *CRC Critical Reviews in Food Science and Nutrition*, 45(3), 165–191.

Moskowitz, H.R. and Gofman, A. (2005) System and method for performing conjoint analysis, U.S. Patent Pending, No. US2005/017388 A1.

Moskowitz, H.R. and Gofman, A. (2007) Selling Blue Elephants: How to make great products that people want before they even know they want them. Upper Saddle River, NJ, Wharton School Publishing.

Moskowitz, H.R., Gofman, A., Beckley, J., and Ashman, H. (2006) Founding a new science: Mind genomics, *Journal of Sensory Studies*, 21, 266–307.

Moskowitz, H.R., Gofman, A., Itty, B., Katz, R., Manchaiah, M., and Ma, Z. (2001) Rapid, inexpensive, actionable concept generation and optimization—the use and promise of self-authoring conjoint analysis for the foodservice industry, *Food Service Technology*, 1, 149–168.

Moskowitz, H.R., Poretta, S., and Silcher, M. (eds.) (2005a) Concept Research in Food Product Design. Ames, IA: Blackwell Publishing Professional.

Moskowitz, H.R., Reisner, M., Krieger, A.M., and Oksendal, K.O. (2005b) Steps towards a consumer-driven concept innovation machine for "ordinary" product categories in their later lifecycle, presented at the *First Innovate! Conference*. ESOMAR (World Society of Market Research), Paris, France and Amsterdam, the Netherlands.

Urban, G.L. and von Hippel, E. (1988) Lead user analyses for the development of new industrial products, *Management Science*, 34(5), 569–582.

Vigneau, E., Qannari, E.M., Punter, P. H., and Knoops, P. (2001) Segmentation of a panel of consumers using clustering of variables around latent directions of preference, *Food Quality and Preference*, 12(5–7), 359–363.

Wittink, D. R. and Cattin, P. (1989) Commercial use of conjoint analysis: An update, *Journal of Marketing*, 53, 91–96.

11 Defining and Meeting Customer Needs: Beyond Hearing the Voice of the Consumer

Carole Schmidt

CONTENTS

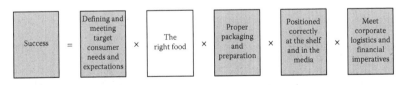

11.1 INTRODUCTION

How often have we heard, "59% of buyers prefer fruit to savory flavors?" "We've got an 18 share." Forty-seven percent chose "with sauce" over "in marinade." "All natural" and "sugar free" are the "top-two box" scores among 33% of women. Useful? Of course. Important to know? Absolutely.

But, do you know *why* your consumer answered the way she did? Do you understand the context of her response? Do you know how she interpreted the question or concept? Sure, you can hear her from the statistics—but are you truly listening to her voice? Some real-life examples:

1. Did "all natural" mean without preservatives or unprocessed or organic?
2. Did she assume that "sugar free" meant calorie free or no high-fructose corn syrup?
3. Was she expecting a separate sauce packet or in-sauce product?
4. Did calling it "energy" mean geared to teens or with caffeine?

The difference in understanding may be top box vs. bottom, trial vs. repeat, and delight vs. disappointment.

Many R&D and marketing people have become so hung up on the quantitative aspects of consumer targeting that they have become deaf to what the consumer is really trying to say. It is important that you push farther, come to understand your consumer's wants and needs from her perspective, and learn to speak the language your consumer uses to explain her attitudes and behaviors.

Why is her voice so important in product development? Because frankly, *she* knows what she needs and wants better than you do. She can tell you why it works for her, or why it does not, because she has no vested interest in the outcome of your product's success or failure. Most importantly, because the cost of ignoring her voice always far exceeds the cost of the most expensive research you will ever do.

By moving beyond hearing—to listening—you will create products that fulfill a physical or emotional need (*motivating*), deliver benefits in a unique and meaningful way (*compelling*), and will communicate directly to her (*relevant*).

OK, I get it...getting inside of my target consumer's head is critical to my success. So what does "beyond hearing the voice of the consumer" really mean? How do I do this?

11.2 WHAT IS QUALITATIVE RESEARCH?

The best way to know your consumer is to get out of the office and meet your consumer in person; in other words, through *qualitative research*. Qualitative research is the careful analysis of unstructured information gained about a select audience to surface attitudinal and behavioral insights.

In my "elevator speech," when asked about what qualitative research is, I tell people: "I ask your customers questions to help you make smarter business decisions with the answers." Quantitative research produces the "what" with its numbers and statistics; qualitative research seeks out the "why" (and often the "how," "when," and "where," too) around a select topic.

On the most *informal* level, you can conduct qualitative research simply by *asking for candid feedback* from anyone you can query along every step of the development process.

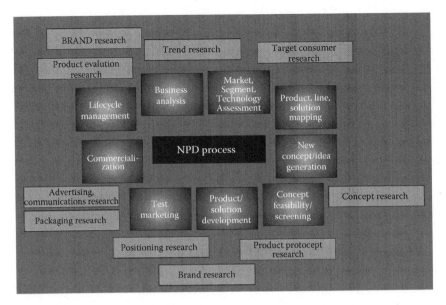

FIGURE 11.1   Qualitative research applications.

Qualitative research can be effectively used to gather key consumer insights and general direction in the following areas. This learning can be used for final decision making or to inform setting a questionnaire for quantitative research (see Figure 11.1).

1. Trend research
   a. *Purpose*: To identify "big picture" themes and influences that will impact your business or create new opportunities.
   b. *Examples*: What are the new food trends? What is happening in the category of interest? How are consumer lifestyles changing? How have health behavior changed? What ingredients are consumers desiring or rejecting? How might food costs affect consumer behaviors?
2. Target consumer research
   a. *Purpose*: To identify and understand the need-states among the potential purchaser/user audience for your proposed product.
   b. *Examples*: What attitudes and behaviors should I understand about my intended audience? What makes them "tick?" What are their values? What or who influences their lives? Who, then, is likely to buy this product and why? Who is my core or "bull's eye?" How might I adapt my product to appeal to other consumers? When and how might my product fit into their consumption patterns?
3. Concept research
   a. *Purpose*: To expose and screen many proposed solutions or concepts to target consumers to assess their reactions, to ensure that the concepts are communicating clearly, and to strengthen the ideas using end-user input.

   b. *Examples*: Do consumers understand this idea as intended? What is appealing and unappealing about each concept/idea? Why does it appeal more to some consumers than to others? What are consumers assuming or expecting about this product, from the concept? Is there anything confusing or hard to understand about this idea? How are consumers likely to use this product? What is the perceived competition?

4. Product protocept research
   a. *Purpose*: To assess target consumers' reactions to the actual product, to determine if the product execution delivers on the concept promise, and to identify any glaring changes needed before production.
   b. *Examples*: Are reactions to my product as enthusiastic as they were to the concept? Do target consumers understand how to open, prepare, use, and serve my product? What are product benefits or drawbacks compared to those of perceived competitors?

5. Brand research
   a. *Purpose*: To gauge how the brand will serve as an asset or barrier to your product, to understand how brand perceptions are affected by your product, or to explore what product opportunities may exist for your brand.
   b. *Examples*: What do target consumers assume or know about my brand? Is there emotional appeal and relevance? What does my brand stand for? How does this product affect consumers' opinions and beliefs about my brand as a whole? Should I consider a sub-brand for this product?

6. Positioning research
   a. *Purpose*: To explain the "location" of a brand in consumers' minds, to aid in defining the product's unique role in the marketplace, to understand how to communicate that positioning.
   b. *Examples*: How does the target consumer segment see my brand relative to that of competitors? What aspects of my product are most compelling to consumers? Is the "angle" I am promoting meaningful to the target consumer?

7. Packaging research
   a. *Purpose*: To explore which potential packaging designs will best communicate the intended positioning and benefits of your product, to ensure that the packaging is customer friendly, that the packaging will compete effectively at shelf.
   b. *Examples*: Does my packaging have breakthrough potential? Do consumers understand what comes in the package? Am I communicating the most important attributes clearly and prominently?

8. Advertising, communications research
   a. *Purpose*: To assess which campaign or promotion is most appealing, motivating, and which best delivers the intended message(s).
   b. *Examples*: Which campaign is most clear in message? Are the proposed campaigns relevant and meaningful to the intended audience? Do my target consumers understand my product from this ad? Does one of these campaigns contemporize my brand better than another?

9. Product evaluation research
   a. To explain acceptance or rejection of commercialized products, and to identify how second generations can be improved.
   b. *Examples*: What stimulated so much trial, yet did not turn into as many repeat purchases as forecast? What is the essence of our product's success that we might be able to leverage into a line?

There is a whole industry with a cadre of "trained brains"—qualitative research professionals—who have studied and refined their psychology and social anthropology skills into a business-practical, creative art form. We can help you to identify who your target consumer is, help you find them so that you can hear from them, uncover rich and revealing insights using careful probing and other discovery techniques, and finally, provide you with careful, deliberate analysis with independent, thoughtful recommendations to help you make smarter business decisions.

11.3   HOW DO YOU START A QUALITATIVE PROJECT?

The most successful qualitative research projects begin with sharing information about the "back story"—the factors that influenced you to commission research in the first place. It is equally important to identify your goals, communicate your preferred working style, and to be clear about how you will use the results and the learning from the research. The bottom line is that the more time you spend up front planning the research effort, the richer the research results.

Here is a "cheat sheet" to help identify what you need, want, and will contribute to the qualitative research project. (It will work for just about any research project.)

- What led to the need for research?
- What is the business decision that needs to be made?
- What three things do you want to get out of this research effort, specifically? (What are the "need-to-knows" and the "nice-to-haves"?)
- Who is your target consumer? What do you already know about your target consumer? What don't you know?
- How do you define a successful project?
- What you *do not* want is…
- What are your hypotheses about what will happen?

11.4   CHOOSING YOUR TARGET CONSUMER

So, who is my consumer? This is a question that you should ask every time you think about new product development (NPD) *before* you start developing products. Then also ask yourself—why would she be interested in what I have to sell?

Let us assume you make high-performance nutrition bars. When you launched your business sales performance was good, but now you are losing shares to competitors in a crowded marketplace. You want to expand your line of products to appeal to a broader audience, and thus, better compete against recent new entries. Furthermore, you have a new technology that allows a true fresh fruit flavor.

Explore the universe of potential customers. Who are all the people who might be interested in your food product? Start with the broadest common denominator shared among all people who eat your product—in the nutrition bar example, this might be "snack eaters...." Spend time really thinking about your target to narrow it to a more productive target. Too often, product developers lump enormous segments of consumers together, e.g., "our target is people 18–64 who like snacks." While on the surface, it may appear that this broad target will yield higher sales volume by sheer numbers, we have observed this way of thinking to be a dangerous practice. Within any large population, there are many segments and subsegments, each with different motivations, different need-states, and different purchase drivers. There are not many successful products that appeal to everyone.

Conversely, just as risky is selecting too narrow of a target (unless you can really afford to be a small, niche marketer). Selecting a consumer segment bull's eye that is actually "a needle in the haystack" will be expensive and time consuming to find and may not give you the volume sales you need.

When choosing your target consumer, look beyond the obvious demographics. The demographics of your target (age, income, family status, ethnicity, education, employment, etc.) are important, of course, but it is equally (and often more) important to look beyond demographics, to attitudes, behaviors, and other factors that shape consumers' beliefs, goals, and purchases. Figure 11.2 presents some of the different considerations.

The key to success in identifying your target segment is to begin asking questions early in the product development process. Returning to the nutrition bar example, let us explore the target "snack eaters" more thoroughly. What behaviors surround snack eaters? People interested in health? What attitudes toward health? What level of health interest? Elite athletes? Health motivated? Those who want to get healthier, but are not doing much yet to get there? Sedentary persons who want a better-for-you snack? What day-parts? What nutritional requirements? What stores should they shop to buy their snacks? Continue exploring and narrowing until you develop concentric circles representing the universe of potential consumers down to your bull's eye target (see Figure 11.3).

11.5 CHOOSING THE RIGHT QUALITATIVE METHOD TO HEAR THE VOICE OF THE CONSUMER

Determine what you really want to learn about your target consumer. The best way I have found to do this is by having your product development team perform a "brain dump" of all the things they want to know. This exercise will identify what you already know, and illustrate what you do not. Next, prioritize the list in terms of most-to-least important. Perform this task up front because it often generates a lively, and perhaps a painful discussion. The attention to this "prep work" will pay off on the back end of the research effort when your goals are met, the results matter, and you have gotten the "biggest bang for your research buck."

Lifestage

- Kids, tweens, teens
- College students
- Young employed singles
- Transitioning, relocating singles
- Married/partner, no kids
- Pregnancy
- Young family (parent as gatekeeper)
- Tween/teen family (parent influenced by children)
- Confident mid-age woman
- Empty nester
- Young senior
- Aged senior
- Needing eldercare

Attitudes

- Evangelists
- Enthusiastic
- Positive/accepting
- Neutral
- Unenthused
- Negative

Behaviors

- Claimed behaviors
- Observed behaviors
- Compensating behaviors
- Early asdoptors
- Users
- Lapsed users
- Rejectors
- Aware nonusers
- Unaware nonusers

Other influences

- Relationships
- Environment
- Competitive set
- Culture, habits, traditions, rituals
- Media consumption
- Economic factors
- Geography
- Day-parts
- Social/political

FIGURE 11.2   Considerations that influence your consumer target(s).

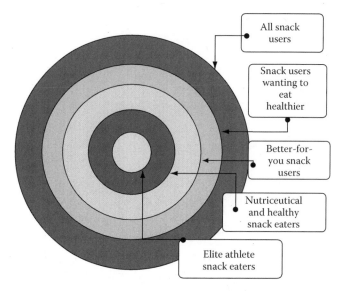

FIGURE 11.3 Concentric circles representing the universe of potential consumers.

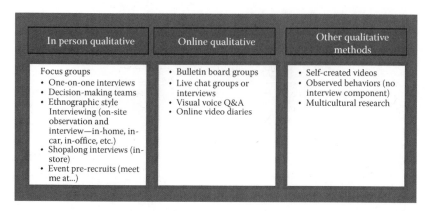

FIGURE 11.4 Key methods used in qualitative research.

There are many resources that provide a guide for how to execute the array of qualitative methods that are available to you. A list of key qualitative methods appears in Figure 11.4.

11.6 WHAT KIND OF LEARNING ABOUT THE CONSUMER WILL HELP DEVELOP BETTER PRODUCTS?

You should be able to define your target consumer's "need-state." Many failed products were created because they could exist, rather than because they should exist. Is there a reason why your target consumer will need or want your product? She must intuitively understand why your product exists (basic needs), understand where the product fits into her life (to engage her, induce trial), and know how to use the product when she gets it home. Finally, she must enjoy the experience with it enough to buy it again (how to turn her from a "trier" into a repeat customer or better, a loyal user).

To know your consumer, you must spend time with your consumer. The rich insights come from moving beyond the surveys and into the dialogue, verbal cues, body language, and expressions. It is critical to explore the open-ended array of emotions and motivations that lie below the surface responses.

- Who is my consumer beyond her physical presence?
- What are her attitudes, beliefs, and values that guide decisions?
- What is the importance of, role of my category, brand, or product in her life?
- Where and how does she shop?
- Which categories, brands, and products does she consider to be the competitive set?
- What are the influences that cause her to consider, buy, use, reject, or buy again?
- What product gaps exist in her life?
- Are there compensating behaviors that fill the product voids now?

Consider the following "typical" case study: Doyle Research conducted a series of in-depth interviews among adults representing a broad range of demographics as well as attitudes toward frozen food. Prior to their interviews, respondents were asked to complete a two-part homework assignment: videotape themselves as they were planning their dinner meals (video diaries), and to attend a personal interview to evaluate some new product protocepts under consideration. As a precursor to their video journalizing, respondents were questioned extensively about their lifestyles, satisfaction levels with current products, and to talk about the role of frozen foods in their lives. Particular attention was paid to their reactions toward particular frozen food categories. The research resulted in a number of distinct consumer profiles (delivered to the client via "posters" along with a presentation summarizing key findings and implications). These consumer profiles continue to be utilized by product development teams as stimulus for NPD.

The challenge for every product developer is to remember that the consumer is not stagnant, but, indeed, a moving target. To be relevant to that consumer, you must frequently pay attention to her, check in with her, stay up-to-date on her changing needs. Be aware that she is flawed and that she can be inconsistent as she is influenced by her environment, her evolving lifestyle and family situation, and, of course, the ever-changing marketplace.

Several savvy clients dedicate a physical, visual space (such as an entire wall in the office) to develop and showcase the research-driven "consumer personas" that represent their target consumer segment(s). The research findings they discover are posted, updated, challenged, and revised as marketers and product developers become more educated about the target consumer's voice. The persona is used as a "check and balance" to ensure that the entire marketing mix (product development, promotions, packaging, communications, and pricing strategies) reflect the needs and wants of the target consumer.

11.7   MOVE BEYOND HEARING: HOW TO REALLY *LISTEN* TO THE CONSUMER

In that first exposure of your product to your target consumer, she will agreeably tell you what she claims she likes and does not like. It is important to hear that voice. As a product developer, your job is not only to accept the consumers' comments at face value, but also to recognize that the true consumer reaction is a more complex puzzle, comprised of many pieces of her life put together to tell a story about her current need-state.

Be accurate when capturing consumer language. When conducting focus groups or interviews, it is important to capture specific words the consumer used as she stated them, rather than trying to summarize what she says. Not only does this demonstrate respect for the consumer's opinions during a focus group or interview, but her description of a new juice product as "health inducing" is more meaningful and telling than presenting her opinion as "the consumer thinks our product is healthy."

Understand and include the context of the consumer's voice. As mentioned in the beginning of this chapter, many products have failed because the client failed to explore what the consumer meant beyond her surface responses. There is a reason why we all have heard the old adage about what happens when we "assume" (it makes an "ass" out of "u" and "me"). Three examples of real-life, yet poor assumptions made by product developers around consumers' comments on new products are

1. Assumed that the term "probiotic" meant advanced technology, forward thinking, the best, when instead, consumers rejected it as "cold," "sterile," "medicinal," and "prescriptive." Reported one consumer, "I would rather eat something that aids digestive health, or keeps me healthy, than something that sounds scary."
2. Assumed that the concept communicated as intended: An enriched juice product expected to perform well in quantitative concept tests repeatedly failed, perplexing the developers. Qualitative research revealed that the consumers' interpretation of the word "shot" was conveying the presence of alcohol, which consumers did not want, and thus, was throwing the concept, "you mean it *does not* have alcohol in it?"
3. Assumed incorrectly what would be the usage occasion: Returning to the nutrition bar example, product developers queried consumers and learned that they associated the product with energy. The product developer assumed the consumer would embrace the product as a breakfast item and launched the product accordingly, with dismal sales. Postmortem qualitative research on the failed launch revealed that the developers did not understand the context behind consumers' interpretation of the word "energy." The consumer envisioned the product as a heavy filler, better served as a post-workout snack, not an everyday breakfast food.

The most revealing qualitative learning emerges when you probe the consumer to explain the words she chooses. My favorite tool to do this is to simply repeat back the word or phrase the consumer has used to express her opinion and wait for her to explain more about what she means. The approach works every time.

Pay attention to expression. Not only is it important to tune in to the consumers' words, but also to tune in to the emotion or emphasis that was used in speaking the words. All too often product developers will relay consumer opinions to those who did not observe the consumer research without providing the context, or the expression behind a consumer's statement that helps reveal what was *meant*, and not just what was said.

In a recent example, a consumer stated: "oh yeah, I really love the smell of that sauce." When this comment appeared on a note from product developer to a marketer it looked very different than what the consumer actually reported, because that particular consumer spoke sarcastically, and was, in fact, complaining about the aroma of the sauce.

Note the presence or lack of passion. Moving the consumer from product awareness to product action (purchase) requires passion. The consumer not only has to understand what the product is and does for her, but she also has to connect with and believe in its promise—product passion. Product passion is evident through expression that demonstrates positive emotion, be it enthusiasm, joy, delight, excitement, intrigue, curiosity, or even skepticism. Skepticism is often "hope" in disguise.

The presence or lack of passion is also necessary to understand when exploring survey responses. To move beyond hearing the consumer means to understand the *relative* measures in full disclosure, and not simply rely on the score of a product or food concept. Qualitative research is used by smart product developers to better understand what is driving food product or concept scores. In example, product developers made a painful discovery when qualitative research revealed that consumers' high absolute rating of a new vegetable soup product was, in fact, simply the best of the presented set. Depth interviews revealed that there really was no passion or motivation to buy the product, despite performing significantly better than any other soup concept tested. While the results meant heading "back to the drawing board" for the product developers, it was clearly a better and much less expensive outcome to have learned this important finding *before* product launch.

Where does the product fit? Addition or replacement? Consumers only have so many dollars to spend and so many minutes in the day. Another cue for gauging consumer passion for the new product is to explore how and where this product fits into her day, her meal repertoire. The goal is to create new products that fulfill an unmet need. Is the consumer willing to make room for this product in her life by, e.g., reducing competitive product purchases, changing her behavior? That is passion. Or, does the target consumer talk about the product as an addition to what she is already buying, meaning she is not willing to replace the product she currently uses. The latter is *not* an example of passion.

Observe nonverbal cues. In conducting good qualitative research, keen observational skills are as important as smart interviewing. Look at the body language. Does the consumer look at the interviewer or look down as she responds? Does she lean forward or sit back? If the consumer says that she would love to buy the new frozen entrée but sits back with arms folded, it is worthwhile to explore this response a bit more. The body language says something different than the verbalized statement.

Four other important nonverbal cues to pay attention to are

1. Facial expressions: Facial expressions are remarkably telling. Make sure to pay attention to consumers as they are exposed to an idea in writing or verbally, to discover these revealing cues. Raised eyebrows without a smile often signal a concern or objection. Raised eyebrows with a smile indicate relief or discovery, usually a positive reaction.
2. Audible cues: While reviewing consumers' faces, also listen for "oohs" (excitement), sighs (typically, a sign of disappointment, indifference, disinterest), or snapping, clicking, or clucking (often reveals a concern or objection).
3. Squirming: Probing this behavior may help reveal barriers to your food product or concept.
4. Looking around at others for agreement: This is a tricky behavior to read. It often means that she recognizes her response is not likely to be a popular choice whether she is interested or not.

Capture compensating behaviors. "Compensating behaviors" refer to the consumer solving a product problem on her own, finding a "workaround." Compensating behaviors are considered a nonverbal cue because consumers often are not really aware of the significance of their solution-finding behaviors, but they are worthy of attention because they are often a rich resource for new product ideas and refinement needs.

For example, I observed families eating a breakfast of proposed prepackaged new products for each of four days in a single week. Interestingly, despite consistent kudos for one of the breakfast sandwich products tested, almost all members of the family would innocently add a condiment to the product every time they consumed it—ketchup, hot sauce, salt and pepper, and Worcestershire sauce—in most cases, without thought or complaint. What emerged was a clear annoyance with the lack of flavor, which was ultimately enhanced. The learning and subsequent flavor enhancement resulted in a successful product launch. Other compensating behavior examples are shaking products, heating/chilling a product not intended to require that action, contorting the product to make it easier to handle or consume, removing some aspect of the product, heating it twice, etc.

Observe the environment. When looking at the information presented during a qualitative session, note the participant's style of dress, quality of attire, attention to detail, and the like. What is the "story" about this person who spends so much money on putting herself together, who insists that her health is priceless, yet insists that the proposed shelf-stable bread product is too expensive? There is a bigger story being told than what the consumer is letting on.

When conducting qualitative research on site, the consumer's need-state story may be further illustrated by conducting spontaneous pantry and refrigerator checks. In revisiting the nutrition bar example, during the course of discussion at various homes, a simple pantry and refrigerator check revealed that the product's "competition" for consumption included *several* product categories, some which clearly lay far outside the range of product types that were originally thought to be the competition. This causal observation generated meaningful news and a strong insight to the development team—that their product was not *perceived* to be as nutritionally rich as claimed. The range of competitor products suggested the potential for a more distinct product line, by offering broader range of flavor profiles and a stronger nutritional story.

Shopping with the consumer. Another nonverbal cue is to note the consumer's pace as she shops certain categories. The amount of time the consumer spends browsing the shelves or case often reveals the passion and interest for certain categories and products.

11.8   WHAT TECHNIQUES WILL HELP MOVE BEYOND SUPERFICIAL OR SURFACE RESPONSES?

When new products are introduced in a crowded or low involvement category, consumers may find it difficult to express their needs, reveal their perceptions, or articulate their interest in a new product, in a way that can be easily understood. Additionally, there may be NPD situations where you want to surface more imagery, emotions, and latent opinions that come from the incubation of thoughts and feelings. As a product developer, you can turn to many smart, helpful "tricks of the trade" to help you move beyond hearing the voice of the customer, toward a real understanding of what she is trying to say.

Projectives is the name given to exercises used to help elicit latent emotional responses from consumers which at the surface level may not seem emotional at all. Projectives work by providing the consumer a "third party" outlet to help express—or project—her own feelings.

For example, two different new yogurt product concepts each scored very high and shared many common diagnostics in a quantitative evaluation, making them seem very much the same. However, when consumers created collages at home to "project" the nature of each concept as they perceived it, the personas of the yogurts were radically different. One persona was a youthful weekend athlete, whereas the other emerged as an introverted health nut. Projectives helped the product developer better select the yogurt concept to develop, and fueled future communication efforts, as well.

Here are eight projective methods, with a short précis of each:

1. Collaging: Collaging requires consumers to reuse existing visual stimulus (magazine images and phrases, Internet photos, color or fabric swatches, natural materials, etc.) to create a visual representation of their emotions and feelings about a particular topic. Collaging was recently used to help consumers express how they felt about cooking on the weekdays vs. the weekends. Findings were used to revise cooking and preparation language used on the packaging, to help make the product better suited to weekday, everyday usage.

2. Personification: Imagine the door opened and in walked the "store brand" (private label) version of a jarred pasta sauce and beside it, a new microwave form of pasta sauce. Do you see the same person? Neither did consumers who explored this concept. The exercise asked consumers to envision each pasta sauce's nature, and dress, car, family status, home style among others, which revealed distinct differences between the images of these products which informed a fresh communication strategy and packaging development.

3. Eulogy: While having target consumers write a eulogy for a product may at first seem somber, even questionable, this is a very effective technique for revealing consumers' passion levels for a product. Eulogies allow the consumer to project deeper emotions in a structured way. Eulogies should reflect feelings about the product that has "died," such as what might be missed if the product did not make it to the shelf or were taken off the shelf, what the product might have accomplished, what the product may have "died" from, and the like. This technique often surprises consumers as they reveal their deep-seated thoughts and feelings about a product or brand.

4. Say/think: This exercise features a character that the consumer uses to represent herself. The exercise encourages consumers to capture both what she might say (the more surface responses) and also provides her a safe forum in which to express what she may really be thinking (her more accurate response) about a new product or concept.

5. Mindmapping: A visual exercise that can quickly raise consumer perceptions, attitudes, and emotions around a product or concept. Each "spoke" of the mind-map worksheet provides a place to capture consumers' stream-of-consciousness associations, thoughts, and feelings as they surface. Mindmapping was recently used to draw out the perceptions, benefits, and drawbacks of canned vs. fresh vegetables to aid in identifying the need-gaps in produce packaging.

6. Product or product-concept sorting: Having consumers physically sort products—either by their own criteria or against an assigned dimension provides a telling story about what gaps might exist in a particular category and consumers' perceptions of competitive offerings. For example, sorting a variety of snack bars (e.g., on the basis of what is "healthy") informed a product development team about consumers' negative associations with several seemingly innocuous ingredients, a contributing factor to infrequent purchases.

7. Value/trade-offs: In this exercise, consumers use paper money to first assign value (level of importance, dollar value, level of convenience, etc.) to certain features or attributes of a new concept or product. Afterward, consumers are instructed to explore and explain how their dollars were "spent" to surface the most salient aspects of a new product or concept.

8. Deprivation: Asking consumers to go without a product or brand they regularly use quickly helps to elicit what is sacred and what is disposable about a product. Deprivation was recently used to identify critical brand elements that could strengthen the product positioning. These brand elements became a requirement for future brand extensions.

The eight tools mentioned above are designed to enhance the fresh and frank dialogue you open with your consumer through qualitative research. Use them judiciously and do not overwhelm your personal discussions with too many of them.

11.9 HOW TO MAKE THE MOST OF THE LEARNING EMERGING FROM THE RESEARCH

Conducting qualitative research with target consumers creates a fertile foundation for developing more meaningful new products. The research provides more than simple statistics with which to create a framework. Qualitative research helps explain the emotional drivers—the whys—behind consumers' attitudes and behaviors toward new products. So, how do you apply the learning you gain? How do you make the findings actionable? Here are seven guidelines to move beyond hearing the voice of customer:

1. Make voice-of-the-consumer research a respected priority in your NPD process.
2. Involve multifunctional teams in the research. Give each team member several opportunities to hear the consumer "straight from the horse's mouth."
3. Embrace, rather than fear, what you may hear. Good or bad, remember that it is far more important and, of course, less expensive to absorb the consumer perspective early in the product development process than to have to explain why you did not ask the consumer her opinion after a failed product launch.
4. Move beyond simply reporting the findings toward more active—and interactive—dialogue around understanding the consumer's voice.
5. Involve all members of the multifunctional team to discuss the findings, and encourage the team members to turn the findings into insights, and insights into action. At the onset of a consumer research project, develop a "next steps contract" with your product development team so that agreed-upon actions are accomplished. The contract is an "if-then" scenario that embodies different outcomes and specific actions to be taken.
6. Keep the voice of the consumer alive and visible where you work. Assign a space for collages, personas, image boards, etc. that reveal who she is, what her values are, what makes her "tick," what you have to do to meet her needs, engage her, and gain or retain her as a customer.
7. Periodically, check in again with your target consumer (the moving target) to ensure you are staying on track with her. Over the course of a consumer's lifetime, many influences will change her needs, her attitudes, and her behaviors.

"Resting on your laurels" is tantamount to the kiss of death in NPD. Key opportunities to enter a consumer's life and gain her loyalty are

1. Solve a problem in her life.
2. Respond to consumers' needs as she experiences inevitable changes in life stage or circumstances: events (weddings, births, employment, home purchase, etc.) and crises (stress, family, crime, medical, death, disaster, divorce, financial, and political).

3. Become a regular part of your target consumer's day-to-day life. Consider how you can fit into her wake up, breakfast, daily nutrition, energy, workout, snacks, lunch at work, away from work, travel, new diet needs, health practices for self or others, relaxation time, on the go, kids, satisfaction/ qualm hunger pangs, dinnertime, dessert, and when entertaining friends, family, and colleagues.

11.10 CONSIDERATIONS

This chapter was written to motivate product developers to look in a new direction, beyond simply the convenient statistics and number crunching that has come to be a part of today's process of NPD. The chapter provides guidelines and ideas about what to do, what works, where the approaches can be applied, and what might emerge. The chapter shows how to move beyond hearing the consumer's voice, really *listening* what is being said.

Parting words: Go now and be fearless as you peer into the lives of your *current* consumers. Perhaps even more importantly, become passionate and assertive about seizing and creating opportunities to understand your potential *target* consumer. The rewards will be fruitful, your products more successful.

12 Observing the Consumer in Context

Jacqueline H. Beckley and Cornelia A. Ramsey

CONTENTS

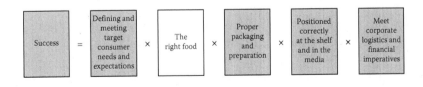

12.1   INTRODUCTION

In our experience, engaging the consumer early in the product development process is critical when developing products that are relevant and meaningful to the consumer. Focus groups, interviews, surveys, and questionnaires are all the techniques we have used to capture the consumers' perspective. Yet, for all this plethora of approaches, what specific techniques are best to "in-context" understanding? What techniques provide the most relevant products for consumers? What techniques authentically incorporate the consumer voice into product design? Is a multimethod approach feasible in consumer research? How does participant selection strategy fit with the data collection methodology?

In the following sections, we suggest how to approach these and similar types of questions for in-context research. We provide examples of research design and a sampling strategy that has produced valuable insights regarding consumer product experience. At a more practical level, the approach has provided strategic direction for product development.

12.2   WHY "IN-CONTEXT"?

We believe that a strength of in-context consumer research is the "open inquiry" philosophy established from the onset of research. This open inquiry contrasts with the frequently preset, fixed ideas that generally direct a lot of traditional consumer research.[12] This does not mean in-context consumer research lacks established goals or objectives; quite the contrary. The in-context approach has specifically outlined goals and objectives for consumer research. Yet, the in-context approach does not engage consumers in a constrained structure of preexisting researcher-defined hypotheses. Nor does it feature preconceived ideas of how consumers will respond to research questions. Within the framework of the design, researchers start with a "clean slate" to be populated by consumers' ideas and expressed experiences.

In-context provides the research subject and the researchers with more normative situations, which help facilitate situations, behaviors, and thinking (by both researcher and subject) which cannot necessarily be replicated in more traditional staged research situations (at least at the beginning of a product development engagement). While some researchers have not found much need for in-context situations,[1] others are proponents.[2] For us, whether it is a staged context in a more central location or the consumers' environment, we find that a more familiar and comforting situation for the research allows the subject to feel more "at ease" and tends to stimulate more naturalistic conversation. This situation then allows for a comfort level for the consumer which provides less guarded discussions and conversations.[13]

12.2.1   Defining an "In-Context" Research Design?

Our approach to understand the consumer in-context is based, in part, on the principle that consumer research conducted for product development can be effective when it is constructed as a modified form of *narrative inquiry*.[3] As applied to consumer research, this means studying the consumer in-context can be a process of using "… an amalgam of interdisciplinary analytic lenses, diverse disciplinary approaches and both traditional and innovative methods… revolving around an interest in … particulars as narrated by the one who lives them."[3] In other words, we as researchers, apply the lenses of many perspectives (e.g., consumer research, sociology, psychology, anthropology, grounded theory, phenomenology, etc.) to the study of consumer experience as told by the consumers themselves. This multidisciplinary, narrative inquiry philosophy naturally leads to the development of multimethod (i.e., multilens) research design adapting modified features of ethnography and phenomenology.

12.2.2 PREPARING FOR IN-CONTEXT RESEARCH

Before entering the field it is always a good policy to think through the steps in the research process, to play the "inner game." What does the researcher want, or better need to achieve via research? What are the specific and measurable objectives for the research? What is the best way to solicit the information from consumers? What specific research design(s) (e.g., online surveys, telephone interviews, etc.) will best meet the goals and objectives with the target population? These are logical steps to create logical, meaningful, and ultimately practical research.

Look at Figure 12.1, "Steps in the mixed methods sampling process."[4] Although the figure was developed specifically to illustrate the steps involved in the mixed method sampling process, Figure 12.1 clearly and simply describes the critical points of a research process. We believe the steps in Figure 12.1 provide a good blueprint for the research process.

12.2.3 FITTING THE "IN-CONTEXT" CONSUMER EXPERIENCE INTO THE RESEARCH PLAN

In our experience, each phase of the research can be answered within the "context" of the consumer experiences. In other words, everything from the goal of the study (step 1) through the individual sampling schemes (e.g., screening) (step 6) can be designed within the parameters of the contextual experiences of the consumer.

---

1. Define the research goals

2. Specify the measurable objectives and timelines

3. Outline the specific research question(s) per population of interest

4. Design the research plan

5. Determine the sampling design

6. Select the participants

FIGURE 12.1 Key steps to develop a research plan. (Adapted from Onwuegbuzie, A.J. and Collins, K.M.T., *Qual. Rep.*, 12, 281, 2007.)

Let us consider step 1, the goal of the study. The goal may be to explore consumers' perceptions of a test product within the context of their experience. Now move to step 2, the research objectives. These may be to understand the patterns of product usage and reasons for product usage within the overall consumer context (e.g., where, when, how, why, who, of overall individual consumer realities of the product experience).

Each subsequent step (e.g., research design and sampling design, etc.) follows this in-context framework as well. Several qualitative research methods provide guidance to researchers trying to understand such contextual phenomena. We adapted principles of ethnography and phenomenology to inform our in-context research.

12.2.4 "Blended Philosophies"—A Productive, Pragmatic Approach to "In-Context" Research

Let us begin with a definition. We define the "hot," fast-growing field of ethnography as follows:

> The study of people in naturally occurring settings or "fields" by methods of data collection which capture their social meanings and ordinary activities, involving the researcher participating *directly in the setting* (emphasis added), if not also in the activities, in order to collect data in a systematic manner but with out meaning being imposed on them directly.[5]

A key feature of traditional ethnography is the location of the research effort "directly in the setting" of the participant. Such involvement in the participant's "natural habitat" requires an investment of time. This time usually includes site selection, travel, and organization of the location and therefore, may run to several weeks, requiring a commitment of time and dollar resources.

Let us look at the world of business and consumer research within this framework. There are limits to the amount of time a consumer is willing to devote, no matter how good the intentions are. Reality intrudes, especially in today's increasingly busy environment. Thus, the time spent with consumers is self-limiting. The interviews, focus groups, one-day shadowing, and the like, reduce the informality and slower pace of an anthropologist/ethnographer into a more compressed period. The result is the limited ability of even the most skilled interviewer to develop a necessary close and trusting relation with the consumer participant. Yet, in tradition anthropology it is just such a relation that is most productive.

With this time limitation in mind, there are other approaches that must be used to shortcut the necessary steps. Recognizing this limitation and reflecting back upon the principles of narrative inquiry, we believe the ethnographic interviewing technique of "storytelling" can be brought into consumer research. Storytelling permits participant consumers to recount their experiences through the use of unstructured interviews conducted in a "conversational" format.[6] Through ethnographic story telling, the field notes, verbatim quotes, and researcher observations are recorded. This recorded information becomes the data of the consumer research ethnographer.

Storytelling cannot take the place of a true ethnographic relationship with our participant consumers, nor does it allows us the "insiders view" of the consumer experience. On the other hand, the conversational approach establishes a friendly, flexible environment for discussion. The unstructured nature of the questions allows the participants to tell their stories, in their own time and weave together the threads of their real-life experiences into a comprehensive tapestry of product experience. In addition, the unstructured nature of the interviews allowed the participants to direct the conversation, to initiate pertinent points of their product experience, and yet, enables the interviewer to probe for further understanding and clarity. The interviewer thus obtains the participants' full perspective of the experience, both positive and negative, based on questioning techniques used, not so much from direct observation as from listening to the stories, empathizing with the emotion, giving "control" to the participant, and hearing with the "third ear."

12.3   PHENOMENOLOGY

To support the research plan, we also implemented what is now referred to as "phenomenology." As we see below, this method applied to consumer research helps to capture the consumer experience in the context where it is experienced.

Phenomenology attempts to go "deeply into the thing itself," and reveal the object or phenomenon to which meaning is attached by the individual.[7] The term phenomenology comes from academic philosophy. Yet, if we think about the meaning of a product experience in consumer research, there are a great many related intricacies involved. A study of the product "as used" provides critical insights into consumers, and uncovers the traditions of product use.[8]

The key point here is that in the interpretivist disciplines such as phenomenology, the words used by the consumers are the source of the data to be analyzed.[9] Taken differently, phenomenology when applied to the world of consumer research is the study of the consumer experience from the point of view of the consumer. Let us contrast this with ethnography. Ethnography attempts to understand the consumer experience by analyzing the interview data, by direct observation, by deeper understanding of field notes. By contrast, the phenomenologist directs immediately to the consumer experience for data collection, without the intermediation of an ethnographic observer.[14]

The multimethod research design blends the ethnographic field notes of the researcher with the phenomenological verbatim words from participants. That is, the researcher explores the meaning of the product experience for the participants as "... result of creation between the researcher and the researched and not just the interpretation of the researcher, who may have different contextual factors or agenda influencing the description."[7]

By blending storytelling (modified ethnography) and modified phenomenology, we captured the context of the consumer experience. Specifically, utilizing the actual language (e.g., real-time words) of consumers (via field notes and interview data), together with observations of researchers gathered, we as researchers gain a far more profound understanding of the product experience from the point of view of the consumer. A somewhat different form of this approach has been adapted by the usability testing world and is called "think aloud protocol."[5]

Moving into actual process, the first step is to compile the observational notes. The notes about the stories are examined to discover those key words, phrases, and expressions, in the consumer's own word, that signify the "product experience" and also define the context. Through this process, the consumers' own voices describe the experience. In other words, the consumer stories provide the context within which the researcher begins to discern common themes and occurrences. Looking across many such stories in a single project reveals important features of the product and makes clear the recurrent themes.

We have just looked at the general approach, the overarching description of what is to be done—combine the consumer voice and the observation to structure the data, and to find emergent understanding of the consumer–product interaction, and thus the experience. But now, the question becomes "how to do it." How does the researcher bring this general way of thinking into action? Are there best practices and if so how are they specified, to what degree, and what do they provide?

The "best way" to solicit the consumer voice, translate and integrate this voice into product development remains a source of great debate. We propose that rather than there being a single best way or single best method, we maintain an adaptive, holistic approach to accomplish three distinct objectives:

1. Maintain the advantages of both qualitative and quantitative techniques.
2. Adhere to the authenticity of the consumer voice throughout the product development process.
3. Maintain the flexibility necessary to be applied within diverse research endeavors (e.g., academic, industry).

With these overarching objectives in mind, let us now see how to do this integrated approach in action.

12.3.1   INTEGRATING QUANTITATIVE AND QUALITATIVE APPROACHES

We begin with the notion of convergence. Both quantitative and qualitative research can provide a lens through which researchers begin to glimpse the consumer voice and experience. Taken alone, however, neither offers the complete vision. One begins to comprehend both the messages of the consumer through the consumer voice, often done by quantitative research methods. At the same time, one begins to understand the "contextual meaning" of what one obtains in these quantitative studies, through qualitative exploration.

Principle of triangulation: Integrating or "triangulating" quantitative and qualitative research methods and techniques permits a completeness of one's understanding. By triangulation we mean applying multiple quantitative and qualitative methods and techniques systematically throughout the research process (e.g., establishing and exploring hypotheses, discovering consumer insights and findings, and empirically testing findings). This triangulation approach reaches beyond traditional paradigms of quantitative and qualitative research. The qualitative findings are neither simply used to inform the development of quantitative instruments, nor are the quantitative results simply used to validate the qualitative findings. The approaches are used together in order to provide both breadth and depth of understanding while also contributing to the validity and reliability of research findings.[15]

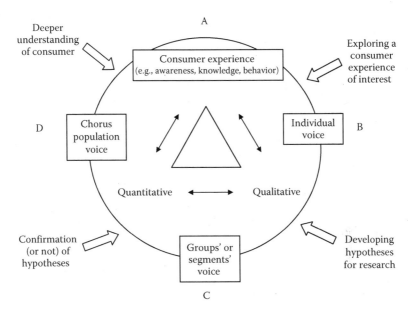

FIGURE 12.2   Principle of triangulation.

Let us look more closely at the process shown in Figure 12.2. The triangulation begins with the consumer experience (Step A). Next, we listen to the individual voice (*idiographic*) (Step B), working with a small sample of participants and qualitative methods. In this way, we explore the consumer experience. Moving on, we use both qualitative and quantitative methods (e.g., card sort, simple surveys or ballots, etc.) working with more participants (Step C). In Step C, we explore hypotheses that emerged and were developed from the individual voice. Finally, in Step D we confirm these insights from the groups or segments, by means of quantitative methods, using a larger sample (nomethetic). This population voice can then deepen the understanding of the consumer experience which we worked with in Step A. Finally, this understanding can be verified by the application of the understanding back to the individual voice via qualitative methods.

An example of the application of this triangulation approach to consumer research for product development is provided in further detail on in this chapter.

How can quantitative and qualitative research be smoothly integrated? The answer is the practical employment of a theoretical framework that allows for the smooth interaction of both approaches. To implement triangulation, both phenomenology and participatory action research (PAR) provide the practical guidance to successful integration of methods. We describe them below and show how they bring the authentic consumer voice into product development and product development research.

12.3.2   AUTHENTICITY OF CONSUMER VOICE: ETHNOGRAPHY, PHENOMENOLOGY, AND PARTICIPATORY ACTION RESEARCH

As described earlier, phenomenology in the context of consumer understanding explores the circumstances, happenings, events, and perspectives of the consumer through the eyes of the person(s) experiencing the phenomenon. The appropriate phenomenological qualitative approaches, as the researcher employs it, should put the consumer back into the situation under study, so that the consumer "relives" the experience directly with the researcher. By doing so, the researcher begins to take the journey with the consumer. The descriptive conversation, the use of props and imagery, and similar techniques allow the researcher to see some of the experience through the eyes of the consumer. The consumer does not simply report answers, but rather describes the experience.

A second way to achieve authenticity of the consumer voice applies PAR. PAR is used quite well in the so-called community development initiatives. In these initiatives, both the researcher and consumers from the target population group work "in tandem" to define the problem(s), identify and implement solutions, and then afterward evaluate results.

Whatever qualitative techniques are used (i.e., interviews, focus group, etc.), there are key elements that identify a PAR approach:

1. *Consumer empowerment*: Consumer is the authority on research topic
2. *Shared control*: Co-ownership of the conversation and situation of data collection
3. *Participant*: Set pace
4. *Active engagement*: Consumers are involved in data collection conversation and props
5. *Participant validation*: Consumers validate researcher interpretations during engagement
6. *Establishment of common language*: Consumers are valued for their creation of language and its meaning
7. Ultimately resulting in the integration of knowledge, beliefs, and behaviors into the understanding of the consumer experience

Using both modified ethnography and a phenomenological philosophy to see the research topic through the eyes of the consumer, and by applying a PAR approach to the qualitative technique of data collection, the consumer voice is preserved and can be infused into the product development process. The consumers and their lives have been appreciated and the end result embodies their framework—not the researchers.

The value of any research approach, method, or technique, such as the "triangulation of quantitative and qualitative approaches" proposed here, is its generalizability of replicability. We believe this approach is applicable in different settings for the following reasons. First, the approach is an amalgam of research methods and experience successfully implemented by the authors in both academic and practice settings. Second, it is based on sound research theory and practice standards (e.g., theoretical basis, design of experiment). Finally, this approach has been successfully implemented across several research initiatives within the industry arena (see Section 12.4).

12.4   PRACTICAL APPLICATION OF METHODS—A CASE STUDY

The triangulation approach described above has been successfully implemented in consumer research industry. The following research outline (description of actual steps taken to achieve a highly favorable outcome) was based upon field research conducted between January 2006 and December 2007.

The consumer research was executed using the integrative, iterative process depicted in Figure 12.1. Although outlined below in a linear format for explanatory purposes, the research process was circular in nature (e.g., step 1, step 2, step 1 and 2 in conjunction with each other, step 1, etc.).

STEP 1: Stage—Exploratory formative research

Type of Research: Qualitative Research
Approach: Phenomenology
Technique: One-on-one, in-context interviews
Purpose: Defining the experience through consumer eyes
        What are the boundaries of the consumer's experience?
        What makes up the consumer experience?
        What is the language describing the experience?
        What is the meaning of the experience?
Outcome: Hypotheses Formation

STEP 2: Stage—Testing hypotheses

Type of Research: Qualitative and quantitative
Approach: Phenomenology & PAR
Technique: Interviews facilitated with Card sort, Ballot scoring
Purpose: Prove/disprove hypotheses
        Initial quantification of consumer insights—forced choice
        Explore consumer priorities and trade-offs, qualitatively
        Include consumer language into quantitative instruments
        Identifying segments of interest
        Identify product attributes
Outcome: Instruments for large-scale research, instruments include consumer driven ballots and statements useful for trade-off or elimination by aspects analysis.

STEP 3: Stage—Quantification of segments, product preferences

Type of Research: Quantitative
Approach: Design of experiment
Techniques: Large-scale survey, conjoint analysis, discrete choice analysis
Purpose: Apply learnings to population
        Verify product design
        Scale understanding (conduct pilot [small scale] then move to larger
        scale testing)
Outcome: Predict consumer choice, consumer acceptance, consumer decision making

STEP 4: Stage—Confirmation of predictions

Type of Research: Qualitative, Quantitative
Approach: Phenomenology, Tracking metrics
Techniques: One-on-one, in-context interviews; tracking studies

Purpose: Confirm consumer engagement is as expected; confirm decisions/ choice is as predicted

Outcome: Quality check-n-balance—consumer voice in product developed

12.4.1 ANALYSIS

Just as the data collection phase of our triangulation approach uses an integrated approach, the analysis and interpretation of data is enriched by the integrated approach. To this end, the results of quantitative research and findings from the qualitative research are blended and interpreted in total. Neither is judged as superior, rather all sets of data are judged to be good and therefore, useful for understanding. This multicomponent approach to data analysis is valuable for several reasons. First, it provides "checks and balances" for validity of the data. For example, if the different data sources are indicating the same results, it strengthens the findings. Second, this process can troubleshoot misunderstandings or prevent misdirections in product development. For example, if the data are indicating conflicting results, it may indicate that more research is needed or measurements are not reliable, and refinements need to be made. Often, it is this area of analysis that will point out small but necessary changes in language[10] or direction in the quantitative component that will lead to superior alignment of results and conclusions. Next, this provides timely breadth and depth of consumer understanding. Not only are the measures (i.e., quantitative results) of consumer experiences established for product development but also the contextual meaning and implications of those measures (i.e., qualitative findings) for product development are understood. Finally, this iterative process provides a means to keep the voice of the consumer "alive" throughout the research and through the product development process. It ensures the consumer language, context, and perspective are ever-present in the sights of the researchers/ product development team.

12.4.2 IMPLICATIONS FOR PRACTICE

In order to develop a context that authentically represents the product experience of the participants, it is important not only to record consumer language and researcher observations, but also to validate that language with the target consumers. This is accomplished through several iterations of research (usually three).

First, as described in "fitting the in-context experience into the research plan," open-ended story telling interviews are conducted. Two types of data are recorded: ethnographic field notes, observations, and notes of interviews and the verbatim stories of the participants. Using constant comparison method of data analysis (reference), the researcher identifies the common themes, words, and phrases across participants. In other words, the product experience is "constructed" as it exists across participants' stories. Second, these common themes are extracted (or deconstructed) from the stories and written on cards. Third, these cards are given to participants in subsequent research and participants are asked to respond to these cards, select those that apply to them, reject those that do not apply to them, link those cards that should be linked to experiences, decouple experiences that do not below, etc. Through this process, participants are, in fact, "reconstructing the product experience." Through this process, new themes may be added, themes may be discarded until there is an agreement among participants in the "deck" of cards and that the meaning of the deck represents their product experience. Through this process, the researchers are incorporating a multimethod design of ethnography/phenomenology with the techniques of interviewing and card sort to establish a common context of consumer experience.[16]

Before leaving the discussion of research design, we believe it is important to mention a critical point in research design that is often overlooked in consumer research: the sampling strategy.

12.5 SELECTING THE BEST PARTICIPANTS: SAMPLING STRATEGY

Since the purpose of the early stage product development research is not to generalize findings to the population (if it were, we would not be conducting smaller scale qualitative research), but rather to understand a particular consumer product experience, it is important to carefully determine which consumers to include in the research. Once research design and data collection procedures are established (above), there is a need to determine the sampling strategy that is consistent with the research design and goals of the research and general business needs. Generally, in consumer research a goal is not to "...generalize to the population but to obtain insights into a phenomenon..." (e.g., specific product-based experiences), we suggest a purposive sampling technique is appropriate for such qualitative product development research.[4] We suggest using a "modified multistage stratified purposive criterion" sampling strategy. Multistage stratified sample means that a researcher looks at the target population in segments of relevance to the product experience. For example, if it is important to understand the differences and similarities of consumers by gender and by the type of product used, stratify first by gender (male and female) and then stratify by the type of product used. So, when the qualitative engagement happens, there will be males and females that represent each type being explored. A *purposive* sampling strategy means that selecting participants who are most pertinent to the research[11] are sampled. Therefore be purposeful in the participants you pick for the study, think about who they should be and how they will advance your understanding of the research engagement. For example, if researching a product, a researcher may select participants with varying use patterns of the product. In the end, the populations explored for the in-context qualitative component will NOT be general populations, but rather will be focused populations that should be expected to have desired traits and experiences. If not, the iterative nature of the testing comes into play, and the sample is adjusted, accordingly.

12.5.1 SAMPLE SIZE

As recommended by Onwuegbuzie and Collins,[4] a phenomenological study should include approximately 10 participants. An ethnographic approach should include 30–50 participants (total), and if using interviews as a data collection method, approximately 12 participants are suggested. We suggest conducting three rounds of interviews per research question with sample sizes varying from 9 to 17 participants per site (average of 14 participants per site)[6] resulting in an average of 42 total participants per research question should be sufficient to meet the objectives.

12.5.2 TOWARD A MORE AUTHENTIC DIALOG WITH PEOPLE

You cannot step in the same river twice

Heraclitus, 400 B.C.

Everything we do to "research" a subject produces change. The goal of the in-context approach is to try to walk gently with the person who is our subject. To value them as people and to create a comfortable and engaging situation that allows them to be close to who they are WHILE we are with them, as they were before we came into their lives, and the way they become when we leave.

For us, the path to the use of in-context approaches reduces much of everything else in market research and consumer engagement to good tries, but not quite there. Try the thinking that is included in this chapter and you will find your research much more enriched.

REFERENCES

1. S. C. King, H. Meiselman, A. Hottenstein, T. Work, and V. Cronk, The effects of contextual variables on food acceptability: A confirmatory study, *Food Quality and Preference* 18 (1), 2007: 58–65.
2. H. H. Kelley, J. G. Holmes, N. L. Kerr, H. T. Reis, C. E. Rusbult, and P. A. M. Van Lange, *An Atlas of Interpersonal Situations* (New York, NY: Cambridge University Press, 2003).
3. S. E. Chase, Narrative inquiry: Multiple lenses, approaches, voices, in: *The Sage Handbook of Qualitative Research*, Eds. N. K. Denzin and Y. S. Lincoln (Thousand Oaks, CA: Sage Publications, 2005), p. 651.
4. A. J. Onwuegbuzie and K. M. T. Collins, A typology of mixed methods sampling designs in social science research, *The Qualitative Report* 12 (2), June 2007: 281–316.
5. J. D. Brewer, *Ethnography* (Buckingham, United Kingdom: Open University Press, 2000); *Think Aloud Protocol* (Seattle, WA: University of Washington, 2008).
6. H. Moskowitz, J. Beckley, and A. V. A. Resurreccion, *Sensory and Consumer Research in Food Product Design and Development* (Ames, IA: IFT Press/Blackwell Publishing, 2006), Chapter 3.
7. P. Wimpenny and J. Gass, Interviewing in phenomenology and grounded theory: Is there a difference? *Journal of Advanced Nursing* 31 (6), 2000: 1485–1492.
8. M. Annells, Triangulation of qualitative approaches: Hermeneutical phenomenology and grounded theory, *Journal of Advanced Nursing* 56 (1), 2006: 55–61.
9. C. Goulding, Grounded theory: The missing methodology on the interpretivist agenda, *Qualitative Market Research: An International Journal* 1 (1), 1998: 50–57.
10. G. Zaltman, *How Customers Think—Essential Insights into the Mind of the Market* (Boston, MA: Harvard Business School Press, 2003), Chapter 1.
11. J. M. Morse and P. A. Field, *Purposive Sampling was Adopted to Select Participants for Field Studies, Qualitative Research Methods for Health Professionals*, 2nd edn. (Ames, IA: Blackwell Synergy, 1995).

12. R. W. Belk, *Handbook of Qualitative Research Methods in Marketing* (Cheltenham, Northampton, UK, MA: Edward Elgar Publishing Inc., 2006).

13. T. Kelly and J. Littman, *The Art of Innovation* (New York, NY: Currency/Doubleday, 2001), Chapter 3.

14. D. W. Smith and A. L. Thomasson, *Phenomenology and Philosophy of Mind* (New York, NY: Oxford University Press, 2005).

15. R. Stake, Qualitative case studies, in: *The Sage Handbook of Qualitative Research*, Eds. N. K. Denzin and Y. S. Lincoln (Thousand Oaks, CA: Sage Publications, 2005), p. 453.

16. K. Charmaz, Grounded theory in the 21st Century: Applications for advancing social justice studies, in: *The Stage Handbook of Qualitative Research*, Eds. N. K. Denzin and Y. S. Lincoln (Thousand Oaks, CA: Sage Publications, 2005), p. 522.

13 Getting the Food Right for Children: How to Win with Kids

Bryan Urbick

CONTENTS

Success	=	Defining and meeting target consumer needs and expectations	×	The right food	×	Proper packaging and preparation	×	Positioned correctly at the shelf and in the media	×	Meet corporate logistics and financial imperatives

13.1 INTRODUCTION

In order to truly win with kids in the development of new food and beverage products, it is critical to understand them. This understanding should include their fast-paced development, their key and underlying drivers, and particularly the factors involved in making their food and beverage choices. A deep understanding of kids and their food also has implications for product testing and then ultimately successful communication of the proposition. With a solid foundation of understanding, success in new product development process becomes far easier and far more likely.

Each year, when working with kids on food and beverage projects in many countries throughout the world, we see a recurring pattern. "Give us something new and different" goes their mantra. Hearing this, we work diligently to deliver. Again and again, when these "new and different ideas" are ultimately tested with consumers, they frequently fail. This is because kids have innate neophobia (fear of the "new"). A true understanding of young people's behavior generally, and specifically neophobia, and how to combat it, give us deeper understanding as to why we must view new product development and particularly new product innovation as "baby steps." It is important to conduct an integrated, incremental approach, rather than an approach that is a giant leap or huge stride, so often desired by companies in an attempt to generate stronger market position.

13.2 CHILD DEVELOPMENT

The first step to understanding children and young people considers the nature of basic child development. Through the years, this knowledge has led us to form better development projects, better research activities, and importantly create better products. This knowledge becomes the foundation and then the first step to better understand ways in which to "get the food right" for kids.

The idea of "kids" or "children" cannot be truly understood without acknowledgment of the pace of change. This is evident in our work with newborns, and continues through projects with older teens. "What a difference a year makes!" is a frequent client comment as we work with one grade or year group and then move to the next. So much happens from birth to teens—probably the greatest natural changes that will occur in a 15-year period in our lives, though arguably one could also say that the end of our natural lives may have a similar pace.

Children's first 15 years are full of physical and psychological milestones. Throughout these stages, they have different needs and wants. Though the study of child development can be involved, and indeed many volumes are written about, it is important to understand some of the highlights of this growth period. In order to create better, more appropriate, more accepted foods and beverages for kids, we now turn to some basic aspects.

In the first year of life, the child spends most of his or her time sleeping, and being fed and nurtured, both physically and emotionally. The baby is completely dependent on its parents. The child's energy is utilized in physical growth; in the first half year of life, the child doubles its size in mass.

During the early weeks, there is relatively little time when the child is awake, but when awake, he or she instinctively observes the world. A key aspect to understanding this developmental stage comes from considering the impact of a child's instinctive observation and imitation. This learning by copying will continue strongly until around the age of 7 or 8, and importantly will become the basis for the learning of new skills all throughout life.

As the child gets a little older, he or she transforms from the helpless, observing, sleeping baby to a more awake, active, then crawling, standing, and walking toddler. As observing adults we see this change as a big milestone. Perhaps the even bigger milestone, and the one most exciting to us, is when the child learns to speak: first in single words (predominantly nouns), and then in simple combinations of nouns and verbs. Eventually, the child begins to use the words "I," "me," and "mine." This is an important sign of psychological function.

With regard to food and drink, in the first couple of years the child is able only to indicate "when" it wants to feed. Over time and with continued development, the child is able to decide what it wants to eat. Importantly, this is the time when the child's influence of food choice begins. Though the child is still completely dependent on the parent, when he or she can indicate like or dislike of what is being fed, parents start taking notice!

With respect to foods and beverages, the child's "copying behavior" can be important. Even if the young child does not appear to like the food offered, the mom usually tries ways to get the child to eat the food. She may eat the food herself in broad, overemphasized manner, uses big, happy expressions and says "Yum! Yum!" Hopefully this behavior is then copied by the child, who then happily eats.

In the early years, before the age of around 9, the mom has much more influence in what her children eat. After this point, the mom begins to lose some control of the child's diet and her strategies change, typically from the original playfulness to a new practicality. In one of her mothering roles, she acts as gatekeeper. She may allow only certain products into the home, and from these gives the child a choice. In other, parallel mothering roles, occasionally conflicting with the role of gatekeeper, the mother may adopt other strategies. The study of mothering is a separate and complex subject, though in brief, she plays numerous roles providing physical and emotional nourishment. Sometimes, we assume that she is only driven by "nutrition" and "healthy foods"—that is not the case. Sometimes, she desires to reward, sometimes treat, sometimes punish, sometimes merely to show love. We need to understand these roles and provide the right products for each role.

As the child further matures, more of the world is discovered and understood. In this process of discovery, play has very important role. Play gives the child an opportunity to experiment with various aspects of the world and try to make some sense of it. Amusingly, until about the age of 4, the child will play with whatever is available and in reach: pots and pans, boxes, scraps of paper, etc. At this young age, up to 4 years, the child depends on what is available in the vicinity and tends to want only the things it can see.

From around the age of 4, play changes quite dramatically with the addition of the child's fantasy and imagination. Suddenly, the child is no longer dependent on that available to generate play, but can play with ideas generated from within: tables covered with blankets become the genie's lamp; large boxes become rocket ships. And though fantasy may seem to the outside observer as random and chaotic, it is not. A child's imagination has a structure and pattern, with "rules" evolving over time and repeated several times.

Throughout the child and teen years, there is a great deal of physical change. For this whole process, the child requires a great deal of energy to evolve and develop. The child also expends a lot of energy in physical manifestations of running around, playing, and making lots of noise! This kind of energy is often a cause of frustration for parents, but all of these aspects are critical for the child in order to learn about himself and discover his place in the world.

From the age of 4–7 (depending on the region and culture), the child goes to school and learns more about social and intellectual aspects of life. The child then learns more about the world and continues to discover and develop his own capabilities, intellectual learning, and social structures.

In that stage of 4–7 years, the baby who had instinctively observed and copied, is now matured to become a child who lives in a very "black and white world." As every parent knows, words like "good" or "bad," "like" or "dislike," "on" or "off," "yes" or "no" are clearly understood. Gray words like "maybe," "perhaps," and "if you are good, then you can have…" are not well understood. The child does not yet have the references to relate the "in-between" or more abstract aspects of life.

At around the age of 8 or 9, though, there is a burst of brain development and a distinctive change in the development of children. The black and white world opens up, and the child will draw attention to details. It is not that these details were not noticed previously, but the child will now comment on them. Those comments are most often sharp and will shock or astonish adults. The child's reaction to his own comments is similar to the parent's, and these observations are disturbing the child's perception of a comfortable, safe world. From these new observations, the young child at this age begins to make new rules for himself (i.e., "… if I lay on the left side of my bed, then the monster won't get me"). Emotions are now beginning to be managed by the intellect.

The important thing about this process is that the child starts to develop the "in-between," the gray area between black and white. As a consequence, the child begins to tell us "why" he likes or why he does not like something. These reasons are critical to understand when testing food and beverage products for kids, and has implications for scales used.

Starting from the age of 5–6, usually when children go to full-time school, it becomes important to distinguish between girls and boys. Though both boys and girls go through roughly the same developmental steps, girls tend to socially mature faster and physically reach puberty approximately 2 years prior to boys. After 11 years, children then move through prepuberty and puberty, leading them to adulthood.

In recent years, the physical development in puberty has moved down toward the age of just over 12 years, with a few girls as young as 8 years beginning to menstruate. There are two major schools of thought: one suggests the high level of estrogen in the "Western diet" as the cause; the other indicates that the increase of estrogen levels in the water supply is a major contributing factor. Whatever the cause, this earlier onset of the physical elements of puberty then adds more stress to young people because it involves drastic physical and emotional changes, now operating concurrently.

How each child progresses through puberty is very individual, but tends to run a wide range of "no problems" to "incomprehensible stress." It is a period of rebuilding, most manifested in the development of the physical body. The body becomes adult much faster than the emotions can often handle. Inside, there rages a struggle for independent expression, and paradoxically, social acceptance. The teen no longer feels the need for any direct attention from parents or other adults. As the child moves to increased independence, the practical control of parent over the child gets looser, although ideally the emotional bond becomes stronger as the child moves to a mature "friend" relationship with the parent. Importantly, the child is very likely to reject some of the parents' ideals and values at this age, and seeks new role models in peers, sports stars, music or movie stars, even older teens.

With regard to food and drink, in the teen years, cautious experimentation creates new favorites, preferred tastes are beginning to be more complex, and simple blends (something known with something new) are enjoyed. The need for acceptance in a social group can drive the desire to like what peers like (to feel that they "fit in").

13.3 THE "BASIC FIVE": AGE BANDS FOR NEW PRODUCT DEVELOPMENT

Based on the general development of children, in new product development, we should look at kids in five different age groups. Even though we may desire a broader target age range for our products, it is wise to consider a "bull's eye" target within the structure of these groups. Various factors in different cultures drive the nature of these ages, such as the age at which children start school or the age in which they advance from a lower school. In general, the principles and order of the group remain the same, and only the age varies at which the child enters a group . Arguably, the first year of life could be a separate group important for infant formulae and weening foods, but for the sake of this discussion we will focus from the age of 1 year. These five new product development groups are

1. 1–3 years (toddlers)
2. 4–7 years (young children)
3. 8–11 years (tweens, i.e., between child and teen)
4. 12–14 years (young teens)
5. 15+ years (teens, this is the age in which out-of-home consumption without parental influence significantly increases, and the products for these ages begin to align with the products selected by young adults)

13.3.1 1–3 YEARS (TODDLERS)

In the 1–3 age range, children are attracted to bright colors and by simple lines and images both new and those they recognize. The marketing is, of course, mostly to the parents. The rule still holds, however, that if the child would not eat it, then the parents probably would not buy it again.

With regard to foods and beverages, in the first year, the child gets nourishment from mother's milk/formula and eventually from "real" food, albeit blended so that the child can consume it. In time, as the child's tastes and "will" develop, he wants to do things more independently, such as get the food separately on a plate rather than all mashed together. At this stage the child starts to feed himself, while the parents desperately try to teach table rules and manners. Note again the importance of the role of imitation with regard to table rules. Parents teach the child to imitate what is said until the imitation has become a conscious act, and then an unconscious habit.

Products for these younger children need to have aspects of built-in imitation and then discovery. Communication needs to reinforce product usage—showing context and occasion can help—to teach the child proper product experience.

A good example, though not in the food realm, is the Procter & Gamble product Pamper's Kandoo (a range of toiletry products for very young children). It clearly shows on-pack (supported on-air), the iconic frog character demonstrating the use of the product—the imagery is important because, of course, the child of the target age cannot read. Food product communication to young children can provide a fun context, but it is important that it shows how the product is consumed, and that it is desired and enjoyed.

13.3.2   4–7 Years (Young Children)

Younger children up to the age of 6 or 7 are particularly influenced by the basic sensory attributes of foods and beverages, namely taste, color, and texture, and in some categories, the sense of smell as well. This is a carryover from the early development stages as the child discovered the world through absorption by the senses. These sensory factors are critical to the young child's enjoyment of foods and beverages. Basic taste profiles, bright colors, and importantly, "smooth" textures tend to be the most popular. These aspects should be carefully considered and tested with young people on a product-by-product basis. This age group is also beginning to be attracted by communication. The child can be motivated by the appearance of the package itself.

Fantasy can be very useful with regard to food and beverages for this age group. Foods and beverages that encourage children to experience different "places" in the comfort and safety of familiar surroundings have great appeal to these young people. Observing children and the way in which they play with certain foods will give good ideas of ways in which to incorporate the "fantasy" qualities in products. The game can be even better than regular toys because the child gets to eat the "toy" afterward!

13.3.3   8–11 Years (Tweens)

As children grow into the next phase, age of 8–11, they are becoming more critical about the world around them and the items with which they interact. The basic sensory attributes are still important, but added to these sensory attributes are the more abstract and less tangible aspects of products and brands.

As the child matures to understand these abstract concepts, he explores these new-found abilities. The exploration manifests itself in the child's increased interest in fads, promotions, TV, Internet, mobile communication, and video games. Interestingly, in some food and drink categories, there is also an increased acceptance of more intensely flavored products. Taste continues to be an important influencing factor—but now we see a desire for not only sweet but also the "game" of the sour and heat sensations.

Peer group is now more important and the child's food choices will be influenced by them. It is also worthwhile to acknowledge the humor of this tween group. They are increasingly attracted by humorous advertising, and often repeat and discuss ads that they see. In most cultures, children begin to understand the humor in irony.

The tween age is a good one to involve the child with cooking things and creating and developing ideas. Activities that promote understanding, and making a food or beverage are excellent ones for the tween. The processes of gathering, mixing, pouring, baking, microwaving, and even cleaning up become very appealing, especially because they are involved with the activity of eating. Products that support this, even simple activities such as making toast or boiling the kettle for tea, benefit the child's development.

With regard to foods and beverages, the complexity of the "goal" is related to the age of the child. The goal can be as simple as pouring out their own cereal from the box to mixing up and baking their own biscuits.

Tween social skills are also in use when young people tell each other about products and other things in their lives. The schoolyard conversation about "the newest…" and "the best…" can be one of the greatest assets to a new food or beverage brand. Sharing the product, or even just talking about it endows the product with credibility, and has therefore makes it desirable. This social aspect can be a powerful tool for a product's success. Brand developers should include the social aspect in their marketing plan to tweens.

13.3.4   12–14 Years (Young Teens)

We move very quickly from our children laughing at humorous advertising targeted at them, to scoffing at it in the 12–14 year old range. The world is now theirs: they own it and, from their perspective, they know everything. Tastes they like from childhood are still liked, but the child's repertoire is increasing. Peer influence is even greater, and the appeal of adult-targeted products starts to emerge. "Health," particularly in relation to appearance and physical prowess, may start to influence choice. The obesity issue starts to emerge as a personal concern or worry. One consequence is that education about health and nutrition begins to make more sense to the child.

Peer pressure is a factor, but peer pressure is often a "catch-all" phrase. The bottom line is that young people want to express themselves as individuals, yet still fit in. Sometimes, the desire to "fit in" becomes very strong, developing in an almost overriding factor in the behavior, and that is when "peer pressure" becomes most effective. Some children find it difficult to fit in with a group for any number of reasons, but importantly the desire to fit in is part of the normal process of maturation.

Music, sport, and fashion are among the ways in which young people express their individuality and therefore, are major influences of the young teen. Incidentally, increasing independence also brings the need to live within a budget, so price promotions exert an increasing effect on choice. Also, in order to keep the young teen interested in a product, it is important to build in elements that drive variety.

The ever-important aspect of taste still influences young teens, yet evolving to include wider choices and even products that take "getting used to." Rather than simply enjoying intense flavored products, they now know that it is relatively safe to try new flavors and blends of simple flavors.

As children reach puberty and their teen years, their friends, music, sport, fashion, independence, and acceptance are all very important. Again, basic sensory attributes are important, although there seems to be a renewed interest in at least being perceived as trying new things. Young people of this age say constantly that they want "more variety," yet most often they choose products that are more basic and "comfortable." To reiterate, simple blends of tastes are very good for this age group, but are best perceived if the taste is predominantly familiar with a hint of a new taste.

13.3.5   15+ YEARS (TEENS)

The teenagers (15+ years) are a healthy blend of adult and child, though very much more toward the adult. Our work with teenagers would, at first glance, not seem that different from the way one would work with adult colleagues, though it does have an important difference. Teenagers find it a lot easier to talk about feelings, fears, and expectations, and share their views of the world, usually quite an optimistic one. In most projects, we neither seek out "trend setter" teenagers, nor do we seek out those teens who are particularly unusual. We work with the average teenager. For the most part, we have these average teens to be creative, articulate, engaging, and very delighted to give their opinions and share their ideas.

13.4   TARGET AGE

The issue of "target age" needs to be specifically addressed. As a rule of thumb, always target the age of a product highest in the age band. Target the communication to the "aspirational age" of children, namely the age at which children wish to be perceived. Although generalizing greatly, the aspirational age is 1–3 years older than the calendar age of the child. If the target age is wide, then maintain the taste to be acceptable to the youngest age in the target. Taste is usually the safest aspect of a product to "down age" because older kids do not reject a product that has a simpler flavor profile.

13.5   SOME IMPORTANT KEY DRIVERS FOR KIDS

Now we have a good understanding of the basic child development aspect, it is important to look at some key drivers of kids through all ages. Though the drivers may manifest differently, depending on the age and stage, the drivers are the same.

13.5.1   CONTROL

The most important driver of kids is "control." Control is about the desire of young people to "be in charge" of both themselves and of their immediate environment.

There are numerous examples of products that give kids control, such as convenience products, products that fit easily in the hands, easy-access/easy-open products, products that give choice, and even simple generally single-serve products to name a few. Giving kids a sense of control to a new product is likely to be a key success factor. It is very important to give the control aspect a great deal of attention in product development and branding.

When children are young, they do not control much of their world, although they want to. In broad generalizations, children are told when to get up, when to go to school, what to eat, when to do homework, and even when to play. The child reacts, striving to acquire control. This striving for control translates into popular food and beverage products. Selecting the flavor of ice cream in a supermarket or ice cream shop is one example. Selecting the topping to go on the ice cream gives a bit more control. Having multiple choices of inclusions for ice cream provides even more control. To use another example, even the simple act of pouring out a bowl of cereal and adding milk gives a child control. In sum, food and beverage products that enhance the target-aged child's perception that he or she is "in control" is more likely to have appeal than those products that do not.

13.5.2 Aspiration and "Creating Excitement and Stretching Boundaries"

Kids aspire to the benefits of being older. They also want to enjoy themselves and say that they want to avoid "boring" and "the same old thing." It is important to note, though, that control is often overriding desire for "new and different," and therefore, tends to be the most important.

Aspiration is usually driven by the child's desire, almost obsession, with being perceived as older: 3-year olds want to reach a higher shelf, 5-year olds want to stay up later like their older brothers or sisters, 8-year olds want to go places with their friends and without the parents, 12-year olds want to stay out late, and 15-year olds want to drive cars. As a result, the drive of aspiration accelerates children through their childhood years.

Food and beverage products need to support this drive. Products that allow children to "feel" older or more adult have greater chance of success. Unless specifically instructed to do so, we find in our work with children that they never design a product that is for "little kids." Rather, and probably not surprising, they always design products that are for older children, teenagers, and adults. In our creativity work with kids throughout the world, we always ask them to design products for "young people their age and older" so they don't only work on ideas for "little kids."

Usually, food and beverage products targeted to very young kids will not appeal to older kids, even though the taste itself might appeal. Even if older kids eat those products, they won't admit that they still like them. While they may watch certain TV shows because they have a younger sibling that does, or eat a certain product because it is in the house for a younger sibling, children would not be seen to choose these things themselves. Underestimating the importance of a young person's aspiration to be "older" can completely undermine the success of a product.

Making things extreme enough to stretch a child's imagination while still letting him or her feel safe is a "sure bet" for product success. This "stretching" is exhibited in different ways depending upon the age of the child. It is exciting for a child to be able to interact with his/her food. And interaction, in this case, does not mean "playing with food." It is better explained by dipping food into sauces, or adding cereal or sprinkles to their own yogurt. Other ways to make food exciting is to provide games on the packages, or to provide new and different shapes of the same product, or to provide interesting dispensers or ways to get the food into one's mouth without simply spooning it there!

Humor, starting particularly in the tween years, is another way that kids make their food products more interesting and stretch the boundaries of adult acceptability. Importantly, humor helps them further assert their independence. Being simultaneously disgusting and funny can be a big winner with younger tween boys. If something is conventionally supposed to be a certain way, isn't it funny to make it another way? Repeating the joke over and over makes it funnier every time. And if one's parents are bothered about it, or there is an adult in the room who does not appreciate the humor, all the better!

A word of caution, however. It is better not to attempt humor than to attempt it and patronize the target audience with "childish" humor or humor that is not up-to-date, "cool," or in any way pokes fun at the child. Making fun of adults may be fun, as long as it isn't perceived to be "mean" and hurtful. Making fun of a child is typically strongly rejected. Also, children as heroes and shown as the ones, in the end, who "save the day" can have great appeal, and also children taking on the role of adults who sometimes "just don't quite get it."

13.5.3 NEOPHOBIA

To understand about the phenomenon of neophobia, we really need to understand *fear*. It is important to be aware of the connection and difference between fear on the one hand and curiosity on the other. It is believed that these two states cannot be concurrent—meaning we cannot be both fearful and curious at the same time. We need to understand, therefore, drivers of fear (and "new" is a driver of "fear") and what generates "curiosity." For a new product development to succeed, we need to reduce kids' fear and enhance their curiosity. Keep in mind the "neophobia equation": familiar = liking.

The paradox of neophobia is that kids may express their desire for "new" and "more exciting," yet, in practice, they continue to eat and drink the same things with which they have become accustomed. This paradox creates a challenge for all food and beverage marketers and developers. There is a strong desire, even a need, to convey a sense of novelty and uniqueness in their products, but must also develop tastes and food or beverage forms that are familiar. Many in the food industry will find inserting the physical stimuli to generate familiarity in new products to be relatively uninspiring and less interesting, but it is important with kid's products.

Kids tend to be *innately* afraid of new foods, and this can be evident at a very young age. This is not just about picky eaters; even kids with wider food repertoires show some aversion to new foods, and when they refuse to eat or drink something, it is very often simply because it is *unfamiliar*. When asked, "why don't you like it?" a common response is "because I've never had it before."

The comfort of familiarity can occasionally mislead, when research is involved. When conducting observational research and accompanied shopping interviews, children as young as 18 months often can be observed pointing to items on the shelves. When asked, the mother regularly interpreted this behavior as meaning that the child liked the item, yet it frequently turned out to be that the child merely "recognized it" rather than trying to communicate liking it.

Adults also exhibit neophobic behavior and resort to familiar, trusted, known brands and products—but are more willing to try new things than children. As we age into our 60s and beyond, we tend to again seek more familiar and comfortable foods, reverting to child-like behavior with regard to food and beverage repertoire and choice.

13.5.4 Dealing with Neophobia*

If kids tend to revert to "familiar" flavors, familiar food forms and familiar brands. How then can we develop new kids' foods and beverages?

1. *To increase liking for a certain food or beverage, you must increase positive exposures to it.* This poses an important challenge to food marketers. Depending on the category, it can take anywhere from 8 to 13 positive exposures for kids to be strongly familiar with a food or beverage (and, therefore, indicate that it is liked). Sampling, therefore, takes on a very important role, and "trial" may continue beyond the first purchase. There are also ramifications in new product testing and sensory evaluation.

2. *Good experiences with new foods decrease neophobia.* This is important in a number of ways. First of all, the opposite can also be true. Bad experiences with new foods can increase neophobia. This means that launching products that kids don't like can damage the brands. People can understand they may not like certain flavor variants, but repeated negative experiences with a brand are likely to make that young person avoid trying future new products with that brand name.

3. *Combining the new with the familiar decreases neophobia.* This is probably the best advice of all. Taking something familiar and adding something new seems to be highly successful in developing new food trends with kids. Change the format or the food, but careful about doing both at once. Too drastic a chance is more likely to cause rejection.

4. *Liking of a product by peers, increases liking.* Moms have been saying this for years about kids. "I have tried for years to get my child to eat (insert food or beverage) and he refused. Suddenly he goes to a friend's house and has it. Now he wants it all the time!" The same can also be true for adults—we

---

* Much of the basic understanding about neophobia was gleaned from work by and listening to Professor P. Pliner, University of Toronto, and D. Mennella, PhD, The Monell Institute. It is strongly recommended to search for and read their work. The specific experiences related in this chapter are pulled from several years' worth of projects carried out by the Consumer Knowledge Centre, all throughout the world.

learn by seeing people we know/respect try something, and we think we will too. This is why "word of mouth" is so powerful with kids, and arguably, powerful with all consumers. It is all about reassurance and comfort, reduction of fear, and the increasing of curiosity. This helps to create a willingness in the child to try something new and make him or her less afraid of it.

5. *Be careful of health and nutrition claims, because they have no effect with many kids, and may decrease liking.* Though more information is available about nutrition and healthy eating, the results to get kids to eat better has not been so encouraging. Being told "it's good for you" will not make kids like your product. Sometimes, they may use nutritional elements as a negotiating tool to be able to rationalize or justify what they want: "Mom, please get it for me. It's got vitamins and minerals!" But in the end, we are not usually willing to eat things we don't like purely because "it's good for you."

Once we understand the meaning of neophobia, the kids' stated desire for "something new and different" or "something less boring" takes on a completely different meaning. We certainly need to deliver to them the perception of "something new," but in a familiar way or with a familiar taste. We need to build new products and flavors from their existing experience, and to move step-by-step, rather than in leaps and bounds. We need to clearly communicate in a familiar way what the product is so that kids can feel more in control, and feel as if they know what they will be getting. We need to remember as our mantra the equation "familiar = liking," and not be ashamed to use this to power our new product development.

13.6   SUMMING UP: SOME DO'S AND ONE DON'T

As you are working through the new product development process, remember some important things:

DO be absolutely clear about who you are talking to! Learn about kids and understand their development and importantly, the key drivers. Kids are different as they grow through the various ages and stages, and there are important physiological and cognitive issues to consider.

DO understand kids' key drivers of control, aspiration, and their stated desire to create excitement and stretch boundaries. Remember that their desire for control is most important, and you should strongly consider ways to make kids feel more in control—especially with regards to packaging (suitable for the size of the target-aged hands, fits in their lives, easy opening and access, etc.) and product (flavor, portion size/bite sizes). Aspiration and creating excitement and stretching boundaries should be more in tonality and personality of the brand.

DO accept that boys and girls are different, and find ways to either target a specific gender or ensure that both are satisfied. When you must target both, always skew the positioning to boys. Girls are more accepting of "boy" products, yet boys are more likely to reject something that is "too girly."

DO see the market and category through kids' eyes. Sometimes, we are so involved in the language and definitions of the food and beverage industry, we forget that consumers don't see things the same way. Kids don't think of foods and beverages in the categories we do—they don't say "I want a carbonated beverage" or "I want a dairy drink," rather they say "I want a drink that tastes good." Maybe they may only say "I want something that tastes good," without certainty that it is a drink at all!

DO involve kids in all aspects of product development and learn from their creativity. To design kids' products without involvement of kids at various stages is folly. The way you work with kids is important (they are not just little adults, and have different ways of expressing themselves), but most important is that you work with them! When you do, use humor and have fun, and where possible, work in environments in which they are most comfortable.

DO use the knowledge of kids and their key drivers in your research as well, and increase the time you spend with them. Too many researchers make excuses and say that kids have a short attention span—they have a short attention span if they are bored or if the process is not engaging. Good researchers can develop a program that can keep kids involved for hours, if necessary (and even days if the project requires it).

DO be holistic in all that you do. Though it may seem logical to work on various aspects of the product, or even the marketing mix separately, with kids (and perhaps most consumers) it is best to develop and test the product holistically. A child has a difficult time to separate product attributes, for example, and younger children often take things very literally.

Above all, DON'T ADULTerate.

Winning with kids need not become a terrifying task. We can build on the understanding of the age and stage of the child, maintain a keen focus to deliver control, instill aspiration, create excitement/stretching boundaries, and build on ways to overcome the innate neophobia that kids have. Following these straightforward guidelines will help the company, whether product developer or marketer, create winning products for kids, faster, with much confidence, and ultimately, with more market success.

SOURCES AND FURTHER READING

L. L. Birch, L. McPhee, B. C. Shoba, E. Pirok, and L. Steinberg, What kind of exposure reduces children's food neophobia? Looking vs. tasting, *Appetite* 9, 1987:171–178.

L. L. Birch and D. W. Marlin, I don't like it; I never tried it: Effects of exposure on two-year-old children's food preferences, *Appetite* 3, 4, 1982:353–360.

B. R. Carruth, P. J. Ziegler, A. Gordon, and S. L. Barr, Prevalence of picky eaters among infants and toddlers and their caregivers' decisions about offering a new food, *Journal of the American Dietetic Association* 104, 2004:S57–S64.

E. Cashdan, A sensitive period for learning about food, *Human Nature* 5, 3, 1994:279–291.

P. Leathwood and A. Maier, Early influences on taste preferences, *Nestlé Nutrition Workshop Ser Pediatric Program* 56, 2005:127–141.

A. Maier, *Influence des pratiques d'allaitement et de sevrage dur l'acceptation de flaveurs nouvelles chez le jeune enfant: variabilité intra- et inter-régionale*. These de Doctorat de l'Universite de Bourgogne, Dijon, France, 2007.

A. S. Maier, C. S. Chabanet, S. Issanchou, P. Leathwood, and B. Schaal, Breastfeeding and experience with a variety of vegetables increases acceptance of new flavours by infants at weaning, *Chemical Senses* 31, 2006:E1–E99.

A. Maier, C. Chabanet, B. Schaal, S. Issanchou, and P. Leathwood, Effects of repeated exposure on acceptance of initially disliked vegetables in 7-month old infants, *Food Quality & Preference* 18, 2007:1023–1032.

A. Maier, C. Chabanet, B. Schaal, P. Leathwood, and S. Issanchou, Food-related sensory experience from birth through weaning: Contrasted patterns in two nearby European regions, *Appetite* 49, 2, 2007:429–440.

J. A. Mennella and G. K. Beauchamp, Flavor experiences during formula feeding are related to preferences during childhood, *Early Human Development* 68, 2002:71–82.

J. A. Mennella, C. E. Griffin, and G. K. Beauchamp, Flavor programming during infancy, *Pediatrics* 113, 2004:840–845.

J. A. Mennella, C. P. Jagnow, and G. K. Beauchamp, Prenatal and postnatal flavor learning by human infants, *Pediatrics* 107, 2001:e88.

P. Pliner and S.-J. Salvy, Food neophobia in humans, in *The Psychology of Food Choice*, R. Shepherd and M. Raats (Eds.), Oxfordshire, UK: CABI Head Office, 2006, pp. 75–92.

Part IV

Proper Packaging and Preparation

| Success | = | Defining and meeting target consumer needs and expectations | × | The right food | × | Proper packaging and preparation | × | Positioned correctly at the shelf and in the media | × | Meet corporate logistics and financial imperatives |

14 Food Packaging Trends

Aaron L. Brody

CONTENTS

		Defining and meeting target consumer needs and expectations		The right food		Proper packaging and preparation		Positioned correctly at the shelf and in the media		Meet corporate logistics and financial imperatives
Success	=		×		×		×		×	

14.1 WHAT IS FOOD PACKAGING?

Food packaging surrounds and envelops the contained food, to protect the contents from the always-hostile natural environment throughout the distribution network. Food packaging prolongs shelf life, i.e., helps reduce food safety incidents, retards spoilages, and retains initial quality.

In industrialized countries, food packaging is indispensable for delivering the safest food supply and the greatest diversity of good tasting food in the world's history. Furthermore, in the United States, food packaging delivers the least expensive food supply in world history, the least food waste in world history, at the lowest packaging cost in world history—less than 10% of the disposable income.

14.2 LEGACY OF FOOD PACKAGING

The more we package food, the less food waste we generate. In developing countries, more than half the food is lost between the local farmer fields and the table. In other industrialized nations, food losses are in 25%–30% range. Due to a highly developed, but still imperfect, distribution infrastructure and the effective application of packaging, food waste in the United States is probably around 20%, a quantity that can be reduced. The results of both growth and reduction of waste have arisen due to the growth of away-from home eating and enormous reductions in the mass of packaging used to contain food.

During the past quarter century in food packaging, we have transitioned from glass to plastic, steel to aluminum, rigid glass and metal to semirigid plastic and flexible materials, paperboard to flexible packaging, flexible laminations to coextruded structures, and semirigid plastic to flexible structures. And flexible materials have reduced their gauges while retaining protective characteristics.

14.3   A BRIEF HISTORY OF FOOD PACKAGING: BEFORE WORLD WAR II

Since the future is always an extension of the continuum from the past and present, a short look back is in order to understand where we are and where we are going. Once upon a time in food packaging, pottery and glass bottles were the benchmarks. These heavy hard structures could be closed and "sealed" mostly to protect against theft, animals and "ether." The resulting wine sometimes tasted even better than the original, and could be drunk later or tasted like vinegar. Milk in animal skins is turned to cheese or yogurt. Packaging was employed to enhance the product contents. Were these the origins of active packaging?

What did the French confectioner, Nicolas Appert, know in 1810 when he drove out the phlogiston or whatever from his pottery bottles and corked them to become the father of canning? Kensett linked processing and canning to drive out the "putrefactants" using steam. Guy-Lussac's oxygen theory was not far from the ultimate reality. They all concluded that removal of something from the container interior was better than leaving "it" in.

Louis Pasteur, the "inventor" of pasteurization demonstrated that food is a nutrient and it supports the growth of microorganisms which can grow on foods and spoil them. Heating can destroy (some or most) microorganisms and "preserve" the food. If the package remained hermetically closed, microorganisms could not enter to spoil the food.

During the 1890s, William Underwood (deviled ham, etc.) and Samuel Cate Prescott (the founder of the food technology profession) determined that spoilage of canned foods was due to (Pasteur's) microorganisms. The quantity of heat required to destroy those destructive creatures could be measured and controlled. Further, heat penetration into the food was relevant.

Drs. C. Olin Ball, Charles Stumbo, and Irving Pflug developed quantitative calculations to enable the operators to optimize the thermal effects within cans. Ball and his associate Bill Martin then conceived a bizarre alternative to canning: processing outside the package and assembly in aseptic packaging, a concept that was applied by Hans Rausing in his Tetra Pak flexible pouch, which required really good barrier plastic, etc., packaging for those funny-looking tetrahedrons (Figure 14.1).

During the beginning of 1900, glass bottle manufacture by machine was instituted to replace human lung power. Then came commercial canning in heavyweight tin-plated

FIGURE 14.1   Tetrahedron packaging.

steel cans. Corrugated fiberboard cases, paperboard, and glassine entered the spotlight, and recycling went mainstream for glass, steel, paper, and paperboard. The evolution of packaging was underway, supported by a super recycling infrastructure. During World War I, cellophane became the first "plastic" material from "natural" sources although in reality it was neither. Alone, cellophane was a "nice" transparent material. When modified, cellophane became a useful food packaging material, great for its day, but its product life cycle was only about 30 years because it is vulnerable to the environment.

Between the two World Wars, aluminum foil became the first real flexible barrier food packaging material. But aluminum foil is fragile, and so it had to be married to other materials such as glassine, paper, and plastic to effect synergies. At the same time, packaging machinery emerged. Electromechanical equipment was developed to effect efficiency, reliability, uniformity, labor savings, and increased output while sorting, filling, unitizing, gluing, cooling, etc. Controls monitored and remediated actions. Machines such as double package makers, vertical form/fill/seal units, cartons, and case packers emerged.

Thermoplastic polymers were invented and developed into potential package materials to enhance paper and cellophane and replace (most) waxes and nitrocellulose. Polyvinylidene chloride (PVDC) (saran) and polyethylene were developed before World War II. Aseptic canning for milk to feed the troops, invented by Drs. William McK. (Bill) Martin and C. Olin Ball, was commercially developed.

During World War II, canning and the infamous canned C rations were inconveniently opened with bayonets. Dehydration led to products such as powdered eggs and powdered milk in No. 10 tin-plated steel cans and the infamous K rations.

Sugar shell-coated chocolate candies were the ancestors of M&M's™ chocolate candies in spiral-wound paperboard canisters. Packaging that protected products in both snowy northern Italy and hot tropical Guadalcanal Island included wax, cosmolene, rubber, and tarpaulin materials and structures.

14.4 "EXPLOSION" OF PACKAGING ALTERNATIVES AFTER WORLD WAR II

After World War II, polypropylene, polyester, and other plastic polymers were commercialized as films, sheet, and bottle materials. Retort pouches—a technology to "replace" the can—were first developed. Bag-in-box from Scholle came forward first for battery acid, and then for fruit and tomato purees and milk. Plastic coatings for paperboard helped to expand the applications of these substrates—and into milk cartons and barrier pouches.

Horizontal form/fill/seal machines such as Bartelt and Jones, and thermoform/fill/ seal machines such as Mahaffy & Harder, Multivac, Filper, and Formseal became standard for food plants seeking to efficiently package foods and beverages.

During the early 1950s, Korean police action, Tetra Pak pioneered the move to ambient temperature shelf stable milk in Europe where a willing absence of refrigerated distribution in the market waited. The "Delaney clause," zero tolerance for some chemicals in packaging, which haunts us to this day, was added to the Food, Drug & Cosmetic Act.

Ionizing radiation, gamma and electron beams to destroy microorganisms without heat, began in earnest for military feeding: the greatest food and packaging research program in food history began during the early 1950s and still waits universal acceptance.

Controlled and modified atmosphere food preservation originated at Whirlpool as the ancestor of Transfresh and then Fresh Express, but required more than 20 years to achieve commercial fruition, as did Cryovac's barrier bags for fresh red meat primal cuts for distribution packaging in place of hanging carcasses (Figure 14.2).

FIGURE 14.2 Barrier bags for fresh red meat primal cuts for distribution packaging.

Microwave food heating began in 1951 by Raytheon, followed by the first microwave susceptors, but the stampede did not begin until the 1980s when compact magnetrons were developed.

During the latter half of the late twentieth century, few of us thought of barrier packaging as "passive." We addressed barrier packaging with all manner of metal, glass, polymer, and combinations. We redefined barrier with water, water vapor, lipid, oxygen, and carbon dioxide, but we were not satisfied with the results, and so we compromised the distribution time and temperature to accommodate the new lighter weight materials and structures.

Forty years ago, military rations were designed in order to have very long lives, 3-year distribution and safety, but biochemical deterioration still adversely affected the quality of products. Package barrier alone was not enough. "Something" had to remove the residual oxygen and so palladium-catalyzed reaction with hydrogen gas in the headspace helped preserve dry milk powder in hermetically sealed steel cans.

Wood crates for distribution packaging began to surrender to corrugated fiberboard cases and wood pallets. During the 1960s, controlled atmosphere was ensuring that the environment in and around fresh food contained the correct quantities of oxygen, carbon dioxide, and water vapor. Temperature was controlled because controlled atmosphere and its direct descendant, modified atmosphere, do not function without temperature control.

From a food packaging perspective, the 1960s also featured the commercial introductions of:

1. Barrier vacuum bags for primal cuts of meat by Cryovac (now Sealed Air)
2. Case-ready fresh red meat—still volatile and evolving (Figure 14.3)
3. Absorbent pads for meat trays
4. Barrier flexible packaging materials and vacuum packaged processed meat

FIGURE 14.3   Case-ready fresh red meat.

5. Oxygen scavengers from Japan's Mitsubishi and Multisorb Technologies in the United States
6. Moisture scavenger sachets from Multisorb Technologies
7. Microwave pasteurization and sterilization: fully developed and commercialized in Europe by Alfa Laval and reinvented in the United States in 2003
8. Zipper reclosures—KCL invention now almost universal in flexible pouches
9. Two-piece aluminum cans—with easy open ends from Coors—now universal in aluminum cans and expanding rapidly in steel cans
10. In-plant extrusion blow molded plastic bottles—for fluid milk, still common practice especially now that plastic bottles have largely captured still fruit beverage and fluid milk packaging in larger sizes
11. Studies and actions on the packaging and its solid waste residual

14.5 FROM THE 1970S UNTIL TODAY

The 1970s began with Earth Day 1970 and the environmental movement and Packaging in Perspective—a coalition that assembled the first positive take on packaging. Ultra high pressure food processing began as a commercial process back then—where is it now? Polypropylene replaced cellophane, and ethylene vinyl alcohol oxygen barrier plastic came across the Pacific from Kuraray, Japan. Stand-up flexible barrier pouches, i.e., Doyen Packs, were introduced for Capri Sun fruit beverages and then cloned by many flexible packaging suppliers to the extent that they are mainstream today. Film metallization and microwave susceptors were commercialized. Plastic bottles for carbonated beverages (originally polyacrylonitrile, and then, polyester, and coextruded barrier plastic) and plastic sheet for thermoforming were launched. During the 1970s, internal moisture control was silica gel desiccant in porous sachets to keep metal products from rusting if the primary package was moisture resistant and the contained product was valuable enough to warrant the added cost, or if the dry military ration was intended to be in distribution channels for years. The sachet contained an admonition, "do NOT eat," but was it obeyed? Purge-absorbent cellulose "diaper" pads in the bottom of what were then pulp or expanded polystyrene trays for meat or poultry became standard and are now extending to active packaging.

In active packaging, Mitsubishi Gas Chemical Ageless and Multisorb Dessicants (now Multisorb Technologies) oxygen scavengers were commercialized using ferrous oxide ensconced in gas permeable Tyvek spun bonded polyolefin sachets. Water vapor activated the ferrous form "rusts" to ferric oxide by reacting with environmental oxygen. In excellent oxygen barrier primary packaging, oxygen scavengers scavenge residual and entering oxygen in dry meat, bakery goods and nuts, portending another slow evolution. Ferrous iron sachets have been followed by ascorbic acid, sulfites, photosensitive dyes, and ligand oxygen scavengers. From the United Kingdom Metal Box Ltd., came Oxbar™ cobalt-catalyzed nylon MXD6 imbedded in the polyester structure of juice bottles during the 1980s, questioned by regulatory authorities, tested and reintroduced in polyester beer bottles in 2000 and in juice bottles more recently.

The 1980s witnessed coextrusion blow molding of polypropylene/ethylene vinyl alcohol/polypropylene barrier (PP/EVOH/PP) bottles. Another 1980s' commercial explosion—25 years after its Swedish debut—arose from aseptic packaging for fruit beverages, and—also 25 years later than its original development—modified atmosphere packaging for fresh cut produce, both of which had been commercial, but quietly, for decades. Retort barrier plastic cans ("buckets") of low acid foods for kids or adults or for microwave convenience were introduced. Ten years after their introduction, polyester bottles extended from carbonated beverages for juices, water, peanut butter, salad dressings and now, 30 years past birth, into virtually every glass, metal, and even other plastic bottles for food packaging.

14.6 PERSPECTIVES ON THE YEAR 2008 AND BEYOND

The new millennium for trends in food packaging can be summarized as

1. Convenience
2. Sustainability
3. Resource utilization
4. Plastics from biological sources
5. An end for hydrocarbon sourced plastics?
6. Energy conservation
7. Dashboard dining (Figure 14.4)
8. Nanotechnologies
9. Temperature signaling (Figure 14.5)
10. Radio frequency identification (RFID)
11. Easy open/reclose
12. Incessant attacks on the wastefulness of food packaging in the media and by uninformed consumers
13. Aluminum bottles for hot filled beverages (Figure 14.6)
14. And more convenience demanded by consumers

FIGURE 14.4 Dashboard dining.

FIGURE 14.5   Temperature signaling.

FIGURE 14.6   Aluminum bottles for hot filled beverages.

Food packaging at this time is far more than containment and protection during distribution. Food packaging has become functional, with the package serving as the usual protection, but also as a cooking and time-saving unit. The "TV dinner" has evolved far beyond a conventional ovenable tray into a device that fits the microwave oven, e.g., popcorn bag, cooks, heats, crisps (e.g., pocket sandwiches), steams, opens when done, and even looks like take-out Chinese food with chopsticks; coffee brewing units; stick sweeteners; thumb-driven candy dispensers; and on and on the trend goes—and where it stops if ever, who knows.

Only the environmental activists, who, coincidentally, are among the major users of these convenience packages, can stem the stampede of convenience packaging that so simplifies eating and life.

14.7 SIGNIFICANCE OF HISTORICAL TRENDS

This brief history demonstrates that there are no guarantees for food packaging innovation success, no matter how good the innovation, how powerful its origins, or how financially sound is the support.

Food packaging development is a long and complex process that must meet a real packager or consumer need and must be skillfully and intelligently implemented. It carries with it a high risk, and is not headline generating. Just as past headlines were hardly a harbinger of success, this analysis suggests that some of today's headliners are not destined for success. And, a similar conclusion will have to be drawn for tomorrow's "innovations" whatever they may be.

14.8 TRENDS TO WATCH FOR IN FOOD PACKAGING

Successful food packaging innovations of recent years that suggest trends for the foreseeable future include

1. Aseptic packaging in paperboard composites and barrier plastic
2. Modified atmosphere packaging
3. High gas transmission package structures for fresh cut produce and especially, fruit
4. Reduced oxygen to prolong quality retention
5. Coextrusion and coinjection of plastics for function: barrier
6. Film: barrier low oxygen packaging for meat
7. Semirigid gas barrier bottles
8. Film metallization for barrier, decoration and microwave heating, especially surface crisping and browning
9. Polyester bottles for carbonated beverages, still beverages, water, and far beyond
10. Easy open and reclose—"creatively swiped from cosmetics" and other personal care items (Figure 14.7)
11. Microwave reheating packages including steam-assisted
12. Stand-up flexible pouches
13. Cook-in barrier bags (Figure 14.8)

These food packaging innovations succeeded because all were conceived and developed by the food or food packaging supply industries with a total commitment to delighting customers. So too were the heralded and occasionally merely forgotten failures, which need not be enumerated here, but which are too numerous.

FIGURE 14.7 Easy open and reclose—"creatively swiped from cosmetics" and other personal care items.

FIGURE 14.8 Cook-in barrier bags.

The origins and development of the successful innovations and trends can be traced to a single dedicated person, team, or company coupled with commitment, but so also that of the heralded failures—through blind commitment to weak notions.

None of these successful food packaging developments was driven by a single supplier material or structure. All were and are elegantly simple in concept. All were fundamentally sound from scientific perspectives. All filled a definite need for the food packager or the consumer. All were introduced complete—little or no after-market development remediation! Most importantly, all were wholly evaluated before their commercial debut.

Still in question commercially are the widely publicized food packaging innovations such as

1. Retort pouch, tray, and carton packaging—still largely off-shore in origin (Figure 14.9)
2. Aseptic packaging of particulates—really not yet commercial
3. Microbiologically safe chilled prepared foods with extended shelf life—everyone is in fear
4. Microbiologically safe fresh cut produce—*Escherichia coli* 0157:H7 remains an issue
5. Microwave susceptors whose application is still limited (Figure 14.10)
6. Glass or silica coatings for plastic films
7. Controlled internal package relative humidity which could extend shelf life
8. Oxygen scavenging in packaging material structures which could extend shelf life
9. Modeling and predicting shelf life which could help extend shelf life

We are awaiting the commercial development of other even more highly publicized technologies and some that are quietly growing in the laboratories and pilot plants:

FIGURE 14.9 Retort pouch and tray packaging—still largely off-shore in origin.

FIGURE 14.10   Microwave susceptors whose application is still limited.

1. Active packaging
2. Effective antimicrobials with no adverse secondary effects
3. Effective, economical oxygen scavengers
4. Multiphase plastics with large internal surface area for active packaging
5. Intelligent packaging—especially that interfaces with consumers
6. Time temperature signaling to consumers
7. Linkages with RFID to control distribution
8. RFID—its role in the packaged food system
9. Supply chain management
10. Food content safety signal
11. Food content quality signal
12. Convenience or even automation in retail check out
13. Monitoring of distribution
14. Information for consumers
15. Nanotechnologies with enormous physical surface area in food packaging suggesting barrier enhancement of biomass materials, to solve the flavor scalping issue of polyolefin plastics and possibly for sensing
16. Reheat packages for the new domestic kitchen food heaters in conjunction with or independent of microwave convection ovens
17. Halogen lamp convection ovens
18. Microwave pasteurization and sterilization
19. Preserve food with minimal heat (Figure 14.11)
20. Quantitative application of hurdle technologies
21. Near zero or micro-oxygen packaging
22. To markedly reduce the biochemical sensory changes in food in distribution channels

FIGURE 14.11   Preserve food with minimal heat.

Some technical challenges to be resolved in the future:

1. Accurately measuring and signaling the real safety and quality of packaged foods
2. How to break out of the 40 × 48 × 52 in. pallet size boundary of distribution packaging
3. Feeding astronauts on long, multiyear space missions
4. Persuading the world, U.S. consumers and legislators that effective packaging of foods will alleviate food waste and shortages

14.9   CONCLUSION

The future is an extension of yesterday and today. When you do not know history, you will certainly revisit it again and again. We can look for intelligence and response to external signals to measure and control food temperature, and measure and signal food content quality. Enhanced gas, water vapor, and flavor barrier will solve the problem of flavor scalping. We must intelligently apply nanotechnology or it will hurt us. Integrated holistic hurdle technology will focus less on specific technology or materials, but more on optimally integrating multiple technologies. And listening to the eating public will be the guideline to enhancing the value of packaged foods for our consumers.

BIBLIOGRAPHY

R. Ahvenainen, *Novel Food Packaging Techniques* (Boca Raton, FL: CRC Press, 2003).
A. L. Brody and J. B. Lord, *Developing New Food Products for a Changing Marketplace*, 2nd edition (Boca Raton, FL: CRC Press, 2007).

A. Brody, E. Strupinsky, and L. Kline, *Active Packaging for Food Applications* (Boca Raton, FL: CRC Press, 2001).

J. H. Han, Antimicrobial food packaging, *Food Technology* 54 (3), 2000: 56–65.

J. H. Han, *Innovations in Food Packaging* (Amsterdam, Holland: Elsevier, 2005).

G. Robertson, *Food Packaging*, 2nd edition (Boca Raton, FL: CRC Press, 2006).

S. Selke, *Understanding Plastics Packaging Technology* (Cincinnati, OH: Hanser, 1997).

15 Evolution of Sensory Evaluation: How Product Research Is Being Integrated into the Product Design Process

Stacey A. Cox and Robert Delaney

CONTENTS

Success	=	Defining and meeting target consumer needs and expectations	×	The right food	×	Proper packaging and preparation	×	Positioned correctly at the shelf and in the media	×	Meet corporate logistics and financial imperatives

15.1 INTRODUCTION: THE VITAL PARTNERSHIP TO GET THE PRODUCT RIGHT

One very important partnership to get today's products "right" is the integration between product developers and their partners. This chapter focuses on one of these partners, the product researcher, and how that researcher provides guidance to developers of a type that could otherwise not be provided by others in the corporation.

Product researchers (i.e., sensory scientists), focusing on how people respond to products at a basic, sensory, subjective level, are by necessity grounded in the measurement of perception. In many companies, the traditional role of product researchers has been as a service, to measure the features of a product in the same way that a clinical technician in a health services laboratory measures the blood, etc. All too often the history of sensory research is a history of a trained technician, who can report to the developer whether the products are different, how they are different, whether the products are acceptable, etc. All these come from the use of research designs, good testing practices, and of course, the appropriate application of statistics. The bottom line is that the product researcher is taught how to "find" the right answer for the developer or marketer, etc.

Developers learn the skill set needed for the formulation, processing packaging and manufacturing great tasting products. Typically developers are not taught the art of food—how to develop and layer flavors; they are not taught how to use complementary and contrasting flavors and textures. To complicate matters farther often, developers are not taught anything about the consumer, who the consumer is or how to best communicate with them. The result is that all too often, developers do not have the necessary conversation with a consumer because they do not have a common language. The developer works in a technical language, foreign to consumers. In food companies, the issue is even bigger, as the developer becomes increasingly remote from the consumer, lacking knowledge about the consumers, and how the developer's product is actually used. Of course the situation is not dire, but the lack of such intimate consumer knowledge remains distressing, even today.

We tend to bring in the consumer only when we think that the consumer response is necessary for a decision, such as when there are concepts to evaluate or products to try. In companies, we often fail to involve the consumer and we thus do not let them help us. The key phrase here is "help." Years of experience working with consumers suggests that the consumers really do want to help make a better product.

The craft of really getting the consumer to tell you his/her needs and let him/her help design a product that is really wanted is more than just a craft alone. It is an art, one that cannot be done without the ongoing interrelationship between R&D who designs the product and the product researcher who understands and translates the mind of the consumer.

15.1.1 Why Don't Companies "Get It Right" Almost All the Time?

If all that is needed to make a good tasting product are the skills and tools we are learning in school, then why do not more products succeed in the marketplace? Even sensory and development departments, not to mention market research, that have been around for 30+ years and have really perfected the skills and tools, still do not succeed every time. Why not?

We believe what is lacking is not an understanding of the tools and one's ability to execute, analyze, or interpret data. Rather, the big gap is how the knowledge gets applied and integrated into the system over time. Companies need to create better systems to understand the customer. It is really not the problem of the tools, but the problem of so what! In fact, what is not being taught or practiced may be the critical part to ensuring a product is successful in the marketplace.

Even if there is a partnership between the two groups, how well and productively does it function? Although it might seem trite in today's objective, bottom-line driven world, the most productive interaction comes when both parties appreciate each other and openly share knowledge. There should be an in-depth understanding of how the two spheres can and should work together. Some of this understanding comes from teaching each group about the tools and expertise of the other. All too often in school and early in one's career, a professional in one discipline misses out on learning how one ought to become half of a true partnership, and how to lead/engage the necessary strategic conversations that drive the business. Professionals schooled in technical services must learn to go beyond that expertise and become a partner at the design table, owning the business and its responsibility as much as do the marketing partner.

With these issues in mind, this chapter presents one suggested process that has been used successfully at Heinz to evolve the relation between the developer and the product researcher, at the early stages of development. We present the vision, the implementation, tool sets, reasons for the change, and most of all how to make it all happen in an organization.

15.2 WHAT CAUSES FAILURE TO OCCUR? CASE STUDY—NEW CHOCOLATE CHIP COOKIE

We begin here with the negative—what we have seen when this partnership was not built and tools were either unavailable, or just a plain wrong to address the problem (or opportunity). The case study highlights what cost it is to the business when products are not designed or made to taste "right."

Let us begin at the outset, with the business picture. The issue was to satisfy a customer request for a new chocolate chip cookie in a line of existing cookies. Concept scores for this new cookie were sufficiently high that the product idea enjoyed, by itself, a reasonable level of success.

Given timelines and budgetary restraints, the team decided to forgo building the product from the ground up and instead, agreed to start the product design using an existing base product. The team decided to use predetermined ingredients, based upon directions that did not emanate from consumers, but were rather dictated from other points of view. (One might say by fiat, but that too would not be totally correct, although might be closer to the truth).

This initial misstep, i.e., ignoring the voice of the consumer, had major ramifications. The team struggled with the "right" formulation. As one might expect, the team members had many different opinions about what the product should be, and even the consumer input through 2 years of the project did not suffice to clarify what to do.

The team tested numerous product options, using sophisticated experimental designs with a thorough list of questions to be answered for each prototype. Products were not the only things tested. The concepts changed during the course of the project, as the team struggled with the benefits, the price, the name, and so on.

Yet, this was not all. In turn, the company invested a large amount of money for a single SKU (stock keeping unit, a facing on the shelf). Money allocated for plant trials grew to $40 K, consumer research cost $80 K, slotting fees to get the product onto the grocer's shelf came to $1.3 K, all in addition to the time and opportunity costs.

So what are we to make of this fiasco? Well, first, it is not rare, but rather all too common. Second, and most distressingly, the loss of the opportunity alone to launch the right product continues to happen to organizations over and over, but not without the strong desire to really follow the new product process with all their hearts and souls.

What is really behind some of the failures, the failure of not getting a product to taste "right?" After, if the job of a food or beverage company is to create and market good tasting products, why do so many of them fail because they do not taste particularly good? One potential cause of failure is really the lack of consumer touch points within many parts of the design process. If one does not get a status check from the consumer, it is likely that the product could be less than optimal, no matter how talented the team members are and no matter how the team feels it is in contact with "today's tastes."

Returning to our case history, consumers were asked their opinion about the proposed product (actually the entire product experience), but probably way too late in the game. Engaging the consumer in product design was used dutifully in the latter stages of development, but in the wrong way. The consumer checkpoint was a numerical-based evaluation, generating a scorecard but really had no deep insight into the tongue and mind of the consumer. The checkpoint was more like a laboratory report from a routine blood test. No "voice of the consumer" had been used in defining the product design at any of the early stages of development. There was no identified consumer need of the product. If at the early stage of the project one were to look at the competitive landscape, the unhappy news would be that in truth there was really no need at all for another chocolate chip cookie in this already-overcrowded category.

One bottom line of the postmortem is that when the initial product design is not right, when the direction is not generated through the consumer, then most likely failure stalks the project and all the remedial flavor modifications in the world will not help. In this case, there was no consumer seat at the design table which later caused the team real difficulty connecting the dots between the product and the concept expectations.

A second cause of failure in our cookie case history was the blind use of current, well-accepted but realistically outdated tools that should have evolved to meet the developer's needs. In actuality, the developer could measure responses, but did not really understand the inside of the consumer mind, namely what was critical, what was nice to have, and what was irrelevant and just along for the ride. The team could not go below the top-of-mind ideas, and so were confined to superficialities that were outdated. It never occurred to the team that the product did not have sufficient "good stuff" so that the consumer would trade current cookies for this new cookie.

Let us go a bit further into this issue of lack of knowledge, because most companies feel strongly that they have the right array of qualitative and quantitative tools for their product research. Part of the problem was that the team was unclear about what they would test, and even importantly, what would they look for. So when working with the concepts, no one thought to give actual products against which the concepts could be anchored. At the back end the team only used prototypes that varied in flavor and nothing else. It was clear that the key problem was limited listening to the consumer. The team heard what it wanted to hear, listened to group speech, and in the end failed to develop the compelling concept and product proposition.

The final cause of failure was the structure of the process, and especially the lack of ownership. Simply stated, there was limited co-ownership throughout the entire process. Some key team members were only engaged as spectators in the process not co-owners. Other key members were not adequately schooled in the process or opportunities emerging from consumer research. No one was the consumer's champion. This limited engagement in the process and tools cracked apart a potentially coherent group, fragmenting into a group of individuals who simply met together and delivered a sub-par proposition.

15.2.1 TODAY'S PRODUCT RESEARCH: THE OPPORTUNITY AND IMPETUS FOR TRUE EVOLUTION

Product research is a tool used by most CPG companies to get the product "tasting right." It is called research guidance, early stage research, sensory analysis, and a host of other names, depending on the company, the intellectual history of the professionals in the company, and the wisdom and guidance of the R&D leader. For many years, the tools of product research were used to create better products, sometimes with success, sometimes with failure, most of the time inexpensively relative to other methods. All these tools are designed to give us the best answer for designing a consumer-wanted product. But what happens? Why are the tests successful sometimes and not successful other times, especially since the tests are run under rigorous conditions, with adequate statistics, by professionals who know how to execute the projects, occasionally, virtually, flawlessly?

The answer may lie within the tool itself, within the use of the tool or within our reluctance at times to change or evolve the tool. No one likes change. Products change, and tools ought to change with them. Are the tools that traditionally have been used for decades still the best? Or do new testing methods, new ways for evaluating consumer responses need to "step up to the plate" and take their proper places? The answer may also lie within the corporate dynamics. Companies, like people, do not like to switch. They would rather fight, relying on what used to work, on norms that are comfortable, and on ways of doing things to which they have become accustomed. The answer may also lie within the strategic conversation of a product researcher and the organization. The answer may lie within the data, but the organization does not want to see it or the researcher is doing a suboptimal job of the conversation to persuade the business to listen to the data.

15.2.1.1 Three Challenges and Opportunities

At Heinz, Product Research and Guidance is one of R&D's key tools for ensuring they are getting the "product right." It is where the tools and minds are housed that focus on translating the consumer's needs and wants into tangible elements that will lead to products and package designs. Product Research provides data to make a decision, but then drives understanding as well about the consumer, as this consumer interacts with product and package. Furthermore, the good product researcher, at Heinz and at other companies, must balance science and art, rigor and vision.

In today's changing world, the product researcher faces three major challenges in the quest to get the product right. The challenges become ever more difficult with changing consumers, changing questions, and changing competitive environments.

15.2.1.1.1 The Ever-Changing Consumer

How does one go about asking questions of the consumer so that we understand what is going to be foundational in the design versus what is wanted just given that moment in time of questioning? As we all know, the consumer in many ways is a moving target for answers. At Heinz, one way to get to the core of consumer behavior is through an evolution of focus groups, which we call Product Labs™. A Product Lab gets to the top of mind, to find the areas of "disconnect" in their conversations, i.e., what they say versus what they do versus what the product is. The Product Lab creates an environment in which the consumers can articulate what they really need or want in the experience. We "break down" the consumer barriers to get at the heart of the consumer need.

Other tools we have added to our process to determine foundation versus moment in time design requirements are the tools such as Value Diagrams™ and Kano Diagrams. We have used these tools to understand design structure and hierarchy, to weed out the must-have from the nice to haves. We have used these tools in numerous cases to understand where to focus resources and what are the elements that consumers will pay full price for in the experience.

In the case of the chocolate chip cookie, the consumer need was not even taken into consideration. There was an assumption on the part of the customer that they knew the desired product design as well as an assumption on the business part. Hence, there ended up numerous iterations of the product tested, because we could never figure out the target or what was missing from what we provided.

15.2.1.1.2 Changing the Question

We have begun to look at how we ask our questions. Instead of asking a whole list of questions, the so-called laundry-list strategy, we now create specific working hypotheses that are meaningful for a developer, and then work with the test stimuli to prove or disprove the hypotheses. Furthermore, we work with prespecified actions to be taken based on the results.

We are working to move the organization from "I have all these prototypes, I need to know which the consumer prefers" to "I believe based on my knowledge, the experience the consumer is looking for is this…let me use some stimuli to see whether or not I get the response I am looking for." That is, over time the Heinz product testing organization and its internal clients are moving from the mindset of prototypes and questions in qualitative research to the discovery of stimulus–response patterns based on team-developed hypotheses. We have also begun to apply the aforementioned tools (Kano Diagrams and Value Diagrams) to provide the product developers watching interviews with the consumers in the "back room" with a structure to help shed light on the conversations with the consumers. The hypothesis generation and diagrams generate more structured and actionable outputs from consumer interactions for not only the developer, but also for the business.

Returning to the case of the chocolate chip cookie, we continued to guess about what would be the best test design. The team was not, in fact, aligned about what we believed to be true for the consumer. Each person had his/her own opinion. Thus, there was no alignment but rather effort, often unproductive effort. We did not test products as stimuli against our hypothesis, only stabs in the black hole of the consumer's psyche. As a team, we were not aligned as to why each product was placed into the research, what was its purpose for being as well as what was the expected outcome. When a team is not aligned on what may occur from the stimuli presented, there is more data confusion and more expensive, perhaps unproductive retesting afterward of yet even more prototypes.

15.2.1.1.3   Changing the Research Tools

The third challenge for researchers is the willingness to change the research tools. A key initial step in being influential in the organization is one's willingness to change/evolve the traditional tools when it becomes obvious that a previous sacrosanct tool is just not working any more. Traditional sensory evaluation tools such as descriptive analysis panels and focus groups have great values in organizations. There are times during which the tool needs to go beyond the current capability, however. The traditional foundations need to be there, but the tool needs to be "updated" to move with the needs. At Heinz, we have done this with the Insight Team, an evolution of descriptive analysis and Product Lab, an evolution of focus groups.

The Insight Team is a self-functioning team built out of the need for a descriptive approach, specifically designed to provide more insights/guidance for the business team at the front end of the new product development (NPD) process. At Heinz, the Insight Team has evolved to become an extension of R&D, comprising more hands and minds to help with the design creation and category understanding. The Insight Team is a way for the business to gain insights not only into product options, but also into the experience in which the consumer engages in the marketplace, and the context in which the consumer makes choices. It is a great tool for finding white space in the market and providing insights into where the team may want to move for a better chance of success. The Product Lab is a R&D owned tool which allows them to move beyond top of mind feedback to in-depth knowledge of consumer trade-offs and perceptions. The tool uses the foundational structures of focus groups with evolved questioning methods and stimulus patterns.

The chocolate chip cookie case history did not use the Insight Team or a Product Lab. Thus as noted before, research was done, but the results failed to reveal what characteristics drove acceptance, and what new opportunities or the so-called category white spaces existed. At that time, R&D used the traditional central location test (CLT) to measure but not to guide. The result was data, but not insight.

Happily, over time the three foregoing issues have been addressed. The outcome, to be explicated below, is truly new, and we feel a better way to drive consumer understanding at the early development stage. At Heinz, the Product Research team continues to work toward understanding the ever-changing consumer, dealing with the ever-changing question, and constantly evaluating and critiquing the ever-changing tools. In doing so, Product Research & Guidance (PR&G) continues to increase its contribution to the business. The skills and tools make the researcher a strong ally and aid to the development organization.

15.3   SECTION 3: RESEARCH & DEVELOPMENT

The face of modern research and development is changing. The competitive environment is vastly different because there are many more options for the consumer, and so R&D has to be sufficiently nimble to identify and incorporate trends. Furthermore, the consumer is more knowledgeable about health, and certainly variety seeking in terms of ingredients and ethnic foods. Consumers' palates are changing. They have and continue to become adventurous and global with their preferences and choices. Consumers are expecting more bold and pronounced flavors.

Experience with foods changes food preferences and consumer requirements. Consumers are inundated with aggressive flavors from restaurants, recipe tabloids, and food shows. Consequently, they are now than ever expecting more from prepared foods. According to B. Joseph Pine II and James H. Gilmore,[1] we are in the experience economy. Consumers, in general, are less satisfied with mediocre experiences. They have so many options these days that almost none of them need to settle for mediocre experiences. It is all about the experience. They want excitement, not just a belly filling meal.

This change in the consumer changes the food industry. Along with that change, the face of R&D has to change as well. Private label sales are capturing an increasing amount of the food market, essentially capitalizing on and cannibalizing the research done by more innovative companies. There needs to be an even larger point of difference between private label and brand name. The role of R&D is also strongly influenced by a saturated marketplace of products. A consumer product (CP) company needs more larger innovations. They need to lead the market with products, not just follow. The company needs to find a balance between mass appeal and niche appeal. The need for truly innovative products creates the need for more innovative research and innovative answers.

With all the focus on innovations, business decisions no longer can be viewed through the marketing or sales lens, but must be looked at through the technical lens as well. In many CP companies, the marketing functions are and should be focused on short-term profitability while the R&D function should be focused on innovative, robust, and sustainable product and package designs.

In a more competitive business landscape, R&D is expected to be equal-share partner in deciding the direction of the business, essentially becoming what we call at Heinz, a Technical Brand Manager (TBM). A TBM is a person who owns and drives the business with a technical focus grounded in the consumer. R&D organizations are expected to know and engage fully in all business aspects of the NPD process. They are expected to provide larger returns on investment, and are responsible for more lines of the P&L than merely delivering a product through the system. In fact, the performance and bonus of many developers are drawn directly from brand

P&Ls. This expectation requires an increase in R&D's ability to influence business strategy and pipeline development.

The challenge with the changing R&D expectations in organization comes from the reality that most developers are not taught a formal development/stage gate process in their schooling. Consequently, developers are not really aware of the critical development steps in the whole business process, specifically those steps that they need to drive, and the other steps that they need to influence even though they are not the primary drivers. Furthermore, when developers are taught the systematic development process all too often they are not told that the process is a fluid one. Rather, they are taught, or perhaps it is their proclivity to believe as technical people, that the development process is fixed in stone. The process should be looked at like a road map. The route is the same but the road conditions will always change. The driver needs to be able to react to these inputs to keep the car on the road. In other words, the process is a process, but the response must be like a fluid, ready to adapt to changing situations.

Going one step further, how should R&D incorporate the so-called voice of the consumer through the process as R&D implements the process? Most product developers are not comfortable dealing with consumers. They lack the necessary skills to facilitate a fruitful discussion with their target. Thus, in the past, consumers have had very few touch point in the front end of the R&D process. We are now moving to maximize the points and give consumers control of design very early in the process. Consumers need to have a seat at the design table, especially if truly innovative, new-to-the-market products are needed. With consumers involved in the early stage, the R&D organization can optimize the product/package design first (fit with general concept and brand) and then in later stages make smaller flavor-based modifications. This stepped process makes the design easier, because it allows for both strategic and tactical inputs. The approach breaks the product and package development process into two clear components… design and then flavor. One may think about the approach like building a house, designs are the walls and flavoring is the paint. If the walls are not in the right place, then it does not matter what color the paint is. Flavor is the last part of the design to fit. Flavor can be refined in the later stages of the process with the more typical quantitative tests. It is easier to get to the flavor once the design is done. The design impacts the flavor.

When the team uses the correct tools, the team gets to the design phase much faster. Let us return for a moment to the cookie issue, armed now with knowledge of how things "should be." The issue facing the team in the case of the chocolate chip cookie was the need to ask the questions on design that were never asked or were asked at the point when "flavor" should have been the focus. Therefore, the team lacked design information. The team talked about the concept and discussed minimal design and flavor questions with the consumer, but did not have the ability to provide the consumer with the stimulus to get to the next level of conversation. Stimuli translate consumer talk into actual consumer attributes. Without the right stimulus, a lot gets lost in the language. There is nothing to refer to. It can be very difficult at time for someone who does not know "food language" to listen to or communicate with the consumer. The stimulus helps R&D to be able to link with the consumer perception.

The stimulus has to be an actual product, and not a product-concept. The stimulus needs to be something that a consumer can play with, can eat, and then react to. Even in concept design, the concept itself is not the appropriate "stimulating" stimulus.

The physical stimulus must embody the on-going hypothesis, or in other terms actually represent it. As an example of best practice, R&D needs to lead more critical conversations with the team around why stimulus has been created for consumer understanding. With everyone aligned to the hypothesis for exploration and stimuli for response, the team is set up for common listening and alignment of understanding after the research is completed.

For design success, in addition to R&D using the right tools, R&D needs to feel the freedom with PR&G to modify the tools to the specific project and product/package needs. The research tools should be treated as fluid. Some of our most valuable insights have come from this "stick and move" mentality with research tools. In addition, for this to work properly, all parties must be present and vested in the process and must have specific knowledge of the tools applied.

15.4   SECTION 4: MAKING THE TOOLS WORK: THE HEINZ MODEL AND APPROACH

What is the unmet consumer need? What is the white space in the marketplace? How can you deliver a product that differentiates itself from those already present? These questions will help lead the way to powerful insights on which great products are designed. When the need or insight is wrong from the beginning, the product will never succeed because the consumer does not really want it.

What follows is a look into some of the key steps to consider when the process involves the design of new products. The steps are those areas to consider when the team must uncover the consumer need, and then design the appropriate product and package to fulfill that need. As we proceed we will look at the steps with the aid of the flowchart in Figure 15.1.

Step 1: Getting the team to focus and become energized: As the marketplace becomes increasingly saturated, need gaps and white space become fewer and farther between. For many years, it was fairly easy for companies to find these gaps and then fill them with mediocre products. Today, the companies have to spend more effort, do more "heavy lifting" to discover and then fill the white space. For the purposes of this chapter, which focuses on CP research, this so-called heavy lifting comprises the consumer research and analysis of the competitive landscape to understand and discover what consumers either cannot find, or do not know yet that they need. Development groups need to dig deeper and deeper to find a deposit of knowledge that will make a sustaining, product success.

Step 2: Homework. For best use of everyone's time in the ideation sessions, the team should conduct a competitive analysis of the marketplace. The homework lets the team understand what is out there, and does so with the different points of view or lenses brought into play by the different team members. The marketplace review should be based on product attributes, attributes that may get to the need. Conducting the competitive review first allows the team to identify potential gaps in the market prior to ideation. As the team works on the homework, it should look at the similarities and differences in the category and begin to investigate what parts of the competitive landscape is relevant or irrelevant to the business. Aligning to the key landscape is critical for a successful ideation process.

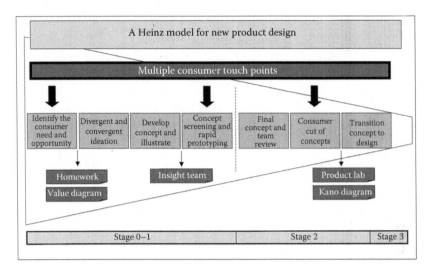

FIGURE 15.1 Flowchart of new product design used by the H.J. Heinz Company, Pittsburgh, PA.

Step 3: Diverge, i.e., focus "widely", and do not let anything escape. Nothing is out of bounds, nothing is too silly or crazy. Cost does not matter at this early stage and everything can be done. All ideas are captured and organized to be revisited later in the session.

Step 4: Reconverge. In convergent ideation, the ideas offered in the divergent ideation session are brought back for the entire cross-functional team to build and elaborate. This reconvergence step offers members the opportunity to better communicate their ideas and perhaps more importantly, take advantage of the input of others. Such confluence of inputs may often be synergistic, so 1 + 1 = 3. In addition, these ideas come from vastly different viewpoints because the ideation team is cross functional, and thus the reconvergence may synthesize new-to-the-world ideas.

Step 5: Create and test in a limited way. Concept screening and rapid protocepting addresses the question "how can we write the concepts that accurately (read better) capture the desired benefits in the correct consumer friendly language?" It is a quick screen to determine which rough concepts show some promise. This is a process by which many concepts are reduced to a few. This stage is the key in whether to scrap a concept or consider it for further testing.

In Step 5, the Heinz's Product Labs offer product developers a venue and tool for testing a developer's initial design hypothesis. This testing is done with protocepts to test the team's hypotheses around what the consumer wants and needs. The systematic approach is one of the best ways we have found to break the concept down into its basic parts and build from the ground up. This stage does not focus on refining flavor as much as constructing the design (the nuts and bolts and general structure). From here, the subsequent development work can be more focused on refining flavor options. Once built, subtle tweaks can occur with minimal effort later in the process. If one does not have the basic structure, no matter how pretty you make the experience, the consumer will not want it.

Step 6: Test the concept. Once the learning from the concept screen is digested, the winning concepts are put into a final concept test to gauge consumer appeal. If the appeal is sufficiently high, then the test estimates the volume sales (called volumetrics). Volumetrics are used to drive the go/no-go decision, and move the project into the next stage of the NPD process. Often volumetric results determine which products move forward and which products do not.

Step 7: Accept the concept for full development effort. Once the concept/product has reached the "transition concept to scope," the concept becomes a full-fledged R&D project. The important aspect of this phase is to keep a very tight connection between the tested concept and the scope. This ensures the development team is working to deliver on the consumer expectations in the design.

15.5 SECTION 5: A PARTNERSHIP TO INCREASE THE SPHERE OF INFLUENCE

In our opinion, in addition to the best tools for getting the "product right," one needs to build a strong and sustainable partnership between R&D and Product Research. This partnership is needed to increase the sphere of influence enjoyed by the technical community in the business. That is, it is important for R&D to become drivers and developers of the strategies and the innovation pipeline. For many years, the role of developers and product researchers has limited to service, often by deliberate choice. The traditional sphere of influence is strictly focused on formulation and process. The developer was influential in the product formulation but not in the design of the experience. The interest simply was not there, or if it was, the circumstances and corporate structures did not allow the interest to be expressed, except perhaps in the perfunctory R&D paid-for MBA.

So why do we want to go beyond our "silos?" With the end goal in mind to deliver the experience consumers want, not what we think we want them to have, R&D and Product Research need to become strategic in the way they work, and not simply responsive to requests. It is important here to lead discussions and deal with points of difference that always occur in conversations dealing with corporate strategy. The consequence is the need to do some "heavy lifting" and contributing on a proactive basis, as we have presented above.

Through this chapter we have presented the notion of food as an experience, not just as a product/package. If the world of food is to become more experimental, then R&D needs to enhance its capabilities and become more of a working, strategic partner than ever before. With this vision, the organization needs to have those with the skills to understand the consumer and to look at food as an experience to design. The developer is becoming more engaged in the entire process, with the consumer and with the business.

With a need for a greater sphere of influence in organization, there is also an evolution of how R&D is engaged in the Product Research process. In the past, R&D

organizations were content to accept incomplete, effectively marginal, and relatively inactionable output from the early development process. Even worse, many times, R&D became accustomed to virtually no real consumer guidance in the early development phase.

Times have changed, competition is fierce, and organizations are becoming increasingly lean, not fat. There is constant and increasing pressure to justify resources. It is a truism, but today anything but right at the first time is no longer an option. The old adage "Not enough time to do it right, but always enough time to do it again" just does not work. The new reality and the new processes require a developer's engagement throughout all of the research because it is the developer's hypothesis that is continually being tested.

We have mentioned above that one example of the R&D evolution into a member of the product research process is the creation of the TBM at Heinz. The position is a hybrid. The R&D person becomes further grounded in and knowledgeable about the consumer. Another form of evolution can be seen at Heinz where the developers are expanding into trend understanding through chef certification. Through the chef certification and TBM, Heinz is marrying R&D skills to consumer and food knowledge. The marriage is becoming ever more stable and rewarding, as increasing number of successes are launched.

Much like R&D, the PR&G sphere is also expanding. At Heinz, we are moving beyond the traditional one, into one of more product and business influence. There is still progress to be made to truly be considered a strategic partner in business decisions. The PR&G group is leading the TBM program. The program educates R&D even more about the consumer, guiding and evolving thought beyond formulation to experience. Product researchers coach and guide the developer through the best solutions and best tools throughout design and execution.

With the partnership and dual ownership of the product design process, both R&D at the development side and Product Research at the "insights" side can increase their contribution to the company, and broaden their sphere of influence. The only way for the sphere to enlarge and be powerful is for both groups to be grounded in the consumer together and to know/own each other's perspective in the process. To have an influential partnership within the team requires both groups to gain a basic understanding of the roles played by each. In order to increase ownership in an area outside one's own expertise, it is important as well that one be willing to give up some ownership of one's specific area of expertise. Cross-functional ownership allows each party to have a fundamental working knowledge of the others "turf," thereby creating a true synergistic collaboration. In order for both parties to commit to, engage in, and influence the process, both parties need to know the process. In order to get the product right by knowledge, and not by guessing, by structure and not by luck, the path needs to be lit by those who know best the consumer and the product.

REFERENCE

1. B. Joseph Pine II and James H. Gilmore, *The Experience Economy* (Cambridge, MA: Harvard Business School Press, 1999).

16 Relevance Regained: Designing More "Democratic" Domestic Kitchen Appliances by Taking into Consideration User Needs and Other Relevant Opinions

Grant Davidson, Tammo de Ligny, and Marco Bevolo

CONTENTS

Success	=	Defining and meeting target consumer needs and expectations	×	The right food	×	Proper packaging and preparation	×	Positioned correctly at the shelf and in the media	×	Meet corporate logistics and financial imperatives

16.1   INTRODUCTION

We do believe the first thing to bear in mind when designing meaningful appliances for food preparation is that you have to think beyond the notion of just developing a new mechanical device. Certainly, factors such as available technology and mechanical construction are of paramount importance; a blender, after all, must be able to blend. But without an intimate knowledge of whom you are designing for and what they expect a product to achieve for them, then you are extremely unlikely to hit the so-called sweet spot.

This chapter presents the way a commercial product designer approaches the problem, and how the design organization within a corporation (Philips Design) works through the issues to go from the concept to the actual product. The chapter deals with five key aspects, as follows:

1. Philips Design reality, or how design is viewed by the company
2. Key fundamentals of High Design, a process for repeatable success
3. Rationale behind design differentiation of appliances for food processing, case created by Philips
4. Case histories of successful products from the recent past
5. A case history referred to a visionary future-oriented program

16.2   COMPANY CONTEXT: PHILIPS DESIGN AND THE HIGH DESIGN APPROACH

Philips Design is the in-house design agency involved in virtually all Philips product-related activities, medical systems for hospitals to urban lighting to consumer lifestyle propositions. Employing more than 500 people located in its studios in Europe, North America, and Asia, Philips Design offers design-related services encompassing everything from experience design for products to integrated communications, identity, and branding as well as consultancy on sociocultural research, strategic futures, and innovation. The agency carries out work for external customers, with a portfolio of past projects ranging from Nike to GM, from P&G to Heineken, from Unilever to Microsoft. Of particular relevance here, Philips Design is responsible for the creation of Philips' kitchen appliances.

In our view, design requires much more knowledge than one might think. In addition to designers and design researchers, Philips Design also employs sociologists, psychologists, trend analysts, anthropologists, and many other experts in "humanistic sciences" in order to develop a more complete understanding of user requirements. This effort is embodied in Philips Design's High Design Process, which is a philosophy of design, as well as an enriched design approach that integrates established design skills with other disciplines in the areas of human sciences, technology, and business.

That being said, where exactly does such an approach fit within the world of food and kitchens? The answer comes from the approaches taken. The backbone to High Design is a proprietary methodology known as Strategic Futures™. Strategic Futures integrates roadmaps of preferable futures extracted from societal, technological, and business forecasts into visions of the future. These forecasts are then translated into concrete applications for system or product solutions addressing the present, emerging, and future needs of target audiences. Thus, any vision of the future of food at home or food as something to be processed by machine can incorporate the Strategic Futures method.

16.3 TWO CONTEXTS: THE GLOBAL MARKET AND THE FOOD INDUSTRY

High Design is founded on the belief that people should be the primary focus of any new proposition. Technology is also obviously important. There's no point in developing high-scoring, groundbreaking ideas if these ideas cannot be put into practice because they are technically unfeasible. Then, there is the consideration of where the design will be implemented. In the case of food and food processing, design is culturally specific to the region where the food processing appliance will be commercialized. A deep understanding of cultural influences is necessary to anticipate emerging or even future needs. Thus, home food preparation requires different types of equipment, depending upon the culture. For example, a family-sized American refrigerator may not fit into an average-sized Tokyo kitchen. There are disparities even in the country-to-country differences for adjacent countries. In Germany, coffee drinking requires different types of equipment than in Italy. And the list could go on.

As with many other consumer appliances, food processing appliances, it was often a case of "one-size-fits-all." The designer in the past had an easy situation, although certainly not the situation that would produce optimal new thinking. Currently, with the increasing affluence of world populations and with increasingly available technologies, designers have become increasingly sophisticated, in step with their customers. Of course, the idea for designs is still the global market and the "one-size," but more and more designers are recognizing that individual regions require tailored solutions.

Let us see how this search for specificity in local markets drives the design of the appliance. In the United States, for example, the coffee machine of the Senseo line accommodates large cups because that is what Americans expect. Another example is a mixer/grinder that Philips sells in India. The physical shape of this product, plus the motor and blades inside, has to be somewhat different than its European counterpart. In India, the machine will be used primarily to grind raw turmeric roots (dry grinding) and viscous batter (wet grinding). Moving to rice cookers marketed in Asia, we note that the products are designed in the Philips Design studio in Hong Kong, to ensure that the specific cultural use is taken into account.

With other studios in places like Amsterdam, India, Singapore, Hong Kong, Atlanta, Seattle, and Boston, the story is essentially the same around the world. Such examples are a perfect way to highlight the strengths of a global design capability. The requirement for the appliance design is to harmonize when appropriate, yet differentiate when necessary. It is particularly critical for customer-used kitchen appliances. It is this understanding of cultural differences that creates the Philips brand promise of "Sense and Simplicity," and its articulation in the specific brand pillar "Designed Around You." The design team acquires insights into the specific habits and lifestyles of the consumer, both in trendsetting urban centers, such as Shanghai or Moscow or Sao Paulo, and in specific regional environments, such as the Indian rural context.

Another crucial factor in the development of more relevant, human-focused appliances for the kitchen is the iterative nature of the design process. By constantly testing the concepts, mock-ups and prototypes with representative users and other stakeholders, the design group follows a step-by-step progression. The goal is to design a relevant, people-appealing, aesthetically pleasing product. The process of incremental refinement extends to the electromechanical performance and quality of the appliance, the resulting quality of prepared food, the usability in terms of pleasure and convenience, the aesthetic appeal, and the overall experience offered.

One of the key elements in this stepwise progression to the optimum is the use of what were once known as "application labs." Starting 50 years ago, at the Philips site in Groningen, the Netherlands, a facility was created to allow people to try out appliances in an authentically recreated domestic kitchen environment. By observing and questioning the participants, designers gleaned valuable information that could be fed back into the creation process. Today, this concept is still strongly in use, with application laboratories in Asia and Europe.

16.4   COMPETITIVE DIFFERENTIATION: A SHORT HISTORY OF FOOD PROCESSING APPLIANCES DESIGN

When designing and developing new products and services, an overriding consideration is that they should be knowledge-based and people-focused. How does that translate to action in Philips Design, and how does the process produce kitchen appliances that are clearly different from the competition, and ideally change the nature of the marketplace? The approach used in our company differs a bit from the new product development procedures used by food companies, primarily because we are dealing with people–machine/person–product interactions, rather than actual consumption of food. Therefore, some of the approaches that we present here might be "new news" to the food industry.

A grand vision of the business context comes first. It is essential to build a complete picture of your business context, starting from what happens in the local markets. Any design organization ought to immerse itself in what is going on. This could mean studying emerging movements around the globe, such as Slow Food or new trends like molecular cooking. Other factors may be biological in nature, such as the rise of obesity.

Experts play a big role in design because they know what is "going on," often more profoundly and far more early than research might discover. Yet, good up-front work does not end with experts. End users, the customers, are important. Thus, in the Philips Design program, the input comes from a wide variety of individuals, each of who provides information in the form of interviews that must be analyzed. For example, for a kitchen appliance it would not be unusual for Philips Design to assemble a great deal of one-on-one interview information from chefs, food producers, nutritional experts, health institutes, supermarket chains, and any others who can help with relevant input. Design, unlike research, does not look for the "right answer," but rather seeks the right pattern, the right guidance.

Once the research teams have identified and validated a trend and discovered the opportunity (or at least ascertained that an opportunity may exist), designers set about coming up with a fitting solution. As an example, the design exploration of the theme "healthy eating" was instrumental in the development of a new Philips line of juicer appliances, which tap into a latent desire for a convenient and no-fuss way of making nutritional shakes and soups.

Let us compare this approach to design in the late twentieth and early twenty-first centuries with the earlier approaches used decades ago. As we mentioned in the chapter introduction, the emphasis on putting the user first is a relatively recent phenomenon. Back in the 1960s–1980s, companies, including Philips, focused very much on the functionality of the appliances. Many new appliances were introduced to simplify the preparation of food or drink, such as coffee machines, electric knives, microwaves, and food processors. These essentially utilitarian products arrived on our worktops because technology enabled them to exist. There was no real notion of how the user would experience them. The thinking was basically that it would now be easier and quicker to make a decent cup of coffee or whip up a bowl of cream, and that was really all that mattered.

Over the decades a turning point was reached, however. The business thinking began to change in the 1990s. It was no longer sufficient to offer improvements solely on a technological level; they had to appeal on a more emotional level too. In other words, it was important to consider *how* a product would be used, *the experience* it provided for the user, and its *relevance* in the context of what the user wanted to get from the time spent in the kitchen.

This trend, perhaps best expressed as "experience," has continued in the new millennium. With celebrity "lifestyle chefs" like Jamie Oliver and Nigella Lawson bringing more color and panache to the art of food preparation and cooking, people give greater thought than ever before to the kind of appliances they want to use and to display in their kitchens. The lesson learned over time, through the decades, is that there should be a greater human—almost anthropological—dimension in appliances, something that enhances the moment when you are entertaining or cooking in a more fundamental manner than just simplifying the chopping of onions.

So what does this mean? Right now, in the early part of this new millennium, people, users of kitchen appliances, for example, want to have more control over the appliances they use. Furthermore, with the rise of "experience" as a key to consumer products, it becomes vital for the designer to envision the features that will lead to a sense of joy and delight when the appliance is used, when it is touched, and when it is shown to others, either deliberately or just in passing.

But how does all the above translate into concrete success? The next subsections will include three case histories demonstrating how the High Design process and its crucial people's insights translated into business success in the market.

16.4.1 CASE HISTORY 1: BREAKING NEW GROUND TOGETHER WITH ALESSI

One of the major milestones in the evolution of kitchen appliances came in September 1994, when Philips and Alessi introduced a groundbreaking new range of high-end electrical kitchen appliances. The four products—a toaster, citrus press, coffee-maker, and kettle—were light years away from anything else available at the time. With their rounded shapes and different colors, they set the trend for kitchen appliances for the rest of the decade (Figure 16.1).

FIGURE 16.1   Breaking new ground: The Philips–Alessi line.

The Alessi range of kitchen appliances marked a shift from rational to emotional, and had a cascade effect on the whole Philips portfolio. In fact, one might actually appreciate how the Philips–Alessi proposition had an enormous impact on the entire kitchen appliances market, generating or at least enhancing the fundamental switch from mechanical food processing to lifestyle appeal. Products like coffeemakers and kettles usually have a permanent spot on the worktop and elsewhere in the kitchen. Why should not they look good as well as provide the required functionality? And although the idea of giving them the appearance that they were all part of the same family seems so obvious now, at the time it was more or less unheard of. Different groups of products did share a superficial "form language," but were not designed to create a coherent range.

The Philips–Alessi line also broke the standard mold because the line allowed the owner to express something of his or her own individuality. We are accustomed to seeing this individuality expressed in one's clothing, furnishings and even cars, but it had never really been evident in the kitchen. By reflecting the cultural values of elegance and humanity, by communicating aesthetic and poetic qualities through friendly, intimate forms, the Philips–Alessi range perfectly captured the revival of interest in cooking, entertaining, and other gastronomic rituals that had started in the early 1990s.

We can go a little further, to the "story behind the story." With the Philips–Alessi line, we can analyze a creative breakthrough. When we deal with design leadership there is always the vision of a person behind the success, regardless of the process employed and how repeatable this process may be. In this case, the vision was shared among three minds. As much as High Design is proven to be the best backbone to Philips aesthetic leadership and business success, the Alessi project came about because, on one hand, Alberto Alessi himself was disgruntled by the homogenous nature of the metal kitchen utensils market, and was prepared to take a strategic move toward something new. Meanwhile, Philips Design CEO and Chief Creative Director Stefano Marzano, as well as Kees Bruinstroop, the head of Philips Domestic

Appliances (now part of Philips Consumer Lifestyle), were convinced that the market needed shake up. Those three men decided to take the lead. Philips wanted to add emotion and Alessi wanted to enter the electronics space for domestic kitchen utensils. And so was born a complementary partnership in the design of kitchen appliances for the expression of oneself.

While the birth of the Philips–Alessi line may seem to the reader like the result of intuition and "gut feelings" among visionaries, it is of course much more than just that. First of all, there was a trendsetting environment, with Alessi having developed some of the best techniques of new marketing of the 1990s and Philips relying on a long track of technological excellence and industrial competence. High Design offered the creative management framework that guided the design teams along the entire journey, from kickoff to the exclusive product launch in the Groninger Museum in the Netherlands. Alessi, Marzano, and Bruinstroop had an affinity to latent developments related to the kitchen and cooking, and could tap into a movement that was just beginning to manifest itself. They had enough confidence in their knowledge of what was going on to realize that they did not have to wait for a new trend to take shape, but indeed they could actually help create it.

16.4.2 CASE HISTORY 2: BILLY, THE BAR BLENDER WITH PERSONALITY

The widespread appeal of the Philips–Alessi line paved the way for kitchen appliances with more "personality." Next to follow in the line of design and commercial success came Billy, the bar blender, launched in 1997 (Figure 16.2). With its bold

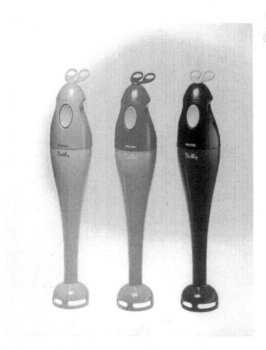

FIGURE 16.2 Billy, the bar blender, a product with personality.

combinations of strong colors, little plastic "ears" so it could be easily hung on the wall, and expressive language form, the bar blender gave a loud and proud style statement that took everyone by surprise. Billy outsold its predecessor by approximately a factor of 7, and the production went into overdrive to catch up on the back orders. This success was, once again, largely due to identifying and tapping into a consumer–market trend, in this case placing the emphasis on the personal touch.

16.4.3   Case History 3: A Juicer with a Difference

A more recent example is the Philips juicer, which appeared on the market in 2006. By carrying out future trends research, the Philips Design teams learned that personal health would continue to be extremely important to people, and that personal health would thus become important when making a decision regarding food and its preparation. The challenge was to translate this information into propositions brought to the market. One way Philips Design discovered to do this was through a new juicer designed according to the new sociocultural framework.

The Philips juicer range was conceived as a means of helping people to maintain a healthy lifestyle. Preparing one's own soup or drinking freshly made juices and smoothies made from fruits, raw vegetables, or a combination of both is a great way to get the daily portion of vitamins and fibers. Of course, this is nothing new; juicers, blenders, and fruit squeezers have been on the market for years. The difference with the Philips juicers is primarily their ease of use. The user simply tosses in entire pieces of a fruit, with skin, pips, seeds, and stems, and the juicer does the rest.

Again, this design was based on the insights which revealed that while many people would like to make fresh juice more often, they inevitably find the preparation messy. Furthermore, afterward the extended cleaning itself becomes an obstacle. The Philips juicer was designed to eliminate these problems. The results speak for themselves. From having approximately 7% of what was a very commoditized and saturated juicer market, Philips increased its share to over 25% in the space of a year.

Another interesting aspect of this juicer range is what we could call a "celebrity marketing" approach to the design and early seminal communication. Philips linked up with one of the most well-known names in the business, Jason Vale a.k.a. the *Juice Master.* Vale, who has written numerous books on the subject and is universally considered to be an expert, provided input on the development of the product and also helped with the product launch. No company can credibly claim to know everything about food and its production and preparation. It is, therefore, often advisable to team up with those who know more about a particular topic to develop meaningful products that have a greater chance of resonating with the general public. This approach can also be extended to include partnerships with companies who have complementary capabilities. Working in this way can lead to the creation of extremely successful joint propositions that simply would not have been possible when going alone.

16.4.4 CASE HISTORY 4: "HARDWARE PLUS CONSUMABLE"

There is more to contemporary design than product design. Systems thinking for new business models have been a growing area of interest for the creative industry in general. Philips Design is definitely no exception to this trend, having actually been a precursor to such approach in the early 1990s. The Philips Perfect Draft is a recent example of this "hardware plus consumable" combination, one of the new frontiers in rethinking about the food appliances business design.

The product is a draft beer tender for homes, which allows users to pour glass after glass of beer from 6 L kegs containing leading Dutch and Belgian brands such as Grolsch, Bavaria, Jupiler, Dommelsch, or Hertog Jan.

Another example of this business model is the aforementioned Senseo coffeemaker, created in cooperation with Douwe Egberts. The notion that you can empower people to make a top-quality cup of coffee at home simply by inserting a pad into a machine is something that has struck a chord in households worldwide. In the Netherlands alone (population approximately 16 million), over 5 million Senseos have been sold since its introduction in 2001. Senseo was a runaway success appealing to people's sense of craftsmanship, yet it took the fusion of Philips' product expertise and Douwe Egberts' long-standing involvement with coffee to make the idea a commercial success.

16.5 A VISIONARY PROGRAM FOCUSING ON THE FUTURE: CULINARY ART

While the Philips–Alessi range, juicer, Perfect Draft, and Senseo have all proved popular in the marketplace, Philips Design does not just limit its activities to developing ideas for new products. A considerable amount of background research takes place in order to build up knowledge and expertise. One such initiative was Culinary Art, a study carried out in 1998 on food and technology trends.

The aim of Culinary Art was to explore new ways to improve people's experience of food and mealtimes, of finding equilibrium between traditional, social, and personal values, and uncovering emerging opportunities afforded by new technologies. Stefano Marzano described it in terms of "rehumanizing the kitchen." "One of the early practical results of our research in this field was the Philips–Alessi line," he said. "We are now taking another step forward, by exploring in greater depth the more emotional side of cooking and bringing color and warmth back into the kitchen."

One proposal generated during the project—intended more as provocative explorations of possible directions rather than as definitive new product concepts—include a wine conditioner which reads the barcode on a wine bottle label and then automatically keeps the wine at the correct temperature (as does the special matching glass) (see Figure 16.3). A second proposal was an interactive tablecloth designed to provide cable-free power to all electric appliances on the table while remaining cool to the touch (see Figure 16.4). A third proposal was kitchen scales that also give detailed nutritional information about the food being weighed (see Figure 16.5), and so on.

FIGURE 16.3 The wine conditioner and glass maintain the temperature of the wine throughout the meal.

FIGURE 16.4 The tablecloth's integrated power circuit provides inductive, cable-free power to all appliances on the table.

With Culinary Art, it was clear that the emphasis is not on the performance of each individual appliance, but about the whole experience, according to true system thinking applied to design concept generation. The Philips Design Strategic Future teams identified a growing tendency to reinstate old rituals in relation to food, such as sharing an evening meal with friends or family, and talking about the day's events. It was an extremely valuable exercise that demonstrated what could be possible if you focused on more than just function.

FIGURE 16.5 The food analyzer weighs food as well as provides nutritional information.

Culinary Art is typical of the breed of major research program conducted by Philips Design because it is extremely wide ranging. Disciplines as varied as sociology, anthropology, trend analysis, materials science, technology, and design were called upon. Numerous experts in Europe and North America were consulted, including representatives and observers of food manufacturing, retail and service, kitchen designers, restaurant owners, food journalists, and leading figures in the fields of food iconography, food marketing, health, and nutrition.

16.6 CONCLUSIONS: THE DEMOCRACY OF DEVELOPMENT

Everything Philips Design professionals do—whether it is observing representative users, interviewing experts, or carrying out extensive trend research—should be seen as part of an almost "political vision of design" that reflects a representation of the future and an ambition for our world today. We might label it "design democracy." Finding the consensus in such a broad and often diverse range of input, from thought leader interviews to cocreative sessions with end users, gives Philips Design teams a much better chance of developing something that will suit many different types of users. Furthermore, when we talk about the user, we talk about more than the "simplicism" of technical roadmaps and aseptic functionalities alone. Rather, we talk about the experience of using the kitchen appliance.

Philips Design professionals, as must all designers, constantly keep people at the center of their focus, and always remain aware that food processing tools, like all lifestyle products, are essentially a means to an end. The "end" is a particular taste, a meal, a quick and convenient way of preparing food and cooking, instilling a feeling of being a good host, or comfortably entertaining friends and family. It is crucial for designers to know what the users are trying to achieve and will aspire to in the first place, in order to realistically create the new, and fulfill their ambition to make a difference to people in the world. Of course, in the context of this chapter, starting from people's kitchens.

17 Chef Formulation and Integration: Ensuring that Great Food and Food Science Work Together

Mario Valdovinos

CONTENTS

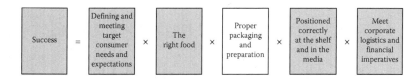

17.1   GENERAL INTRODUCTION: WHY I WROTE THIS CHAPTER

I wrote this chapter to explain the union of traditional food science fundamentals and the integration of culinary art into food product development. The chapter delineates the process from the chef's point of view. We start from customer insight, move to the creation of the chef "gold standard," and on to initial protocept, and then to production in the form of refrigerated and frozen foods. The chapter also provides a chef's viewpoint on the areas of general food product design, culinary formulation, as well as touching on the basics of ingredient functions, manufacturing product evaluation, food safety, and label regulation.

17.2   VIEW FROM THE INSIDE

17.2.1   ANOTHER INTRODUCTION: THE CHEF'S VIEW!

A chef's life usually involves a combination of long hours in the kitchen, guiding a series of people to help execute a food "vision," and relentless drive for precision and flawlessness in the face of, and in spite of challenging conditions. Every so often there is a little time for family and friends. When a chef gets a chance to change how someone eats in the evening, or in my case helping change how America eats for the better, the sacrifice and effort are truly gratifying.

The truth is food is one of the very few substances that touch all people in many ways. Food simply brings people together during moments in life. These are such life cycle situations such as making a business deal, a special occasion such as a wedding, a celebration for a recent graduate, a festive winter holiday and sadly, funerals.

Today's American chef can be found in many diverse settings. The chef's charisma and passion live in the kitchen, in many of the five star hotels in America, the mess halls of our armed forces, the teeming commissary of a university, amusement parks like Disneyland, and finally the neighborhood restaurant. Our food culture has been built around the heart and nerve of the culinarian, sometimes called a "foodie." The highly successful chef must be a multiple-talented individual. Since we are not cooking for simply our own enjoyment, the chef has evolved beyond creativity and must merge keen business sense with the innate desire to create sensory pleasure. The chef must go to a place where the existence of a business person begins.

Recipes are products that are derived by the "craft" of cooking and the persistence of creating a perfect pair of foods for the meal. The art of cooking comes down to respecting a craft that is so unusual, so intense that it can consume the life of a chef when lived properly. Indeed, it is a consuming passion that most people are only modestly aware of. In my case, the real test of "chefhood" starts when I am or my colleague chefs are placed in a culinary race, instructed to "make" a test sample of a desired food or beverage, and continue to race against time, not knowing where the finish line is, when it will be reached, and sometimes whether it will be reached at all. A recipe or sample may take up to 20 tries before it is accepted.

Food culture in America and its importance should never be underestimated. Many people give multiple hours a week in the foodservice industry and cook daily in the preparation of meals in the kitchens across our great country. These types of duties were just considered "jobs" for many years. For some time, however, a food movement has been building in this country. Now, with the demand for better food and the supply being as big as it is, the demand for expert culinarians has increased dramatically. The spectrum is broad and the likes of the manufacturer, the processor, the broker, the consultant, the chain restaurant, the cruise ship, the hotel, the hospital, the sports stadium, and many other channels benefit from the talents and creative nature of the corporate chef.

I have been a cook in some sort of fashion for half my life. No matter how things go or where my career has taken me, one thing stays constant; that love is my love for food. The passion for food, that runs through me, never changes. More importantly, neither does my appetite to understand more completely how I will be able to make food taste and look better. When it comes to cooking, I think the greater part of the chef community puts a lot of emphasis on the "wow" factor. The "wow" is not a singular element in the creation of incredible food. It is a combination of appropriate cooking techniques, proper execution, and the right ingredients. Those are all the parts to create the "wow" in food. On top of my patient, positive outlook on food, I am always pushing myself to help new product developer's progress into the new century. I share this passion with many of America's cooks, chefs, and R&D professionals. Some may say the test kitchen may lack the excitement of a restaurant kitchen. The knowledge I have gained, working with most experienced new product developers in retail and foodservice, is every bit exciting. I have asked myself many times: are "Test Kitchens" as important as "Food Labs?" Is being a new product developer all art or science?

It is easy to claim that both chefs and technical food developers are the dominating force behind a great-tasting, well-received prepared food. However, at the recent national Research Chef's Association meeting in Seattle, food professionals agreed that it takes both disciplines to make products that consumers find inviting, and perceive to have a great value.

In my restaurant life, I learned to make food taste great in the relatively few seconds it took to get it from a hot wood-fired oven onto my patrons' plate. There is no doubt that the satisfaction for a restaurant's chef occurs rapidly in that instant after preparing a dish in that intense moment, artfully arranging a plate before the server picks up and sets it in front of a guest. That harmony of components that seems like minutes, actually may have taken months prior to get the food right.

17.2.2 CHEF LOOKS AT THE DEVELOPMENT PROCESS

There are many approaches to new product development. In order to taste success, the testing process has to be well defined. Any "new food product development" process includes the overall process of strategy, organization, concept or idea generation, product and marketing plan creation and evaluation. The final step of the process is the commercialization of the new food product.

In the most general terms, there are two elements needed for new product development: a process and the people to participate in it. Figure 17.1 presents a generic new product development process model. The model is based on the analysis process by which food companies undergo for new food product development. I will look at it from the point of the creative chef, rather than the R&D product developer, the product testing specialist, or the brand manager in the marketing department. The chef looks at the world differently, even though we all may be talking about the same product, to be served to the same people, or packaged in a container.

Chefs work for both small and large companies. From my experience, one thing keeps coming out. Processes for new product development vary by company. A few processes may have similar steps, but the underlying function of the process is controlled by the specific culture of the company.

In general, the new product development process can be broken down into four general categories:

1. Idea generation
2. Concept development
3. Plan and design
4. Launch and produce

The foregoing four steps are, by now, well understood by companies. What is not understood, or perhaps not well recorded, is how a research chef sees these steps, interacts with others, and on a personal level how the chef feels about what goes on in the corporation.

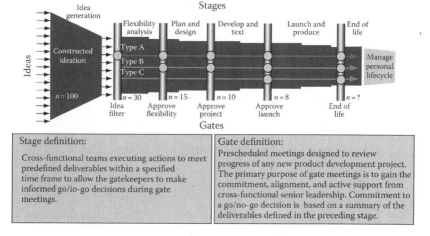

FIGURE 17.1   A prototypical development process. (Courtesy of Deloitte Consulting Group.)

17.2.3 Chef Looks at Corporate Processes

17.2.3.1 Idea Generation

Most American food companies and organizations must develop new products, and at times, product or process improvement ideas for existing product lines. The creation and development of sources for potential ideas is vital to the company's health. There are specific sources that generate great ideas. In larger companies, they may include market research, focus groups, third-party consultants or vendor partners, in field sales teams, technical R&D groups, competitive intelligence, business unit brainstorm sessions, team member suggestion boxes, and internal culinary teams, etc.

When representing the food product design world, culinary R&D teams are great experimenters to work out "40,000 foot" ideas and on the creation of gold standard development. We are creativity hubs with a unique perspective on the emerging trends and broader adoption of food news. In some cases, the R&D chef does not have the appropriate skill-set needed for all the engineering-type tasks that are demanded by that food product development. An example of this would be the chef who has little or no practical experience in commercializing new products. This does not mean that such a chef should not have input—that would stifle creativity and innovation—but at the later stages of development there must be a realistic consideration of attainability by someone familiar with and skilled in the area of commercialization, etc.

Larger food companies face many challenges in the continuous idea creation phase. One is the ability to reduce the large pool of thoughts and platform of ideas. Many food companies have lots of unused ideas. Some may choose to store ideas in a database and revisit them at a later date. Or in some cases, a company may choose to discard them. Food companies should not be replicating new product project work; different products driven by need-states need different development approaches. The food company needs to reflect on what really matters for its food development agenda. Continuous monitoring and evaluation of ideas have to be an ongoing effort. I feel this process of reflection should actually be built into the new product development process. Any new food product development's funding can be scaled back by 5% or 10%, and that time and/or money can be used to think and reflect more closely about the impact of those large pool of ideas.

Over time, I have developed a strong belief that there should be more tolerance within the process to risk-taking. New food product failures need to be built into the learning process. I feel both risk-taking and reflection are necessary to extract great food ideas. This may even change a company's culture of handling a large reservoir of idea platforms. Risk can also be mitigated through small pilot projects. Reflection is necessary, and simply needs time and encouragement.

17.2.3.2 Concept Development

Once the sorted and feasible ideas are selected for additional consideration, they must be further developed, scrutinized, and prioritized before the select few concepts proceed to the prototype phase. The prototype phase requires a more prescribed review and a thorough planning process. This process ensures the feasible concepts are technically viable and will make a commercially executable and profitable product. During this phase, a scan of ingredients and equipment must be performed by a knowledgeable professional.

Here is where the chef can shine. Product viability assessments can be considered a key aspect of this phase. This phase can be performed in the test kitchen or food laboratory, and it provides the initial arena for creativity. It is from this phase that an enthusiastic chef can bring creativity to life. From here comes imaginative and innovative new ideas. This step usually is performed throughout the front end of the new food product development process, from ideas to prototype.

Personal experience is also important, and perhaps even critical. A lot of what goes into my ideas and concepts comes from what I want when I go out to eat or when grocery shopping, and I cannot find a solution for my home food experiences. Once people discover that I am a chef, they enjoy telling me about their interaction with food. I listen to them as they tell me about how they would like a certain flavor or form produced, so, I often bring that insight to this phase. For me, live research also plays a big role here. Eating out and scanning cookbooks, lifestyle magazines, and trend-right food news help me create initial concepts.

Let us look at an example, specifically a foodservice item for a restaurant chain. The chef's product creations must go to the laboratory for development and scale-up. All the many technical issues for scale-up do not have to be worked out in this phase. From here, and after multiple samples have been generated, the final product must go into the production environment and return back to the kitchen, for operational validation. Every step of the sample creation process must travel down the same course meeting a delicious, visually appealing, and cost-effective product. Many issues and changes may have to be addressed. These issues may require specific equipment or functional ingredients. Once completed, the new food product still has to "work" in a real kitchen situation, in the home of a consumer, or in a foodservice operation.

It is here that the chef and the food technologist really interact. My past successes in developing a great tasting well-received foodservice product, modeled on successful new "limited time offers," proved to me again and again that one truly needs both viewpoints—the chef for design and creation, and the food technologist for scalability, economics, and "reality." One thing stands true, however, the chef's point of view comes first and last, always with attention paid to the technologist's science.

17.2.3.3 Plan and Design

During the planning and design phase of a new food product, an ideal cross-functional development team comprises research chefs, food scientists, packaging specialists, process engineers, nutritionists, sensory scientists, and as well as fundamental scientists such as protein scientists, rheologists, microbiologists and, of course, flavor chemists.

During this phase a few things occur. Typically, the construction of a formal strategic brief or business plan begins with initial forecasts. These forecasts may be derived by a BASES test, as well as from a clearer and more defined product description, along with the first stages of establishing an initial product specification. One of many responsibilities of the marketing group is to build a project timeline and a preliminary marketing strategy. If there are limited resources, then a reallocation may be necessary to produce a new food product. While predictability of success for new products is highly desired, never lose sight for the need for "speed to market." A rigorously researched new product that reaches the market, after the consumer has moved on to other needs is worse than never introducing the product at all.

Other activities may include an executive review. During the review, additional resource allocation may be established and/or a general approval for the product may be given to design. This review stage involves turning the concept into a real-life product. During the design process a working prototype or protocept is generally used as a bench mark, sometimes called the "gold standard." Product eating quality and raw material changes should be monitored to determine whether the quality of the product has changed, and whether or not customer requirements might be better met with product changes.

If a change is identified and vital to fulfilling customer needs and expectations, then the change should be completed in this design phase. A few raw materials may be specified if the supply is short. Such specification allows the design team to test and verify the ability of the product to deliver the original performance expectations and eating quality.

The basic building blocks of food development should be considered during this design phase. However, the more important action is to use great cooking techniques and high-quality ingredients that elevate normal food to something visually extraordinary and great tasting. At this point in the process, the chef can become personally attached to an idea. It is important to recognize this, and to be open-minded to market realities that may emerge and bring into question the viability of the project. At the same time, the chef can also become the informal champion for a new food idea that he or she feels passionate about.

Passion is important. Research is an integral component of the information to be considered, it is not, however, enough for making a decision regarding new products— although, of course, research is an important component of the information to be considered. The research chef should never become bull-headed or unrealistic when supporting a particular project, but at the same time, the chef must be persuasive when he or she believes strongly in the opportunity.

17.2.3.4 Launch and Produce

Once the final food item is designed and verified through a series of product cuttings, the product is launched into a real production environment. This step is known as the scale-up period. This has to be verified before full production can commence. Most of the individuals involved from the initial concept work are needed as functional parts to ensure a smooth transition between concept design and food manufacturing.

Multiple pieces are at play here and are required before production can begin. In my experience, these may require the official production plan including all product requirements and specifications. A detailed HACCP plan and a quality control diagram/outline are needed in order to show that the food conforms to the original culinary specifications and customer needs.

17.3 FOOD FORMULATION

When formulating foods a few considerations should be made. One, is the rapidly changing diets and tastes of the American palate. Moreover, there are many more choices that consumers demand and crave. A new-product developer should focus on what consumers want and how consumers use food products. Fact-based information is necessary to fulfill attitudes and barriers.

The surge of health and wellness claims in food has influenced how America eats. This movement also changes how consumers want to see food. Packaging plays a crucial role in food development. Garnishes and serving suggestions are also key in a good food experience. For example, it is difficult to incorporate freshly chopped herbs in a frozen food dish and yet, maintaining the sensory integrity. Aroma and texture vary once the herb is frozen, so another delivery system for chopped herbs may be in the form of a sauce, compound butter, or an individually quick frozen (IQF, herb system). The immediate product visual allows people to get exactly what they need. Dedication to that focus and attention to small details are what make the food better. All this shows that a new food product is best designed by a well-educated, well-trained team of food scientists and culinary chefs.

17.4   TYPES OF FOODS AND THEIR RELATION TO THE PRODUCT DEVELOPMENT PROCESS

There are many types of food products that sit on your local grocer's shelf, in the freezer of your favorite chain restaurant and maybe in your refrigerator. They range from dry goods, to frozen pizza, to shelf-stable pasta sauce, to double glazed chicken wings. These foods require different levels of experience for development. They also require a series of proper food equipment to replicate a finished-plate experience. Sometimes a chef, product developer, or food scientist can identify how these foods are produced. Some approaches to development may be modeled on an ingredient basis. Others may be modeled on the execution of a complex dish or menu item with a wide variety of preparation and cooking platforms.

A well-equipped laboratory or test kitchen is necessary for the food developer who will manufacture afterward. Often, most of these products follow an assembly line-like flow. They start as an idea, go to R&D, and then to a food laboratory, on to industrial engineering and production planning and so forth. A good marketing group inserts itself at various stages during the flow. Unfortunately, all too often, like a seashell lost in the sand of a beach, the original product is shaped, refined, and distinguished until, and finally, it no longer resembles the gold standard originally created by the chef. In many cases, this has happened to all great food companies. Perhaps, there is a better way to balance the art and science of food development and make both disciplines equally important.

Regardless of the many faces of a product, one consideration rises to the top. That is, of course, the opinion of the customer/consumer. The final end user makes what is truly the final judgment—will the new product succeed? Like the case with all products, the consumer decides. I do not think a consumer would ever stop purchasing a food item because it tasted "too good," or because it looked "too pleasing." The goal in developing any new food product is to prepare and present a product that consistently delivers satisfaction to the consumer, one that is convenient, safe and wholesome.

The goal is a food product designed to taste great and offered at a cost that consumers are willing to pay.

17.5 CONSIDERATIONS

17.5.1 COMPONENT OR BATCH COOKING OF FROZEN FOODS

Many manufacturers have numerous ways of organizing food production. One of these methods is called a batch production process. This is when, instead of manufacturing things singly or by continuous production, the items are manufactured in batches.

A specific process for each item takes place at the same time with a batch of items. That batch does not move onto the next stage of production or inspection until the whole batch is done. For example, in some cooked facilities, chicken breasts are oven baked in batches. You first must trim your chicken, make your marinade, and then tumble the meat with the solution. After those steps are complete, you then can start oven cooking, chilling, freezing, and packaging. Production facilities are limited on how many filets can be produced at one time by the number of tumblers and ovens that are available.

There are some culinary limitations as to what large food processors can do in big facilities. For instance, they cannot perform some of the flavor-enhancing cooking procedures that can be done in traditional foodservice kitchens. This is a batch production process, since a facility cooks a large number of chicken filets at the same time. Here you simply cannot skip from one process to the next until each process is complete. One way out of the dilemma is to use high-quality commercially prepared ingredients as a starting point. They enable you to offer more variety, while saving time and money.

17.5.2 FURTHER PROCESS AND VALUE-ADDED MEATS

When developing any value-added product, the product developer should fully utilize technology and creativity to efficiently produce a restaurant quality, value-added product that meets today's consumer demand. Some processing techniques can be simulated well in a facility similar to a foodservice environment, including curing, brining, smoking, sausage making, old world style butchering, and drying meat.

Processed meats, such as franks, luncheon meat, and fermented products like pepperoni take longer to test, not because of the sensory evaluation process, but due to the real time it may take to produce a sample. Here a chef may not necessarily understand the technical process, but can provide visual and flavor guidance to the food technologist.

The evaluation of sensory aspects should always include the testing of individual differences in flavor and odor perception potentially associated with differences in food and flavor or aroma. Also, perceptual models and mapping of flavors in further processed meats is important because of the amount of time that a food item may be kept in a freezer. Some foods attract odors and may pick up some off flavors associated with oxidation. Key ingredients may mask these types of unpleasant flavors.

17.5.3 TASTE OF FROZEN FOODS: FLAVOR EFFECTS OF FREEZING

In my experience, food technologists and scientists often do not have the same artistic skills or culinary expertise as the chef does. Yet, both have the same objective. They both want to make great food. In the science world, the discipline is a much more serious-minded task. The new food product must be finished in such a way that it can be distributed by a broadliner truck, locally and nationally.

The food item needs to be commercialized for production at volumes that would cause local restaurant or commissary kitchen to stop functioning. Regardless of role, both face very real concerns. There is no question that frozen food is chemically and sensorially perishable. Because of the nature of large-scale manufacturing, it often is not possible to achieve the fresh made-to-order meals in the same way the local restaurant may prepare food. Solving for the situation often involves some compromises. The ones for consideration are the use of boutique ingredients. Like any foodservice operation, there has to be a profit made. However, unlike a gourmet restaurant, final product shelf price often is a far more serious consideration. The price of ingredients is especially important when large quantities of specialty ingredients have to be purchased. An everyday price has to be established, attractive to the buyer, and reasonable to the seller.

17.5.4 EXTENDED SHELF-LIFE VS. SHORT SHELF-LIFE REFRIGERATED FOODS

If you ask any microbiologist about the behavior of a refrigerated food product, one thing is certain, microbiological organisms love refrigerated food as much as we do! In this case, the food product designer has to develop the products in a stable format appropriate for distribution and marketing. For example, an inhibitor is necessary to satisfy the USDA regulation on fully cooked foods for the prevention of Listeria.

There are many of these antimicrobial agents out in the market place. However, most of these agents significantly alter the flavor of food. The challenge is determining the correct percentage level for an effective kill, while at a level that a consumer cannot detect. This is one area that I would rather use a postpasteurization technology to maintain the original integrity of the food item, rather than introduce a foreign flavor to a formulation. There are some cases, in which only the inhibitor will work, such as in applications that involve highly seasoned fresh meat.

17.6 EFFECTS OF PACKAGING: VACUUM PACK

There are hundreds of companies that have used vacuum sealing equipment for just about every food product on the grocer's shelf. Types of products range from coffee beans to fresh steaks. The vacuum sealing process has clearly benefited the food industry products that need an extended shelf life. It is widely known that food products can greatly benefit from a controlled environment. In an ambient environment, a product can decay quickly. Products like coffee grounds become stale. Other foods in the presence of moisture lose their shelf life.

There are other secondary benefits of vacuum packaging food. Some food items can shift aggressively during the shipment process and cause the products to be damaged. Vacuum packaging secures the product in place, preventing any movement. When designing food, packaging plays a very important role in the preservation or presentation. Simply stated, the package should fit the type of product that is developed. If the product is for foodservice, then a bulk package is necessary to ensure the food and package are suitable for storage, handling, preparation, and serving in the foodservice environment. If a retail product is designed, then a single serve package maybe the obvious choice.

17.7   COMMERCIALIZATION PROCESS: KNOW YOUR CAPABILITY

A piece of advice I can give to any product development team is the need to understand the company's capabilities in terms of the processes involved in food production. Such knowledge ensures that cooking technique, quality, and safety parameters are identified and can be measured. A pilot plant facility is usually required to work out the difficulties of scale-up for production.

The culinary test kitchen augments product development and allows new food product developers to test the capacity and limits of the products in a "home-cooking" environment. Small-scale equipment such as ovens, marinators, and blast freezers, along with a pilot facility mimic production, and help scale-up products. Precise analytical instruments are used to measure critical parameters for the flavor specification (such as refractive index, specific gravity, moisture percentage, acidity levels, salt, etc.).

Specialized skills and equipment are vital for identifying the critical points of the process for scale-up, not only for the value-added meat industry but also to ensure the safety of any final cooked product. The critical control points could include monitoring the process time, essential temperatures (storage as well as processing or shipping), moisture, metal detection, foreign object control, and sanitation validation—all of which require the support of a laboratory and accurately calibrated equipment. This level of sophisticated new product development would be impossible using a chef's kitchen alone or noncommercial commissary.

17.8   COOKING TECHNIQUES AND REPLICATION
OF FLAVORS IN FOODS

Some say refrigerated food is like eating leftover. The truth is that to develop a refrigerated food product is at times more difficult than a frozen food item because of the acceleration in shelf life. Flavors are difficult to replicate due to the fact that microbes eat away some of the sugars and proteins in food, while at the same time creating their own by-product and therefore, creating competing flavors.

For an example of difficult flavor replication, consider the simple tomato-based pasta sauce. Whether looking in a cookbook or replicating a handed down family recipe, it is safe to say that almost every recipe starts with a "sauté" step. This step includes the use of fresh onions and garlic in cooking oil. This is virtually impossible to do in a manufacturing facility. Though some processes may get to a "sweating" step, high heat caramelization is difficult to replicate in the way it occurs in a sauté pan. However, the use of a "sautéed" onion flavor or caramelized flavor to give it a homemade taste maybe the best route. There are some value-added vegetable manufacturers that offer fire-roasted garlic and onions that come close, but may not be identical to those produced in a foodservice kitchen.

17.9 REPLICATING FLAVORS CONSUMERS CRAVE: THE EXAMPLE OF ROTISSERIE CHICKEN

In the late 1980s, oven-roasted chicken was a big hit in the retail segment. Many deli operators made investments in large rotisserie ovens which provided a small "food theatre." As the demand grew, this opportunity evolved into filling the need to provide a part of a meal solution. Home consumers were shifting away from making whole chickens from scratch which required time and many ingredients, and toward visiting the local grocer less frequently. The need then became to purchase prepared chicken with longer shelf life. Afterward, by adding few strategic ingredients—like olive oil, fresh garlic, fresh lemon zest, and herbs—consumers could then "make a meal" in almost no time, replacing the need to buy a hot prepared chicken. One of the great advantages of a manufactured roasted chicken with a commercially prepared approach to food production is the consistency of the finished product. By starting with the same prepared products, home consumers were assured of a consistent taste. In turn, that reliability provided its own consistency on meals that were served. The American appetite is still strong for the taste of "rotisserie".

17.10 SATISFYING THE TASTES OF MORE THAN JUST ONE

After the preamble of the brainstorm session, an idea has to be written out in the form of a concept. A series of tests can be conducted asking consumers which of the culinary concepts they are likely to purchase. Once the results are tallied and the goal concept is selected, such as teriyaki flavored pork tenderloin that is ready to cook in peel-away film, the research chef should find several recipe variations from cookbooks, restaurants, and online resources. The idea is to gather a range of possible flavors. From this point, 4–6 samples should be created to try, and for which to get feedback. After the submission of samples, tasting, and modifications are produced, the cycles occurs again until a final version emerges.

Key steps focus on certain effects of food, one at a time. One frustration that may occur is coping with the speed of progress. New product development is not linear. Advancement may not be a step-by-step process. Often a chef has to reverse track and take highlights from previous versions, and try to balance them with new variations. For example, the sequence might begin with the raw material, then move on to say a desired color in a raw state, then develop to a salt level that the consumer can trust.

Continuing, on to appearance which may include nice particulate definition, and then finally to a desired length of cook time. Once all these steps are completed, a validation step is necessary to determine whether or not the finished products flavor profile and appearance are acceptable to more than 60% of the targeted consumer group.

17.11   FOOD RETHERMALIZATION AND HOW THE CONSUMER TREATS YOUR PRODUCT

One assumption here is that both consumers and foodservice operators follow the instructions provided with the food. In my experience of formulating a recipe, most cooks do not follow the rules. It is important that we understand how people will treat food. On most occasions cooks substitute ingredients, take shortcuts on recommended cook times, and may not have the cooking equipment that was used in the initial product development phase. Therefore, it is safe to assume that overcooking food products is common. For any product developer, the desired outcome should be to create a high-quality product that can be replicated consistently in manufacturing, in light of the expected, standard abuse both by the consumer and by the foodservice operator. The food should have staying power on the shelf. The food should attempt to be dramatically different than anything else in its category. This is important when creating true classic food in a restaurant environment vs. "inspired by" flavors and forms that are sold in the freezer or refrigerated case.

17.12   FUTURE POSSIBILITIES OF COMMERCIAL FOOD DEVELOPMENT

Specific tools and activities are used throughout the new product development process in order to enable organizations to better understand the markets they wish to enter, the products they wish to produce, and the means they need to develop the finished goods. Food has so many aspects and dimensions that the real art of creating a commercially viable food product is to deeply understand the food, packaging, and the consumer. Food changes, as the consumer and technology evolve.

Many organizations use documentation tools and methods that capture new product development information. These tools have a variety of names and activities. Key measures are important elements of any new product development process. They enable practitioners to make decisions about the viability of new food product ideas, monitor those new ideas, that have been selected to enter the new food product development process, and make necessary adjustments to a food product development effort as it moves through the process. Such metrics and activities help create the success or failure of the food product after development. These metrics cover topics such as risk, corporate strategy, market penetration, technology, production, finance, cost, and time.

To end this chapter, let me give a piece of advice. I think that some food products and solutions can be ahead of their time. Companies should periodically revisit old ideas. An idea rejected yesterday may have not been fully fleshed out, and may not have incorporated the voice of the consumer or fit into a specific need-state "at that time." Companies should revisit these ideas at some stage in the product development process to determine whether this cast-off idea now has an opportunity to fit into a food product for tomorrow's lifestyle. The pursuit of creating a food item that people do not even know whether they want yet, by extrapolating old ideas and engaging habits and practices is an activity, and even a major opportunity for future product developers to consider.

17.13   SUMMING UP

The fusion of food science and culinary development can have a positive effect on New Food Product Development if together both potentials and limitations of the disciplines are understood. The food laboratory vs. test kitchen is not art or science, rather Culinology. It is tempting in food development to claim that either chefs or food scientists are the dominating force behind a good-tasting prepared food "hit." However, today's food professionals are in agreement that it takes both specialties to make a product that consumers find appealing.

18 Gastronomic Engineering

Jose M. Aguilera

CONTENTS

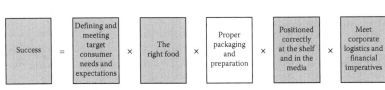

$$\text{Success} = \begin{array}{c}\text{Defining and}\\\text{meeting}\\\text{target}\\\text{consumer}\\\text{needs and}\\\text{expectations}\end{array} \times \begin{array}{c}\text{The}\\\text{right food}\end{array} \times \begin{array}{c}\text{Proper}\\\text{packaging}\\\text{and}\\\text{preparation}\end{array} \times \begin{array}{c}\text{Positioned}\\\text{correctly}\\\text{at the shelf}\\\text{and in the}\\\text{media}\end{array} \times \begin{array}{c}\text{Meet}\\\text{corporate}\\\text{logistics and}\\\text{financial}\\\text{imperatives}\end{array}$$

18.1 INTRODUCTION—FACTS ABOUT THE RESTAURANT INDUSTRY

Higher incomes, increased urbanization, and less time for food preparation cause consumers who live in modern societies to spend an increasing percentage of their food expenditures eating away from home. This number is reaching almost 30% in the United Kingdom, and 42% in the United States. According to *Forbes* magazine, worldwide annual sales of the restaurant industry are estimated to be around $1.5 to $2 trillion. In the United States alone, the restaurant industry sales were estimated to have reached $537 billion by 2007. In the period 2000–2020, per capita expenditures in restaurants are expected to grow by 18%, while those in fast-food outlets only by 6%. Furthermore, there is an emotional component to this growth. A sizable

	Restaurant	Chef	Country
1	elBulli	Ferrán Adrià	Spain
2	The Fat Duck	Heston Blumenthal	United Kingdom
3	Pierre Gagnaire	Pierre Gagnaire	France
4	Mugaritz	Andoni Luis Aduriz	Spain
5	The French Laundry	Thomas Keller	USA
6	Per Se	Jonathan Benno	USA
7	Bras	Michel and Sébastien Bras	France
8	Arzak	Juan Mari and Elena Arzak	Spain
9	Tetsuya's	Tetsuya Wakuda	Australia
10	Noma	René Redzepi	Denmark

FIGURE 18.1 Top 10 restaurants in the world and famous chefs. (Adapted from The S. Pellegrino World's 50 Best Restaurants, *Restaurant Magazine*, 2008, http://www.theworlds-50best.com/2008_list.html, accessed July 7, 2008.)

proportion of consumers—almost 40% in the United States—prefer restaurants for an enjoyable dining experience (Stewart et al., 2006).

Gastronomy and the industry of fine dining are becoming major drivers for food innovation. Top chefs, often the owners of high-quality restaurants, are evolving into major protagonists of a paradigm shift in modern cuisine characterized by the increased use of scientific knowledge for the development of new dishes (Vega and Ubbink, 2008). Several of these high-ranked chefs are celebrities as well as multimillionaires, their empires are based on the restaurants, cookbook sales, endorsement of product lines, and TV performances (Figure 18.1). According to a salary survey done in 2005 by http://www.starchefs.com, 15% of executive chefs in fine dining establishments in America earned annually over $100,000.

Enthusiasm for fine dining does not stop with celebrities. According to Cowen (2006), expenditures in the highest-quality restaurants in Paris (i.e., Michelin dining guide categories from "very comfortable" to "luxury") tripled between 1950 and 2005, adjusted for inflation. Today, a meal in a Michelin three-star restaurant in Paris can cost $300 or $400 a person, not including wine. By contrast, the cost of eating in nonluxury restaurants became cheaper in period. The economic consequences of this trend has led to the formation of The Society of Quantitative Gastronomy (http://www.gastronometrica.org) described as "a scientific organization where mainly economists, econometricians, and managers bring their professional know-how and knowledge to gastronomy."

18.2 GASTRONOMY: ART OF COOKING AND PLEASURE OF EATING

La cuisine est le plus ancien des arts

Antoine Brillat Savarin, 1840.

The *Larousse Gastronomique* defines gastronomy as "… the art of good eating." In France, the word gastronomy came in general use around 1800 and was accepted by the *Academie Francaise* only in 1835 (Montagne, 2003). Another definition of gastronomy is "... the art or science of good eating" (Gillespie, 2001). While gastronomers are referees of taste and define what may be considered gastronomic, a gourmet is someone interested in and has learned the connoisseur of table delicacies (Gillespie, 2001).

Traditional gastronomy is based on recipes passed from one generation to the next. These recipes embody all the information needed to construct gastronomic dishes, albeit in an imprecise way. According to This (2005), recipes comprise two parts, the first having to do with the ingredients and the second with the succession of processes and devices necessary to achieve the expected results. While ingredients are more or less well defined (but now in constant evolution), it is the process that needs "precision." Recipe nomenclature such as "stir slowly," "bake with oven door open," "add drop by drop," and so forth often represent 80% of the text. The steps in preparation are crucial to the success or failure of the preparation. In a positive way, this vagueness permits experimentation by cooks and is a source of variability, creativity, and differentiation.

The pleasure of eating is largely a perception of breakdown of food structures in the mouth. Enjoyment of a meal would not be the same if its contents were reduced to purees, even though all chemical components would be the same. Indeed, some foods such as meat, corn flakes, and cucumber are actually not recognized by most consumers when their structure is obliterated into purees, with flavor becoming the only attribute for identification (Bourne, 2002). Although it is proverbial that "food enters by the eyes," most desirable structural attributes of foods are at a level beyond the resolution of our vision (approximately 100 µm). Texture perception and flavor release, among other properties, depend on food microstructures formed by nature or created during processing and cooking (Aguilera, 2005).

18.2.1 ORIGINS OF SCIENTIFIC CUISINE

When it comes to the art of cooking, great chefs in the late 1980s must have felt like painters at the end of the nineteenth century. These artists realized that reproductions of the common life by the masters of the Renaissance could not be surpassed, perhaps only imitated or duplicated. Precise imaging of the visual world, the new technology eventually to be called photography, was becoming popular. Painters at that time turned to explorations with colors, shapes, and subjects adopting new materials and techniques. Almost in the same period, architects discovered that their trade and engineering need not remain estranged from each other but on the contrary, both could benefit by using new materials and techniques to built astounding structures, for example, skyscrapers (Gombrich, 1995).

A similar evolution occurred in the 1980s in the field of *haute cuisine*. Local ingredients and traditional cooking had been exploited to their full extent for a couple of centuries since the restaurant became established in Paris around 1765. The ambiguity of most recipes alluded before provided the opportunity for experimentation and change. Some of the prefixes ascribed to this modern type of cuisine are fusion, nouvelle, and author's, hyper, experimental, and techno-emotional. All these terms attempt to capture the efforts by modern chefs to break up with the rigidity imposed by traditional ingredients, cooking methods, and tastes of classical gastronomy. Like architects and engineers in the past, chefs and scientists have realized the enormous potential of combining creativity and scientific knowledge to develop new foods and new sensations. In this context, the book of McGee (1984) became a classic in describing the scientific understanding of science-based cooking.

The term "molecular gastronomy" was coined by Herve This and the physicist Nicolas Kurti in 1988 to refer to the "scientific" aspects of culinary transformations and the sensory phenomena associated with eating (This, 2006). According to them, molecular gastronomy is not molecular cooking, because cooking is a craft or an art, and not a science. They argued that scientific knowledge was needed to explain why some recipes work and some did not, to give chefs and cooks (including amateurs cooks) a basis for innovation, and to convey personal imprints in the foods they prepared. Although the term molecular was very fashionable in the 1990s, some chefs were not at ease in giving such a chemical meaning to a science associated with gastronomy. The Research Chefs Association (RCA) in the United States introduced the term "Culinology" (now a registered trademark) to describe "… the blending of culinary arts and the science of food."

18.2.2 GASTRONOMIC ENGINEERING

Only in the last 20 years has the study of foods as materials become a field in its own. This maturation has been fostered by integrating progress in related areas, most notably, polymer science, colloidal science, mesoscopic physics, microscopy, and other advanced instrumental techniques, and by applying it to food structuring (Aguilera and Lillford, 2008).

The battlefield of definitions and scopes in modern gastronomy is expanding. Food material scientists have created the term gastronomic engineering (GE) which is "… applying concepts of food materials science as well as methods and tools of food engineering to guide the physical and chemical transformations of culinary practice and design novel structures that, according to chefs, are uniquely tasty and palatable" (Aguilera and Lillford, 2008). It is clear that chefs are central in the process of creating food structures. They actively participate in the generation of ideas, process development, and final applications. Examples of novel engineered structures already established in modern cuisine are edible films, the airs or three-dimensional sauces, gel beads, and textures derived from the use of cryogenic freezing, just to name a few.

GE also encompasses the principles of engineering that lead to the formation of food structures. Cooking is largely transferring energy to heat a food so it changes in texture and flavor, becoming edible. Broiling, boiling, baking, steaming, and frying are the terms used by cooks that describe different methods to affect heat transfer. Cutting, dicing, grinding, and mashing are all size reduction operations that are well defined in engineering texts. Stirring, shearing, and mixing are based on the transfer of momentum to liquids or particulate materials, and so on. Heat and mass transfer concepts as well as the physical chemistry behind food processing and microstructural changes in foods are presented in the book by Aguilera and Stanley (1999).

18.2.3 Ingredients for Structure Design

GE rapidly and efficiently takes advantage of the ingredients that are new to the gastronomy trade (even though they have been around the food industry for many years) and of novel raw materials and ingredients that have been consumed by ancient cultures (e.g., the Inca's cereal quinoa, edible flowers, berries, etc.).

For example, while traditional gastronomy for the most part used one thickener (starch) and one gelling agent (gelatin), today's chefs use a myriad of hydrocolloids and refined proteins in order to control the viscosity and mouth feel of their dishes. Commercial kits containing standardized thickeners and gelling agents for culinary applications based on hydrocolloids are already in the market (e.g., http://www.texturepro.de). Among these different products we find

1. Agar agar: A polysaccharide extracted from the algae *Gracilaria*, traditionally used in Japan (known as *kanten*) for molded jellies and in microbiology laboratories. Hot solutions at very low concentrations form strong gels when cooled to room temperature.
2. Carrageenans (kappa, iota, and lambda): A family of linear sulfated polysaccharides extracted from red seaweeds (e.g., *Chondrus crispus*) that have been used as food additives for hundreds of years. Used as a thickening, stabilizing, or emulsifying agent in dairy products, reprocessed meats, and to make puddings, salad dressings, etc.
3. Egg white powder: Spray dried egg whites that can be used in the same preparations as regular egg whites to achieve more concentrated flavors, new textures, and foams.
4. Gellan gum: A bacterial exopolysaccharide that may form hard and brittle gels that crumble in the mouth, giving a "melting in the mouth" sensation.
5. Guar gum: Primarily, the ground endosperm of the seeds from *Cyamopsis tetragonolobus* (L.) Taub. (Fam. *Leguminosae*) mainly consisting of high molecular weight (50,000–8,000,000) polysaccharides composed of galactomannans.
6. Methyl celluloses: Water-soluble cellulose ethers derived from cellulose. They have been used as binders, emulsifiers, stabilizers, suspension agents, protective colloids, thickeners, and film-forming agents for many years. Form gels upon heating and return to solutions after cooling.
7. Sodium alginate: A seaweed extract from the giant kelp *Macrocystis pyrifera*. It is a cold gelling agent in the presence of calcium (i.e., a solution of calcium chloride), and is used to make artificial caviar ("spherification") and other gels.
8. Whey protein isolate: Manufactured from sweet dairy whey and spray dried. Used as a protein source and to form gels by heating.
9. Xanthan gum: Polysaccharide gum made by a fermentation process using *Xanthomonas campestris*, widely used as a natural thickener and emulsifier, and a substitute for eggs and gluten. It reduces the thinning of liquid films in foams.

As food scientists will recognize, some of these ingredients have been used for centuries. Their traditional applications are well established and most of them have been amply researched by academia and the food industry. The point is that their adoption by chefs has been fast and steady. An excellent practical example of converting ideas from the chef's mind into actual products using physicochemical principles and some of these ingredients is presented by Arboleya et al. (2008). A bubbly beetroot juice (thin foam) was stabilized with a mixture of egg white powder and xanthan gum in order to trap the aroma of the food. An edible film was obtained by the casting method, using a solution of gelatin extracted from cod skins. The idea was to fool the sense of vision and taste by covering the pieces of white fish with an artificial skin.

18.2.4   FOOD STRUCTURING: KNOWLEDGE OF MOLECULES TO PRODUCTS

Our understanding of how food structures are built from the molecular level to the macrolevel has advanced steadily in recent years. For a more detailed description of the science behind food structuring, the reader is referred to the book published by Aguilera and Lillford (2008). It is a characteristic of foods that several structure-building phenomena occur simultaneously at different length scales from the molecular level to the macrodomain (Figure 18.2). Furthermore, thermal gradients build up during heating that lead to heterogeneous and composite structures. Examples of the former are bubble growth and gluten setting during baking, and of the latter, the formation of crust in fried and baked products.

Some specific phenomena leading to structure formation in foods are

1. Stabilization of interfaces by small molecules (e.g., emulsions)
2. Formation of liquid crystalline phases
3. Denaturation of proteins
4. Aggregation and phase separation of macromolecules

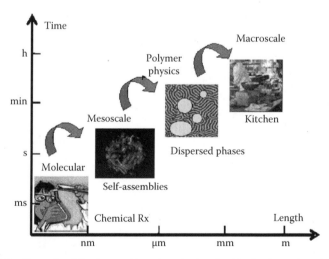

FIGURE 18.2   Example of the time and length scales involved in the formation of gastronomic structures.

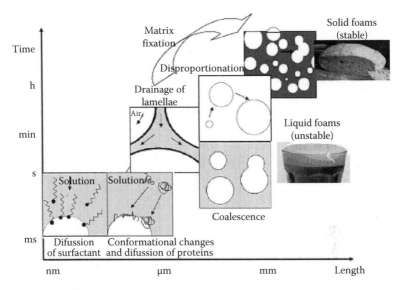

FIGURE 18.3 Time–length diagram of events leading to the formation of stable or unstable food foams.

5. Hydration and swelling of starch granules
6. Assembly of colloidal networks of proteins and polysaccharides (e.g., gels)
7. Formation of fat crystal networks

Control of food microstructure starts at the molecular level. Food chemists continually modify the molecules for added functionality (e.g., modified starches) and of course, safety. The next level of organization is the assemblage of molecules at the mesoscale (10 nm–1 μm). Further up in the scale is the realm of polymer physics where interactions of molecules give rise to phenomena such as growth of separated phases (crystals, bubbles, droplets) and formation of amorphous matrices that stabilize the product in a metastable state. The last level is performed directly in the kitchen. Figure 18.3 depicts some of these events in the case of the creation of solid and liquid food foams.

18.3 WHAT SEDUCES INNOVATIVE CHEFS?

Top-level chefs continually generate new ideas that must be converted into dishes. In this endeavor, empiricism and the trial-and-error approach are replaced by the application of scientific knowledge to achieve the desired results, in an effective and efficient way.

Nine of the more popular concepts that attract the attention of modern chefs are

1. Converting liquids into light three-dimensional structures, e.g., foams
2. Entrapping natural aromas distilled in the laboratory and smoke from indigenous woods
3. New gases and aromas to aerate solutions and solids

4. Entrapping solutions in the solid state as gels
5. Forming transparent films
6. Structures formed by freezing with liquid nitrogen
7. Cooking meats and fish under vacuum at low temperatures and for extended periods of time
8. Development of crispy, crunchy, and crackly textures
9. Teasing and astonishing people, with new sensations (e.g., explosive candy)

As pointed out by Vega and Ubbink (2008), it is ironic that both scientific knowledge and technological concepts originally developed for industrial food production, including those described in patents, are adapted and used to their advantage by the practitioners of scientific cuisine.

18.3.1 SOME FAMOUS COMMERCIAL FOOD STRUCTURES

Several of the most popular food structures are relatively new in the history of foods, considering the fact that Western gastronomy has been traced back several thousand years, to at least several centuries BC (Gillespie, 2001). Figure 18.4 presents some commercial products with their typical structures and the approximate date when they were well known or commercially available.

It is interesting that from a material's viewpoint these foods span a wide range of structures. The point is that there are ample opportunities for creating novel food structures that satisfy specific needs and tastes.

18.3.2 CHEFS AND THEIR LABORATORIES

Most reputed chefs involved in scientific cuisine have their own laboratories, or command outsourcing from independent entities (e.g., AZTI-Technalia). For

Food	Structure	Origin and Date[a]
Breakfast cereals	"Crunchy" solid	Kellogg (1884)
Freeze dried coffee	Porous solid	Nestle (1938)
French fries	Soft composite	After WW I
Potato and corn chips	Brittle solid	Around 1930
Puffed snacks	Solid foam	Around 1950
Milk chocolate	Particulated solid	Around 1818
Cotton sugar	Strings of amorphous sugar	Around 1900
Ice cream	Frozen aerated emulsion	Around 1832
Mayonnaise	Emulsion	Late 1700s
Ketchup	Structured fluid	H.J. Heinz (1876)
Chantilly cream	Liquid foam	Nineteenth century

[a] Date refers to that of extended use or commercial production.

FIGURE 18.4 Some of the most popular "traditional" food structures.

example, top chefs Ferrán Adrià and Heston Blumenthal have research laboratories and staff dedicated almost entirely to the creation of new dishes and techniques. It is well known that scientist Hervé This and chef Pierre Gagnaire have been collaborating for years.

Sophisticated techniques such as rheometry to characterize semisolid foods or gas chromatography-mass spectrometry for aroma research are not uncommon in well-equipped laboratories. Some sort of pilot facility to mimic reactions, extraction processes, blending and drying operations is now necessary for gastronomic research. Precise analytical instruments are compulsory to measure critical parameters that define product specifications, such as refractive index, specific gravity, moisture, acidity, etc. (McEvoy, 2005).

The Arzak Laboratory, located near the restaurant and operational since March 2001, combines modern equipment and a team of "alchemists" who are indispensable to generate the dishes for a kitchen that does not abandon tradition. The "bank of flavors" includes more than 1600 flavors from around the world that are carefully sorted and classified, normally applied to some dishes or ready to be introduced into new developments (Gastronomía and Cía, 2008).

At The Fat Duck, they have developed an internal computer-based encyclopedia (*duckopedia*) that serves as a virtual laboratory notebook where all the procedures, results, and conclusions of their culinary experiments are registered. Blumenthal has published a book where he explains the chemistry and physics behind cooking to college students and the general public. At elBulli, all creations are systematically cataloged by grouping them into families, in order to better understand the fast evolution of their innovative cuisine (Vega and Ubbink, 2008). Adrià is also the president of the *Alicia* Foundation created by the *Generalitat de Catalunya* and *CaixaManresa*, which owns a modern laboratory installation located in *Sant Benet de Bages* since October 2007.

18.3.3 NEW TOOLS AND EQUIPMENT

The *batterie de cuisine* of the scientific chef now includes new equipment, some of them more typical of the development laboratory of a food company than of a restaurant. Among the many new apparatus are pipettes, siphons, syringes, refractometers, thermocouples, pH meters, and thermoregulated baths (Kingston, 2006; elBulli, 2008; van der Linden et al., 2008). Process equipment such as homogenizers, rotary evaporators, centrifuges, dryers, and freeze-dryers are not uncommon. The use of mechanical devices to create new or unfamiliar textures and consistencies (e.g., aerated, thick, emulsified, and crispy) also distinguish traditional cooking from scientific cooking (Ruhlman, 2007). Commercial appliances to cook under vacuum (*sous vide*) and at low temperature are becoming standard equipment in restaurants to tenderize meat and fish in unique ways while intensifying flavors as the food cooks in its own juices. The same device can be used to impregnate solid foods with the surrounding solution and to achieve uncommon combinations of familiar textures and exotic flavors.

Chef Homaro Cantu of Moto restaurant in Chicago is recognized by his creativity in implementing new devices. He has modified an inkjet printer to print the menu on an edible paper with various tastes using vegetable inks (that is later eaten by the customer; Lee Allen, 2005). Another favorite tool used by Homaro Cantu is a laser, similar to the type used for eye surgery, which he uses to vaporize bits of cinnamon or vanilla (Thorn, 2007).

18.3.4 WORKING WITH CHEFS

The author has had the experience of working with a young chef for the last 3 years. Chef Rodolfo Guzmán, trained at the Mugaritz in the Basque country (number 4 restaurant in the world, see Figure 18.1), uses the laboratory to develop some of his creations. Winner of numerous national awards in gastronomy, including Most Innovative Chef 2007, after opening Boragó (http://www.borago.cl) in 2007 he has been involved in extracting aromas from flowers, structuring with liquid nitrogen (Figure 18.5) popping out of Andean cereals quinoa and *cañihua*, and toying with edible films.

  Working with chefs presents several advantages for the food technologist, and no disadvantages, as least thus far. In this unique experience of product development we have realized that modern chefs are eager to know the scientific and engineering concepts behind their creations, even the technical vocabulary (apparently to make a distinction with traditional chefs, making them a "breed apart"). Once results are attained, their innovations are rapidly converted into dishes. The market is well defined (customers of the restaurant), and cost is not a major restriction since these costs are passed on to customers. No major scale-up into industrial production levels is needed. For example, on only a couple of dozen portions may be prepared daily with laboratory-size equipment in the kitchen. Also important is that restaurants have been free till now of some of the regulations (e.g., labeling ingredients) and inspections required for industrial production of foods.

FIGURE 18.5 Chocolate coolant having a solid outer shell rapidly formed by immersion in liquid nitrogen and interior of hot chocolate, standing over "desert sand" made from ground cereals and sugar. (From Chef Guzmán, 2008, Restaurant Boragó, http://www.borago.cl.)

Another major plus of working with chefs is that they are very credible to the general public. Examples of this can be seen from the numerous TV shows all around the world. Chefs can pass on to people the science behind cooking with minimum distrust, which is not the case of the food industry, always suspected of deceiving consumers by using unhealthy additives and unidentified generic ingredients (e.g., ground meat from unknown species and GM foods).

18.4 SEIZING THE MOMENTUM OF SCIENTIFIC CUISINE

It is obvious that a limited proportion of consumers have access to top restaurants and to experience the novelty of scientific cuisine. In time it is expected that lesser quality restaurants and perhaps, food chains will benefit from the "trickle down" of these innovations. Meanwhile, product development centers of food multinationals are listening to chefs and bringing them into their quarters to profit from their approaches. As the use of new products become more popular and disseminated in the general public, ingredient companies will find expanding niches for their products. Manufacturers of kitchen appliances are likely to see expanded opportunities as the precise control of heating, cooling, mixing, and shearing during cooking becomes a must in achieving unique gastronomic results.

Food product developers should be alert about this top–down trend of innovation as it may lead to improved goods with added novelty. Top chefs excel at disseminating their creations in books, Web pages, and TV shows. At the end of the day, most of the developments of scientific cuisine have been picked up from the published scientific literature.

18.5 CONCLUSION

The turnover of the restaurant industry (eating away from home) is almost one-third that of the food industry. Chefs of top restaurants all over the world are becoming the foremost innovators in terms of using new ingredients and creating novel dishes, much like small companies in the high-tech industries. For the first time in the history of foods and gastronomy, we have the scientific knowledge to design food structures with specific goals, whether by creating astounding textures and flavor sensations or, hopefully, contributing to improved nutrition. This is the ambit of GE. Working with chefs may be a unique opportunity for food technologists and food engineers to see their efforts in research reaching the consumer. Results from the marriage between chefs and scientists will expand the variety of foods, introduce new functional ingredients to our diets, and increase the understanding of the science behind cooking.

ACKNOWLEDGEMENT

Contributions by Mr. J.P. Vivanco in some Internet searches and revising the final manuscript are appreciated.

REFERENCES

J.M. Aguilera and D.W. Stanley, *Microstructural Principles of Food Processing and Engineering*, 2nd edn., Gaitersburg, MD: Aspen Publishers Inc., 1999.

J.M. Aguilera, Why food microstructure?, *Journal of Food Engineering* 67 (1–2), 2005:3–11.

J.M. Aguilera and P.J. Lillford, Structure-property relations in foods, in *Food Materials Science: Principles and Practice*, J.M. Aguilera and P.J. Lillford, Eds., New York: Springer, 2008, pp. 229–253.

J.C. Arboleya, I. Olabarrieta, A.L. Aduriz, D. Lasa, J. Vergara, E. Sanmartín, L. Iturriaga, A. Duch, and I. Martínez de Marañón, From the chef's mind to the dish: How scientific approaches facilitate the creative process, *Food Biophysics* 3 (2), 2008:261–268.

M. Bourne, *Food Texture and Viscosity: Concept and Measurement,* 2nd edn., San Diego, CA: Academic Press, 2002.

T. Cowen, In the language of gastronomy, those Michelin stars translate as dollar signs, *The New York Times*, July 13, 2006.

elBulli, July 2, 2008. The Story of elBulli: Our story from 1961 to today, http://www.elbulli.com/historia/version_imprimible/1961–2006_en.pdf (accessed July 2, 2008).

Gastronomía & Cía, February 6, 2008. Laboratorio Arzak, cocina de investigación, http://www.gastronomiaycia.com/2008/02/06/laboratorio-arzak-cocina-de-investigacion (accessed July 4, 2008).

C. Gillespie, *European Gastronomy into the 21st Century*, Oxford: Butterworth-Heinemann, 2001.

E.H. Gombrich, *The History of Art*, 16th edn., London: Phaidon Press Ltd., 1995.

R. Guzmán, 2008. Restaurant Boragó, http://www.borago.cl (accessed July 1, 2008).

A. Kingston, Atomic chefs, *Maclean's* 119, 7, 2006:44–49.

R. Lee Allen, New 'molecular gastronomy' trend raises doubts: Is it fine dining or Frankenfood?, *Nation's Restaurant News* 39, 2005:27.

J.H. McEvoy, Ingredient challenges: The chef's edge: Food lab vs. test kitchen: Art or science?, *Prepared Foods* 5, February 1, 2005.

H. McGee, *On Food and Cooking: The Science and Lore of the Kitchen*, New York: Fireside, 1984.

P. Montagne, *The Concise Larousse Gastronomique: The World's Greatest Cookery Encyclopedia*, London: Hamlyn, 2003.

The S. Pellegrino World's 50 Best Restaurants, *Restaurant Magazine*, 2008. http://www.theworlds50best.com/2008_list.html (accessed July 7, 2008).

M. Ruhlman, The new new cuisine, *Restaurant Hospitality* 91 (5), 2007:26–28.

The magazine for culinary insiders, *StarChefs*, 2008. http://www.starchefs.com (accessed April 17, 2008).

H. Stewart, N. Blisard, and D. Jolliffe, Let's eat out: Americans weigh taste, convenience, and nutrition, *USDA-ERS Economic Information Bulletin* 19, October 2006.

H. This, Molecular gastronomy, *Nature Materials* 4 (1), 2005:5–7.

H. This, *Molecular Gastronomy: Exploring the Science of Flavor*, New York: Columbia University Press, 2006.

B. Thorn, New cookery becomes old hat, *Nation's Restaurant News* 41 (12), 2007:36–37.

E. van der Linden, D.J. McClements, and J. Ubbink, Molecular gastronomy: A food fad or an interface for science based cooking?, *Food Biophysics* 3 (2), 2008:246–254.

C. Vega and J. Ubbink, Molecular gastronomy: A food fad or science supporting innovative cuisine?, *Trends in Food Science & Technology* 19 (7), 2008:372–382.

19 The Right Preparation Technique

Dave Zino

CONTENTS

Success	=	Defining and meeting target consumer needs and expectations	×	The right food	×	Proper packaging and preparation	×	Positioned correctly at the shelf and in the media	×	Meet corporate logistics and financial imperatives

Contribution funded by The Beef Checkoff

19.1   INTRODUCTION

This chapter continues the focus on—what we call in the beef industry—"creating crave." We understand that consumers eat for a variety of reasons and seek food products that satisfy one or more needs—flavor, fuel, convenience, or nutritional requirements, to name a few.

But we also know—in fact, research has proven it—that beef eaters tend to *crave* our products. And no matter how brilliant a new product idea is, if end users cannot recreate it in their own kitchens, we have failed them, and we have lost a customer.

Getting consumers to go beyond *wanting* to try a new product because they are curious or think it is nutritious, to *craving* a product involves a variety of factors. One of the more important components is how that food is prepared.

The folks on the marketing or product development teams, who have identified a consumer need or expectation, hand off their product concept to the food professionals in the research test kitchens. These professionals do their part by figuring out the best way to meet this consumer need or expectation through extensive testing.

Product developers need to not only create a great product to put on the shelves, but must also consider how the end user will treat the product before it reaches the plate. Even though it is the last step, developing the right preparation technique should be considered at the beginning of the development process.

One of the most challenging phases in the development process may be creating simple yet versatile package instructions that help consumers create the same "craveable" flavor, whether at home or in food service establishments. Remember, it does not do anyone any good to have professionals create a masterpiece in the laboratory that average consumers or food service employees cannot replicate in their kitchens.

With that in mind, we will take a look at what goes into creating the optimal preparation options for new food products and how to translate those findings into foolproof package directions that make consumers feel like master chefs.

We will discuss

1. Optimizing flavor
2. Balancing product quality and convenience
3. Satisfying consumers' need for preparation choices
4. Keeping "healthy" meals fit
5. Testing your product to ensure consistency
6. Maintaining food safety standards at home

It is important to note that simply creating package instructions is no longer enough. Consumers want more than just how to cook the product, they want the product to

1. Taste great
2. Cook in minutes
3. Offer a choice of cooking methods
4. Come with serving ideas
5. Include nutrition information beyond the Nutrition Facts panel
6. Be used as an add-in for other recipes so they can customize their own meal ideas

19.2 CREATING CRAVE—THE ART (AND SCIENCE) OF FLAVOR OPTIMIZATION

Optimizing the flavor of any product is not an accident, at least for the most part. Flavor optimization begins in product development and is central throughout every step of the process. It is just as important to consider when developing the package directions as it is at the first brainstorming meeting. Most of us use the terms taste and flavor interchangeably, but they are actually different (Figure 19.1).

When the beef test kitchens work on new product development, flavor optimization is the goal. Two ways to help reach that goal are umami, which we will go in-depth here, and the Maillard reaction, which should be on top of everyone's mind in the test kitchens.

19.2.1 UMAMI

Umami (oo-MOM-ee) is not a preparation technique but may be the reason why some classic beef flavor combinations are so popular. Understanding umami can greatly influence new product development.

Umami is the fifth taste (along with salty, sour, bitter, and sweet). Umami is described as "meaty and savory" and is derived from *umai,* the Japanese word for delicious. It is the taste of glutamates—the salts of an amino acid—and other small molecules called nucleotides. Although umami has been known for quite awhile, umami receptors on the tongue have been clearly identified only recently, making it a bona fide fifth taste.

Umami was discovered in the early 1900s by a Japanese scientist, Dr. Kikunae Ikeda, at Tokyo's Imperial University. He undertook research to ascertain the true nature of the "deliciousness" of *konbu,* or kelp, an indispensable part of Japanese cuisine. He succeeded when he extracted glutamate from the *konbu,* discovering that it was the main active ingredient and the key to its delicious taste. He coined the term "umami" to describe this taste.

Many years later, in 1997, taste researchers Stephen D. Roper and Nirupa Chaudhari of the University of Miami Medical School clearly identified taste buds on the tongue, and have since cloned receptors that respond to umami. Their discovery and demonstration solidified umami as the fifth taste, and even put it on a firmer footing than some of our original four tastes. Note that the underlying physiology for the sour and salty receptors is well understood, but there is still controversy about bitter and sweet, which appear to be more complex.

Taste (five basic receptors)	Flavor (taste plus the other sensations)
Sweet	Tastes
Salty	Aroma
Sour	Texture
Bitter	Juiciness
Umami	Mouthfeel
	Color

FIGURE 19.1 Taste versus flavor. (Courtesy of the National Cattlemen's Beef Association.)

The science behind umami is becoming increasingly clear. The umami taste is produced by a number of naturally occurring compounds; amino acids like glutamic acid, and nucleotides like guanosine 5'-monophosphate (GMP), inosine 5'-monophosphate (IMP), and adenosine 5'-monophosphate (AMP). Amino acids are the basic units that link together to form proteins. Nucleotides are composed of a piece of protein and a nonprotein part.

Some basic umami foods are rich in "free amino acids" (i.e., lots of umami when raw), while others have "bound amino acids." If these bound amino acids are to contribute umami taste, then first they must be freed from their protein molecules through cooking or aging/fermenting.

Dr. Harold McGee, author of *On Food and Cooking*, points out that big food compounds like carbohydrates, proteins, and fats are not very flavorful, but their small component parts such as sugars (from carbohydrates), amino acids (from proteins), fatty acids (from fats), are in fact, extremely flavorful. Sources of umami (amino acids, nucleotides) are these small, intensely flavorful compounds. Again, aging, fermenting, ripening, or cooking are all great ways to break big compounds down into these flavorful gems. Think about great flavor enhancers like mushrooms, soy sauce, Worcestershire sauce, aged cheese, or wine. They are all rich in umami.

Beef comes equipped with three natural sources of umami–glutamic acid, glutamate (simply the salts of glutamic acid), and nucleotides, all of which account for its great natural flavor. Since umami compounds are synergistic, layering umami ingredients gives the product many times the flavor. To punch up the already delicious flavor of beef, we only have to add some other natural sources of umami (e.g., mushrooms, tomatoes, or an aged cheese).

19.2.1.1 Putting Umami to Work

Part of umami's great flavor power comes from the fact that the whole is greater than the sum of its parts. When individual umami compounds are combined, they have a magnifying effect on each other. This explains the delicious pairings of mushrooms and steaks, and wine or tomato sauces with beef. A 50–50 mixture of two umami compounds can produce eight times as much flavor as either of the compounds alone!

As mentioned previously, ripening, aging, and fermenting foods can dramatically increase their umami. That is why a truly ripe tomato (Figure 19.2), aged Parmigiano-Reggiano, and fermented foods, such as wine and soy sauce, possess enticing, complex flavors and also pair well with beef dishes. Figures 19.3 and 19.4 present lists of umami-rich foods.

There are two key ways to take advantage of umami while developing preparation directions.

1. Ensure the product itself takes advantage of umami.
2. Offer serving suggestions on the package to help consumers pair accompaniments to your product that help create the umami phenomenon. When featuring a burger show it with cheese and tomato, or with a marinated steak product include a serving suggestion: "sprinkle with grated Parmesan or blue cheese." For ideas, see Figure 19.3.

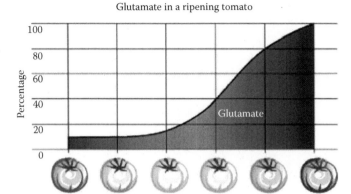

FIGURE 19.2 Glutamate in a ripening tomato. (Courtesy of International Glutamate Information Service. With permission.)

Natural Sources of Glutamic acid and Glutamates*	Natural Sources of 5'-guanylate (a nucleotide-small part of protein compound)*	Natural Sources of 5'-inosinate (a nucleotide—small part protein compound)*
Kelp	Dried mushroons	Bonito
Cheeses	Mushrooms	Mackerel
Green tea	Beef	Sardines
Seaweed	Chicken	Tuna
Sardines		Beef
Fresh tomato juice		Prawns
Peas		Chicken
Corn		Cod
Mushroom		
Tomatoes		
Oysters		
Potatoes		
Chinese cabbage		
Duck		
Soybeans		
Chicken		
Spinach		
Mackerel		
Carrots		
Beef		
Beets		
Milk		

* In descending order for each group.

FIGURE 19.3 Umami-rich foods. (Courtesy of the National Cattlemen's Beef Association.)

The Essential Umami-Rich Pantry

Sauces (soy, worcestershire, Asian fish)
Balsamic vinegar
Canned tomato-based products
Aged cheese, e.g., Parmigiano-Reggiano
Dried mushrooms (Shiitake, Porcini)

FIGURE 19.4 The essential umami-rich pantry. (Courtesy of the National Cattlemen's Beef Association.)

19.2.2 MAILLARD REACTION

In the culinary world, no chef would ever consider skipping the preparation technique commonly known as browning. Browning creates beef flavors that can only be produced through dry heat. These are unique flavors and aromas that are not intrinsic to the beef itself. During browning, temperatures of 350°F or higher on the surface of the beef cause proteins (amino acids) and carbohydrates (sugars) to caramelize, creating intense flavors and aromas. The very limited number of carbohydrates in meat is just enough for the browning reaction.

This browning process is called the Maillard (may-YARD) reaction, named after a French scientist, Louis Camille Maillard, who discovered and described it. Everything from baked goods to coffee beans to beef, foods that are not primarily sugar, benefit from this complex reaction of sugars and amino acids caused by higher heat. When developing preparation directions, take advantage of the Maillard reaction.

19.3 CREATING CRAVE—CONSUMER-CENTRIC PREPARATION

Throughout the process of creating preparation directions, the consumers' needs and wants must remain central. We have already discussed flavor optimization, probably the most important factor to product developers and consumers. Now we cover five others:

1. Convenience
2. Cooking options
3. Nutrition
4. Cooking as science—The art of testing
5. Food safety

19.3.1 CONVENIENCE—KEEPING IT SIMPLE

If the consumer wants a convenient meal, you do not want to complicate things in the kitchen. Consider self-serve gas stations that do not allow for at-the-pump payment. Pumping your own gas becomes less of a burden and more of a convenience once you could pay at the pump. It is the same with meals prepared for consumers' convenience. From the decision to buy the product, to the purchase, preparation, and clean-up, it all has to be convenient. Not just one part in the process—every part.

Let us try to understand consumers for a moment. We know they are strapped for time, especially when it comes to preparing dinner. (Breakfast too is no easy feat for most families). But we also need to know:

1. Who are they?
2. What cooking equipment are they using?
3. What are their time constraints?

19.3.1.1   They Are Women, Mostly

According to The NPD Group's Annual Food for Thought Study, women continue to be responsible for the planning, shopping, and cleaning up of meals in America. But unlike several generations ago, the majority of today's women are also working outside the home.

The U.S. Department of Labor reports that:

1. The female labor force has doubled in the last 50 years with 60% of all women now employed
2. Women have increased their work hours by 50% since 1976
3. There is an increase in single parent households.

When you also consider that most women have fewer cooking skills than previous generations and more demands on their time, yet still have a strong desire to satisfy their families, you have the recipe for a dinnertime dilemma!

19.3.1.2   They Are Not Just Hungry—They Are Time Starved

Because these consumers—female or male—are time starved, the buzzwords for meal preparation are simplicity and convenience (Source: The NPD Group's NET Research and Wirthlin Beef Category Architecture Study, both studies funded by the Beef Checkoff) are:

1. Supermarket customers are cutting back on meal preplanning
   a. About 2/3 of all dinner decisions are made on the same day
   b. Of those, 1/3 do not know what they are going to have "just before" dinnertime
2. The focus is on streamlining meal preparation
   a. During the week, almost half of all dinner meals are prepared in less than 30 min
   b. Dinner meals are becoming more basic. The number of ingredients used is at an all-time low and fewer dishes are served
   c. Scratch meals have declined and assembled meals have increased.

Fewer meals are prepared at home. Instead, more meals are purchased in restaurants and supermarket deli take-out.

19.3.1.3 They Prefer Simple Cooking

When meals are prepared at home, the traditional slow cooking methods, such as braising or roasting, do not fit with the time constraints of many consumers. Between 1985 and 2001:

1. Stovetop cooking has declined from 67% to 52%
2. Use of slow cookers has slightly increased
3. Casseroles and other unattended oven meals grew in popularity
4. Microwave cooking has maintained its place and is mainly used for reheating

When asked how consumers prepare meals using beef at home, we learned they are more likely to prepare simple meals, although two in five do prepare complex meals.

19.3.2 COOKING OPTIONS: HOW THEY ARE COOKING

When it came to cooking dinner 60 years ago, consumers had few choices. They could bake it in the oven or cook it on the stovetop. Today, we have nearly endless options. Here are a few of the more popular cooking options consumers have at their disposal:

1. Outdoor grill—charcoal, gas, and electric
2. Indoor grill—tabletop (i.e., George Foreman®), stovetops
3. Conventional oven—gas or electric
4. Convection oven—gas or electric
5. Broiler—gas or electric
6. Stovetop—gas or electric
7. Microwave
8. Slow cookers
9. Deep fryers
10. Smokers
11. Rotisseries

19.3.2.1 A Convenient Truth—Consumers Love Choices

Today's abundance of cooking options means one needs to consider the available options and their market penetration when developing preparation directions. If the product is appropriate for an outdoor grill, then consider testing it in a skillet or broiler. If the preferred method of preparation is on a stovetop, then is it possible to achieve similar flavor results in a microwave? Remember package directions can state a "preferred" method of preparation, but alternatives should be offered as long as they do not jeopardize product quality and safety.

As an example, research shows that beef consumers prefer a steak cooked on an outdoor grill, but they understand that at times they may need to use a skillet or broiler. The consumers are willing to sacrifice some flavor for speed, but that is a choice consumers wish to make for themselves.

Another example is "beef in a bag" products. Food professionals know that the best quality comes from simmering the bag on the stovetop. However, research shows that consumers have reservations about handling the hot bag. They find it easier to handle when heated in the microwave. Thus, manufacturers need to offer more than one preparation method. The preferred set of directions should appear first on the label.

Given the desire to merge convenience with good taste, here are four examples of convenient meal solutions:

1. Beef's "heat and eat" products are a delicious example of creating a product to satisfy consumers' need for convenience. With these products, all the work has been done. Convenient, versatile, and delicious, these beef products get heated in no time and need only a simple salad or vegetable to complete the meal. Alternatively, with a simple twist or two, they can jump-start favorite family soups, sandwiches, pizzas, and pasta dishes. An excellent meal solution for busy households!

2. Offer meal ideas on the package. These ideas enhance package directions that will help create repeat purchases. Here is another opportunity to take advantage of umami—pairing side dishes with the product that will enhance the flavor and ultimate enjoyment for the consumer.

3. Add copy and visuals to package directions to complement the product offering. The more direction provided to consumers the better, especially while at the grocery store. For example, if consumers pick up a product and read on the package that it can be cooked from a thawed or frozen state, then they will consider it as a meal option tonight or maybe two nights from now. Then, add that couscous and a side salad which complements the product well. Now the consumer can walk right over to the appropriate aisles in the store and purchase them, knowing their entire meal—for tonight or later in the week—is in the bag.

4. Offer preparation directions from both the frozen state as well as thawed state. The option of two preparations makes the product more convenient to consumers. Like the cooking options, one should mention that it makes a difference to quality or flavor when it is prepared from frozen or thawed state. By offering both options, however, the product becomes twice as attractive.

19.3.3 NUTRITION—KEEPING IT FIT

The 2005 Dietary Guidelines for Americans emphasizes the basics: eat foods that are rich in nutrients first and choose foods from all food groups while balancing portions and caloric needs, especially since so many Americans are struggling with overweight or obesity. The nutritional makeup of new products is often an important factor for consumers. The marketing team should have already decided whether or not the product is going to be positioned as a "diet" or a "healthy living" product.

To meet the criteria to be considered "healthy," the developers would have squeezed as much flavor into the product without adding unnecessary calories, fat, sodium, and so on. Keep in mind that the flavor combines taste, texture, aroma, juiciness, mouthfeel, and color.

The current trend is to focus on the positive nutrition attributes such as "excellent source of protein, good source of fiber."

1. Excellent sources are 20% or more of the daily value of a nutrient.
2. Good sources are 10%–19% of the daily value of a nutrient.

Some important vitamins and minerals are not required to appear on the nutrition facts label, although manufacturers can choose to list them. For example, a serving of beef stew is an excellent source of vitamin B12 and zinc, but the consumer may not find that on the label unless the manufacturer chooses to provide the information.

While Americans are eating more, research shows we are getting less from the foods we eat. Choosing naturally nutrient-rich foods first, from each food group, can help people make each calorie they eat count more. Work is underway in the scientific community to develop a quantitative nutrient-rich foods index or "score" to help consumers select foods with the highest nutrients per calorie.

19.3.4   COOKING AS SCIENCE—THE ART OF TESTING

It is imperative to test and then retest package directions. Test in all kinds of conditions. Provide temperature end points as well as visual descriptions so that consumers, no matter what their kitchen expertise is, will know when the dish is ready to be served. Once the product has moved into full-scale production, do not forget to do confirmatory testing to recheck your recommendations (Figure 19.5).

19.3.4.1   The Testing Protocol Checklist

Objectives
- How will this product be used

Product
- Product description
- Product state—frozen, refrigerated, shelf-stable

Cooking
- See cooking, planning matrix specifics, Figures 19.6 and 19.7
- Identify cooking methods and techniques—gas and electric ranges, model, etc.
- Cooking equipment that will be used
- See testing sheet and testing summary, Figures 19.8 and 19.9

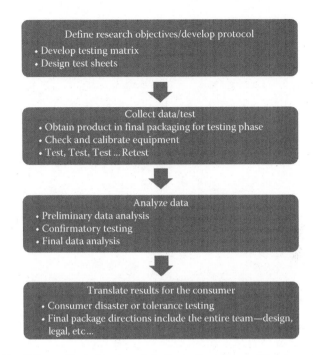

FIGURE 19.5   Culinary Center's approach to the testing process. (Courtesy of the National Cattlemen's Beef Association.)

Equipment	Grills • Gas–natural • Weber genesis silver B 2005 model year

Methodology	• Preheating of grill
	    • Gas grills
	        • Preheat on high with lid closed 10 minutes until grill thermometer registers 500°F–550°F
	          • Adjust burner control knobs to medium temperature.
	• Meat preparation
	    • Use number of steaks/kabobs/patties as specified on chart for each trial.
	    • Arrange meat evenly on grid.
	    • Position thermocouples nearest to center of meat as possible, especially in terms
	      of center thickness point.
	• Testing
	    • Arrange meat evenly on grid of grill.
	    • Cover grill.
	    • Turn meat once. Record time of turning.**
	    • Use tongs, not fork, to turn meat.
	    • Monitor grill temperature occasionally during grilling.
	    • Remove steaks/kabobs/patties from grill at specified temperature. Record end time.
	    • During standing time, observe and record temperature rise for 5 min.
	    • Cut meat in half: Observe doneness visually.

Special notes	** Steaks and kabobs may need to be turned more than once to prevent charring, turn as needed; record number (and time) of turns.

FIGURE 19.6   Cooking specifics. (Courtesy of the National Cattlemen's Beef Association.)

Beef gas grilling trials

Cut	Thickness or weight	Cuts/trial	Grill #1 Medium rare (145°F)	Grill #1 Medium (160°F)	Grill #2 Medium rare (145°F)	Grill #2 Medium (160°F)	Grill #3 Medium rare (145°F)	Grill #3 Medium (160°F)	Tests
Tenderloin Steak	1 in. thick	4 steaks/trial	X	X	X	X	X	X	30
	1-1/2 in. thick	4 steaks/trial	X	X	X	X	X	X	30
Top loin (Strip) steak	3/4 in. thick	2 steaks/trial	X	X	X	X	X	X	30
Ground beef patties	1/2 × 4 in. (4 oz)	4/trial		X		X		X	15
	3/4 × 4 in. (6 oz)	4/trial		X		X		X	15

Total: 120

FIGURE 19.7 Planning matrix. (Courtesy of the National Cattlemen's Beef Association.)

Meat cut	Thickness/weight	Heat level	Desired doneness
Ground beef	4 patties per pound 1/2"×4"	Medium	Medium 160°F

Testing data
Outside temp:
Climatic conditions: (wind/sunny/etc.):
Brand/trial:
Time required for heating:

Pretrial Data									
#/trial: 4 burgers	Grid Temp(°F)		Time (min)	Meat Temp. (°F)				Final Doneness		
Raw weight: 1 lb = 4 oz each	T1	T2		Instant Read		Professional Scientific		Stand Temp.	Stand Time	Visual
Edible weight:				#1	#2	#1	#2	↑	at peak	
Raw illustration:										

Comments/recommendations:

Trial Data									
#/trial:	Grid Temp(°F)		Time (min.)	Meat Temp. (°F)				Final Doneness		
Raw weight:	T1	T2		Instant Read		Professional Scientific		Stand Temp.	Stand Time	Visual
Edible weight:				#1	#2	#1	#2	↑	at peak	
Raw illustration:										

FIGURE 19.8 Testing sheet (beef gas grilling direct). (Courtesy of the National Cattlemen's Beef Association.)

Product: Ground Beef Patties
Equipment Manufacturer: WEBER Genesis Silver B – natural gas – 2005 model

Test #	Grill Type	Raw Wt.	Cooked Wt.	Total Cook Time	Final Temps.	Comments
1						
2						
3						
4						
5						

Comments and questions in general:

FIGURE 19.9 Testing summary. (Courtesy of the National Cattlemen's Beef Association.)

Data Collection

- Weights and dimensions
- Temperatures
  - Probes/thermometers—All probes/thermometers should be tested daily with boiling water or ice water to assure accuracy
    (Ice water test—Fill 1-quart glass measuring cup with ice. Add water just to cover ice; stir slightly. Place probe in mixture; temperature should register $31°F–33°F$. Probes not registering accurate temperature should not be used.)
- Initial temperatures
  - Starting temperatures should be recorded on data time sheets
  - This information is not required in final report, but must be monitored for data analysis
- Internal temperatures
  - To be recorded as indicated on data sheet from start of cooking until the final end point temperature is reached. Consult the appropriate credible source, such as the Model Food Code, when setting end point temperatures
- Standing temperatures
  - To be recorded as indicated on data sheet for a period of 5 min after removing from heat, for the purpose of observing:
    - Any rise in temperature
    - The peak temperature
    - Time when temperature peaks
- Cooking times
  - To be recorded as indicated on data sheet from the start of cooking:
    - Until $145°F$ or $160°F$ is reached in steaks/kabobs
    - Until $160°F$ is reached for ground patties

Determination of Acceptability

- A determination of the appropriateness/effectiveness of the cooking method used needs to be made for each trial. Examples of unacceptable tests:
  - Steak is charred
  - Meat is gray in color; meat "steamed"
  - Indirect grilling recommended
  - Meat was very charred on outside; frequent turning was required

19.3.5 Food Safety

When it comes to food safety, you cannot overcommunicate. After doing your homework—thoroughly testing package directions to ensure the cooking times are accurate—complete the food safety chain by including tips to eliminate food safety issues at home. These tips should be on the packaging where they cannot be missed, as well as peppered throughout the package directions. For instance, when the preparation has the consumer working with raw meat or poultry and produce, clearly tell them to use two different cutting boards, or tell them to wash the board or counter with warm soapy water before working with vegetables.

By integrating food safety tips throughout your preparation directions, you can ensure your messages get across. Following are the examples of on-package directions. (See Figures 19.10 and 19.11.)

Example 1:

Cooking instructions for beef shoulder center steak

Beef shoulder center steaks can be grilled, skillet cooked, or broiled, cut into strips, and stir-fried, or cut into pieces for kabobs.

Grilling
Grill, covered, over medium, ash-covered coals according to table, for medium rare (145°F) to medium (160°F) doneness, turning once

3/4 in. thick	9–11 min
1 in. thick	11–14 min

Skillet
Heat nonstick skillet over medium heat until hot. Place steaks in skillet (do not crowd). Cook, uncovered, according to table, for medium rare (145°F) to medium (160°F) doneness, turning twice.

3/4 in. thick	9–12 min
1 in. thick	13–16 min

FIGURE 19.10 Example of on-package directions. (Courtesy of the National Cattlemen's Beef Association.)

Example 2:

Cooking instructions for fully cooked ground beef crumbles

Microwave oven

Remove desired amount of frozen ground beef crumbles from package and place in microwave-safe container. Microwave on high according to chart or until hot. (Microwave ovens vary; adjust times as needed.)

Amount	Cooking time
½ package (12 oz) about 2⅔ cups	3½–6 min, stirring once
1 package (24 oz)	6–11 min, stirring every 3 min

Skillet

Remove desired amount of frozen ground beef crumbles from package and place in large nonstick skillet. Heat over medium heat according to chart or until hot, stirring frequently.

Amount	Cooking time
½ package (12 oz) about 2⅔ cups	5–7 min
1 package (24 oz)	7–11 min

FIGURE 19.11 Example of on-package directions. (Courtesy of the National Cattlemen's Beef Association.)

19.3.5.1 Additional Package Copy Points—Ground Beef Crumbles

Ground beef crumbles are fully cooked and can be used in your favorite recipes; just omit the browning step and proceed with recipe directions.

- Stir frozen ground beef crumbles into your favorite prepared spaghetti sauce, barbecue sauce, or prepared Sloppy Joe sauce. Bring to a boil over medium-high heat; reduce heat to medium-low and simmer until hot, about 5 min.
- Change ordinary mac and cheese into cheeseburger mac. Stir heated ground beef crumbles into prepared macaroni and cheese.
- Turn a side dish into a main dish—add ground beef crumbles to the packaged rice, noodles, or potato mixes (i.e., Spanish rice into Spanish rice and beef skillet)
- Add ground beef crumbles to cheese nachos for a quick after-school snack.
- Sloppy Joes in a snap—heat ground beef crumbles in your favorite barbecue sauce and serve in a roll.

Consumer tips to avoid cross contamination:

- Keep raw meats, poultry, and seafood and their juices separate from ready-to-eat foods, both in the refrigerator and during meal preparation.
- Do not place cooked foods on the same plate that held raw meats, poultry, or seafood.

- Use separate cutting boards for raw animal products and ready-to-eat foods, such as fruits and vegetables.
- Always wash hands thoroughly in hot, soapy water *before* preparing or eating food and *after* handling raw meats, poultry, and seafood.
- Clean all cooking utensils and countertops that have come in contact with raw meats, poultry, and seafood, with hot, soapy water immediately after use.

19.4   CREATING CRAVE—NOW AND TOMORROW

The effort in creating crave is more than just putting a product in a package. It should be your focus throughout the product development process—from initial concept to delivery. If you are able to create a product that consumers crave, that they are able to create in their kitchens, and one that satisfies one or more of their objectives for a meal, you have succeeded.

20 Recent Developments in Consumer Research of Food

Herbert L. Meiselman

CONTENTS

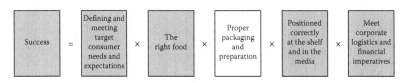

A lot has happened in consumer research since it began to grow rapidly in the 1960s. A number of texts have reviewed the development, and sometimes the decline, of particular data, methods designed to collect them, and statistics designed to analyze them. Behind all of this development have been theories of human behavior and consumer behavior, and from the theories, different concepts of what variables direct consumer choices.

As we enter the twenty-first century, this development within consumer research continues. There have been major developments in the past 10 years, and new developments are now on the horizon promising different ways to collect and analyze consumer data. This chapter presents selected new techniques in experimental design and methodology. The new techniques reflect a growing sense that we have not been going about our business in the right way. This has been noted in recent reviews.[1,2] Other recent reviews of consumer methodology can be found in MacFie,[3] Schifferstein and Hekkert,[4] and Frewer and van Trijp.[5]

The sections of this chapter review, in turn, selected new developments in consumer research methods.

20.1 PSYCHOGRAPHICS: CONSUMER SEGMENTATION USING PSYCHOLOGICAL ATTITUDE AND TRAIT MEASURES

During the course of consumer research, large samples are often divided or segmented into smaller groups for more detailed analysis. Consumer segmentation is not new, but simple segmentation based on demographics (age, gender, socioeconomic variables, etc.) is still the norm.[1,6] Wansink[6] has commented that "Using demographic information to describe a consumer is like using a yardstick to describe a beach ball."[6] In many instances, segmentation of the data set based on gender, age, and income and other demographics is worthwhile, but these types of data are very limited. Very few consumer phenomena divide so nicely between such obvious criteria as age and gender. One example of another criterion for segmenting consumers, noted below, is their degree of concern over dieting and health, which is almost always found to be stronger in women.

Consumer segmentation is often conducted based on the distinction between product users and product nonusers. This distinction requires the development of criteria of product use and decisions regarding whether or not product usage must be long term and regular. A new technique for segmentation only recently available to consumer researchers and product developers are the new attitude and trait questionnaires designed to measure fundamental views of consumers toward many issues. These questionnaire techniques, suited to in-person research, mail research, and Internet research, can uncover important issues of consumer motivation and behavior related to categories of products as well as specific products within categories.

Beginning in the 1990s, a new group of attitude and trait techniques have been developed, making new resources available for segmentation analyses. It would be appropriate to begin with brief definitions of attitude and trait. An attitude is an evaluative statement about a product, person, or event. Attitudes usually contain the notion of liking–disliking about a category of goods, such as "I like apples." Traits are the longer lasting building blocks of personality, and traits represent the ways people differ from each other. For example, passivity and aggressiveness are two (opposing) traits, representing two different personality types. Tests for attitudes and traits can be referred to as psychographics. Many trait and attitude tests are developed within fields of study different from product development and its supporting technologies.

Tests for attitudes and traits have been widely used within some aspects of research on eating, but little has been published applying these techniques to food product development. With the growth of available techniques it seems appropriate to consider application of existing tests to product development. Tests can be used in two different ways:

1. Describe a test population. Samples of consumers are often described with standard demographic criteria such as age and gender, but these new tests provide more insightful descriptions of differences which might impact consumer decisions.
2. Use as a test variable. Studies can be designed in which psychographics can be an independent variable which is manipulated in the study design.

20.2 MEASURES OF VARIETY SEEKING AND RELUCTANCE TO TRY NEW PRODUCT

Two different measures of variety seeking and variety avoidance appeared in 1992. They were born of two different influences on consumer research on food. van Trijp and Steenkamp[7] introduced a scale of variety seeking (VARSEEK), which is a variant on general scales of variety seeking which identify people who prefer to choose from many products including new products. The Variety Seeking Scale appears in Figure 20.1.

At the same time, Pliner and Hobden[8] published a Food Neophobia Scale (FNS), designed to measure consumer reluctance to try novel foods. The FNS is shown in Figure 20.2. This scale comes from the social psychology and abnormal psychology literature.

Pliner and Hobden's original paper[8] lays out the steps they used to develop the scale (Figure 20.3). Pliner and Hobden's original paper[8] measured the food neophobia of a student sample and obtained a range of 10–68, almost the entire theoretical range. More recent testing cross-cultural study of the FNS has shown that items 1, 3, 4, 6, 7, and 10 form a unidimensional group of questions that can be used as a shorter questionnaire.[9] Items 5 and 9 have been criticized on other measurement grounds as well.

The two scales, variety seeking and food neophobia, were developed in the same year. One might have expected variety seeking to become the more widely used scale, given its roots in more general consumer scales. But the FNS has seen much

| 1. When I eat out, I like to try the most unusual items, even if I am not sure I would like them. |
| 2. While preparing foods or snacks, I like to try out new recipes. |
| 3. I think it is fun to try out food items one is not familiar with. |
| 4. I am eager to know what kind of foods people from other countries eat. |
| 5. I like to eat exotic foods. |
| 6. Items on the menu that I am unfamiliar with make me curious. |
| 7. I prefer to eat food products I am used to ®. |
| 8. I am curious about food products I am not familiar with. |

FIGURE 20.1 Variety Seeking Scale (VARSEEK). (From van Trijp, H.C.M. and Steenkamp, J.-B.E.M., *Eur. Rev. Agric. Econ.*, 19, 181, 1992.)

1. I am constantly sampling new and different foods. (R)
2. I do not trust new foods.
3. If I do not know what is in a food, I will not try it.
4. I like foods from different countries. (R)
5. Ethnic food looks too weird to eat.
6. At dinner parties, I will try a new food. (R)
7. I am afraid to eat things I know never had before.
8. I am very particular about the foods I will eat.
9. I will eat almost anything. (R)
10. I like to try new ethnic restaurants. (R)

FIGURE 20.2 Food Neophobia Scale. (From Pliner, P. and Hobden, K., *Appetite*, 19, 105, 1992.)

1. The scale contains 10 items each judged on a 7-point agree–disagree scale; half are stated positively and half are stated negatively (Figure 20.3).
2. Five items require that their scale values be reversed (shown by R), such that a scale rating of 6/7 becomes a rating of 2/7. In this way, all items show increasing neophobia with increasing score.
3. The highest level of neophobia would be a score of 70, the theoretical limit.
4. People with higher scores are referred to as neophobics and people with lower scores neophilics, meaning people who like to try new foods.

FIGURE 20.3 Steps to develop the FNS.

more attention and publication. In fact, food neophobia has been shown to be implicated in many important food-related behaviors in children and adults.[10]

Attitude and trait scales can be used to define a test population or used as an independent variable in research studies. The FNS is beginning to underscore an important issue about recruited test panels. When consumers are needed to test products, usually they are recruited by advertisements calling for people to test products for a modest fee. Many people accept those individuals who respond to such advertisements as representative consumers, or at least representative of consumers who use the product in question or a related product or a class of products.

Typically, investigators, especially those working in commercial, that is, "applied research" studies, define the test panel using demographics, and sometimes using product usage statistics.[11] Recent analyses of the composition of test panels has demonstrated that recruited test panels are not representative of all consumers with respect to food neophobia. Specifically, in recruited panels, the entire top of the distribution, the high neophobics are missing (Figure 20.4). The upper end of the scale (i.e., the neophobic area) is not observed when samples are recruited for academic studies or commercial tests. Finally, the large range observed by Pliner and Hobden[8] has never been reported again in published research.

The "food testing" community is just beginning to discuss the implications of the fact that those reluctant to try new foods are, in fact, not testing new foods.

Pliner and Hobden[8]	Theoretical range of scores	10–70
Pliner and Hobden[8]	Actual range of student scores	10–68 (>35)
Tuorila et al. (1994)[11]	Natick panel range of scores	10–54 (>30)
Tuorila et al. (1998)[12]	Young and old ranges of scores	
	Young	20–53 (>39)
	Elderly	17–58 (>39)
Arvola et al. (1999)[13]	Female range of scores	10–49 (>29)

FIGURE 20.4   Do neophobics volunteer for taste tests? The data show the range of scores obtained. The neophobia score cutoff is shown in brackets, for example (>35).

Those testing our foods are neophilics; they like new things. Does this selection of neophilics thus bias our testing toward newness? Will neophobics like the new products after exposure to them, because repeated exposure is known to counteract neophobia?

Food neophobia has also been used as a variable in research. The results show a good consistency of findings, and importantly implications for testing and for generalizability of the results of current tests. Food neophilics, people who like to try new foods, assign higher ratings to food acceptance. Or conversely, food neophobics rate products slightly lower. Thus, not only are neophobics underrepresented in the test populations, but when they do volunteer, they rate products differently.[11]

20.3   RESTRAINED EATING

One of the main features of recent food product development has been the emphasis on products with a healthy product image. Many products are classified as "functional foods," meaning that an ingredient has a specific benefit.[14] Some products are aimed at the growing obesity problem in many countries. Of interest here are the scales of restrained eating, introduced to measure the attempt to control intake in order to control body weight. It is important to understand that restrained eating does not equate with dieting but refers to a broad and not always successful attempt to control eating. In many types of product development research, it would be useful to test products on people who we know are trying to control their intake.

Scales of restrained eating are the most widely used of all of the attitude and trait scales mentioned in this chapter. Their wide use in the health fields has promoted a great deal of research. A Web search of "restrained eating" generates hundreds of thousands of hits! Recent research has shown the complex nature of the restraint concept.[15] This field began in the 1970s with the first development of a scale of restrained eating.[16] The Revised Restraint Scale (RS) of Herman and Polivy[16] consists of 10 items (e.g., "How often are you dieting?") that are rated on 4- or 5-point Likert scales (Figure 20.5). Scores range from 0 to 35, with high scores reflecting a high degree of restraint.

A longer three-factor questionnaire including restrained eating was developed by Stunkard and Messick[17] in 1985. The restrained eating portion of the questionnaire

1. How often are you dieting?
Never, rarely, sometimes, often, always (scored 0–4) CD
2. What is the maximum amount of weight (in kilos) you have ever lost within 1 month?
0–2.5, 2.5–5, 5–7.5, 7.5–10, 10 + (scored 0–4) WF
3. What is the maximum amount of weight gain (in kilos) within a week?
0–0.5, 0.5–1, 1–1.5, 1.5–2.5, 2.5 + (scored 0–4) WF
4. In a typical week, how much does your weight fluctuate?
0–0.5, 0.5–1, 1–1.5, 1.5–2.5, 2.5 + (scored 0–4) WF
5. Would a weight fluctuation of 2.5 kilos affect the way you live your life?
Not at all, slightly, moderately, very much (scored 0–3) CD
6. Do you eat sensibly in front of others and splurge alone?
Never, rarely, often, always (scored 0–3) CD
7. Do you give too much time and thought to food?
Never, rarely, often, always (scored 0–3) CD
8. Do you have feelings of guilt after overeating?
Never, rarely, often, always (scored 0–3) CD
9. How conscious are you what you are eating?
Not at all, slightly, moderately, extremely (scored 0–3) CD
10. How many kilos over your desired weight were you at your maximum weight?
0–0.5, 0.5–3, 3–5, 5–10. 10 + (scored 0–4) WF

FIGURE 20.5 Restraint scale. *Note*: CD, concern for dieting; WF, weight fluctuation.

contains 12 true–false items in Part I and 9 scaled items in Part II. Stunkard and Messick noted that people with high cognitive restraint eat less but weigh the same as do as people with low scores on cognitive restraint.

Another widely used restraint scale was developed by van Strien et al.[18] The restraint scale (Figure 20.6) is part of a longer questionnaire referred to as the Dutch

Response scale: never (1), seldom (2), sometimes (3), often (4), very often (5)
1. If you have put on weight, do you eat less than you usually do?
2. Do you try to eat less at mealtimes than you would like to eat?
3. How often do you refuse food or drink offered because you are concerned about your weight?
4. Do you watch exactly what you eat?
5. Do you deliberately eat foods that are slimming?
6. When you have eaten too much, do you eat less than usual the following day?
7. Do you deliberately eat less in order not to become heavier?
8. How often do you try not to eat between meals because you are watching your weight?
9. How often in the evening do you try not to eat because you are watching your weight?
10. Do you take into account your weight with what you eat?

FIGURE 20.6 Restrained eating scale from the DEBQ. (From van Strien, T et al. *Int. J. Eat. Disord.*, 5, 293, 1986, Table 2, p. 304.)

Eating Behavior Questionnaire (DEBQ). The Dutch restraint scale has 10 items, and uses a temporal response format as follows: never (1), seldom (2), sometimes (3), often (4), and very often (5). The overall average score was 2.2, with average scores of 1.84 for males and 2.49 for females.

Finally, a 28 item questionnaire on Rigid and Flexible control was introduced by Westenhoefer, Stunkard, and Pudel.[19] Rigid control of eating is characterized by an all-or-nothing approach to dieting, whereas flexible control is characterized by a more moderate approach in which decisions can be traded in order to affect overall restraint.

The questionnaire includes 12 items to assess flexible control (FC12; Figure 20.7) and 16 items to assess rigid control (RC16; Figure 20.8). The first seven items of each subscale are from the Eating Inventory[17] and represent the short forms FC7 and RC7. The original item numbers were 4, 6, 18, 28, 35, 42, and 48 for flexible control and 14, 32, 37, 38, 40, 43, and 44 for rigid control. Scored alternatives are underlined.

1. When I have eaten my quota of calories, I am usually good about not eating any more.
(true–false)
2. I deliberately take small helpings as a means of weight control.
(true–false)
3. While on a diet, if I eat food that is not allowed, I consciously eat less for a period of time to make up for it.
(true–false)
4. I consciously hold back at meals in order not to gain weight.
(true–false)
5. I pay a great deal of attention to changes in my figure.
(true–false)
6. How conscious are you of what you are eating?
(not at all–slightly–moderately–extremely)
7. How likely are you to consciously eat less than you want?
(unlikely–slightly unlikely–moderately likely–very likely)
8. If I eat a little bit more on one day, I make up for it the next day.
(true–false)
9. I pay attention to my figure, but I still enjoy a variety of foods.
(true–false)
10. I prefer light foods that are not fattening.
(true–false)
11. If I eat a little bit more during one meal, I make up for it at the next meal.
(true–false)
12. Do you deliberately restrict you intake during meals even though you would like to eat more?
(always–often–rarely–never)

FIGURE 20.7 Scale for flexible control (FC12). (From Westenhoefer, J., et al., *Int. J. Eat. Disord.*, 26, 53, 1999.)

| 1. I have a pretty good idea of the number of calories in common food. |
| (true–false) |
| 2. I count calories as a conscious means of controlling my weight. |
| (true–false) |
| 3. How often are you dieting in a conscious effort to control your weight? |
| (rarely–sometimes–usually–always) |
| 4. Would a weight fluctuation of 5 lb affect the way you live your life? |
| (not at all–slightly–moderately–very much) |
| 5. Do feelings of guilt about overeating help you to control your food intake? |
| (never–rarely–often–always) |
| 6. How frequently do you avoid "stocking up" on tempting foods? |
| (almost never–seldom–usually–almost always) |
| 7. How likely are you to shop for low calorie foods? |
| (unlikely–slightly unlikely–moderately likely–very likely) |
| 8. I eat diet foods, even if they do not taste very good. |
| (true–false) |
| 9. A diet would be too boring a way for me to lose weight. |
| (true–false) |
| 10. I would rather skip a meal than stop eating in the middle of one. |
| (true–false) |
| 11. I alternate between times when I diet strictly and times when I do not pay much attention to what and how much I eat. |
| (true–false) |
| 12. Sometimes I skip meals to avoid gaining weight. |
| (true–false) |
| 13. I avoid some foods on principle even though I like them. |
| (true–false) |
| 14. I try to stick to a plan when I lose weight. |
| (true–false) |
| 15. Without a diet plan I would not know how to control my weight. |
| (true–false) |
| 16. Quick success is most important for me during a diet. |
| (true–false) |

FIGURE 20.8 Scale for rigid control (RC16). (From Westenhoefer, J., et al., *Int. J. Eat. Disord.*, 26, 53, 1999.)

20.4 FOOD INVOLVEMENT

The notion of involvement came from the service literature rather than the product literature. Bell and Marshall[20] applied involvement to food, measuring the importance of food to the individual. The Food Involvement Scale comprises 12 items, 6 positive and 6 negative, relating to the 5 stages of the food cycle from acquisition to disposal. Items are rated on a 7-point agree–disagree scale (Figure 20.9). The potential range of scores is 12–84, and the average score is 45.

Although currently there appear to be no published food involvement data from a commercial product development setting, preliminary reports indicate that those

1. I do not think much about food every day. (R)
2. Cooking or barbequing is not much fun. (R)
3. Talking about what I ate or am going to eat is something I like to do.
4. Compared with other daily decisions, my food choices are not very important. (R)
5. When I travel, one of the things I anticipate most is eating the food there.
6. I do most or all of the clean up after eating.
7. I enjoy cooking for others and myself.
8. When I eat out, I don't think or talk much about how the food tastes. (R)
9. I do not like to mix or chop food.
10. I do most or all of my own food shopping.
11. I do not wash dishes or clean the table. (R)
12. I care whether or not a table is nicely set.

FIGURE 20.9 Food Involvement Scale. (From Bell, R. and Marshall, D.W., *Appetite*, 40, 235, 2003.)

recruited for commercial product testing score on the upper end of food involvement. If this pattern holds up with further testing, then this imbalance of panelists on food-relevant issues would add to the neophobia findings. That is, consumer panelists participating in studies run by companies tend to be more neophilic and more product-involved than one might expect from average consumers.

Questionnaire development to measure attitudes and traits continues today. Those in need of different techniques should search the literature to determine whether such a technique exits (e.g., search the *Handbook of Marketing Scales*).[21] Another question to ask is whether a technique has been applied to your product category which helps to see norms of data.

20.5 PRODUCT TESTING IN CONTEXT

Product development as well as testing divides between the laboratory and nonlaboratory spaces (e.g., the market and the home). Much of the actual design and development of new products takes place in the laboratory. Yet concept development by companies and postdevelopment product testing occur outside the laboratory. In recent years, it has becoming increasingly clear that where products are tested influences the results. Some of the critical variables are beginning to be seriously explored, as factors that determine these context effects. Contextual and environmental effects have been recently reviewed by Meiselman,[11,22–24] Wansink,[6] and Stroebele and de Castro.[25]

20.5.1 LOCATION: LABORATORY, HOME, AND CENTRAL LOCATION

One approach to assess the impact of eating location on food ratings evaluates identically prepared foods in a broad range of institutional, social, and commercial settings. The multilocation method involves central preparation of the food and distribution to the different locations for eating and rating.[26]

Location/Situation	Mean	N
Army camp	6.63	42
University staff refectory	6.64	36
Private boarding school	6.66	88
Fresher's week buffet	6.69	83
Private party	6.99	77
Elderly residential home	7.05	43
Student refectory	7.08	33
Elderly day care center	7.09	33
THR Patrons	7.58	19
4–5 Star restaurant	7.63	32

FIGURE 20.10 Ratings of overall acceptability in different venues. (From Edwards, J.S.A., et al., *Food Qual. Pref.*, 14, 647, 2003.)

Edwards et al.[27] presented extensive data about ratings assigned to the same food in 10 different meal settings in the United Kingdom (Figure 20.10). The participants in the study comprised a range of age, socioeconomic level, and gender. Thus the data could be looked at as the assessment of the product by different populations. The results were clear, and in line with previous research. A product tested in the institutional locations was rated lower than the same product tested in more home-like or restaurant-like settings. This effect of testing venue on rating accords with data showing that people expect lower quality food from institutional settings.

In the same spirit, but with a different approach, Cardello et al.[28] asked people to say how good they expect the food to be in a number of different locations in a questionnaire study. They observed consistent differences in expected liking in the following order: home > restaurant > fast food > school > military > airline > hospital.

The weight of the data suggests that perception of a product is affected by expectations and that product ratings move in the direction of expectations.[29,30] In the Edwards study and in the prior research, laboratory ratings of food fall above the institutional ratings but below the ratings of restaurants and home.

The picture is not, however, totally clear. Hersleth et al.[31] found conflicting results when varying study locations. They served eight Chardonnay wines differing in three product characteristics, either in a sensory laboratory or in a reception room (capacity of eight people), either with or without food. This makes four test conditions (location, food available). The 55 consumers rated the wines on a 9-point scale, with each consumer participating in the four conditions. The key finding was that context effects were as large as product effects. The presence of food and the enhanced reception room raised hedonic scores 0.3–0.5 scale points, just as the product factors did, with one of the product factors actually reducing liking. The presence of food was a more effective enhancer in the reception room than food in the laboratory.

Even the same researcher can report different effects, however. Thus 2 years later, Hersleth et al.[32] reported no differences in liking scores when six cheeses were tested

| 1. Desire to eat (choice/no choice) |
| 2. Amount of product consumed |
| 3. Number of contacts with sample (several/one) |
| 4. Appropriate time of consumption |
| 5. Environment (comfortable, social) |
| 6. Combination of several food items |
| 7. Sequential versus monadic presentation |

FIGURE 20.11 Seven ways in which HUT differs from laboratory testing. (From Boutrolle, I., et al., *Food Qual. Pref.*, 16, 704, 2005.)

at a laboratory, at a central location test (CLT), and at home using different orders of testing. Furthermore, the degree of social interaction had no effects on cheese ratings. There are at least two possible explanations for the lack of an effect of context:

1. Cheeses are highly familiar, so context plays a smaller role.
2. Meal context, not testing context, is the more effective influence, as we will see below. The study here dealt with evaluation outside of a meal.

Whereas the Edwards et al.[27] and Hersleth et al.[32] studies both included the home environment as one of the various test conditions, Boutrolle et al.[33,34] focused specifically on the differences between laboratory and home testing. The possible difference in scores as a function of laboratory versus home is critical to the food and beverage industries. The typical path is to create the products in the laboratory, test the prototypes in the laboratory, and then select winning products to test at home for longer time periods and more natural conditions, through the so-called home use tests (HUTs).

Laboratory testing differs from HUT in at least seven ways (Figure 20.11). We saw above that laboratory tests of a product generates ratings below those obtained for the same product test in restaurant, but higher than institutional ratings. In a series of studies, Boutrolle and colleagues[32,33] confirmed that laboratory ratings are lower than home ratings.

Therefore, the sum of this work is that *laboratory testing underestimates the product ratings that would be obtained from HUT.*[35] Koster and Mojet[2] have recommended extended HUT as a more valid measure of product acceptability.

20.6 EATING SOCIALLY VERSUS EATING ALONE

Eating is a social activity. This is referred to by the term "commensality," which means eating meals with other people.[36] Commensality refers to two important factors in studying foods. The first factor is the role of meals and meal structure (see Section 20.7). The second factor is the role of socialization or the presence of other people while eating. Other people present at a meal impact what one eats, how one eats, and how one appreciates the food eaten.[37]

The relationship between commensality and food intake was first reported by deCastro and deCastro[38] in a series of papers. They used a food diary, training participants for one day in maintaining a diet diary. People kept detailed records for the next week including with whom they ate. When people ate alone they had fewer daily meals (1.6) than did people who ate with others (2.1). When people ate alone they consumed less food per meal (410 kcal) than when they ate with others (591 kcal). This effect has been called social facilitation of eating, which was recently reviewed by Herman et al.[39] In many replications, deCastro and deCastro found that the amount consumed increased with the number of people present.

The effect of social facilitation of eating has generated many studies that focus on effects of other variables on intake. There are confounding factors, however. Variables that should increase eating also increased the number of people present. For example, dinner is the biggest meal and has the most people present; restaurant meals typically have more people present and are bigger, and meals with alcohol have even more people present and are even bigger.[40]

For some years now, researchers have suggested that the length of the meal, that is, eating duration, might be critical to the social facilitation of eating effects.[41] In several natural eating situations, Bell and Pliner[42] reported that the duration of eating increased with the number of people present. Furthermore, Pliner et al.[43] reported results from another study that independently varied both eating duration and group size. Increased intake was related to eating duration, but not related to group size. This result needs to be replicated in a variety of eating environments, since it reveals a possible mechanism by which to increase or decrease eating and thus a way to indirectly control one's health.

Focusing attention on social effects and eating should also cast attention on eating alone, the absence of socialization and commensality. But eating alone has received very little attention. Eating alone is important for both research and practical reasons. There is a widespread belief that the frequency of eating alone is increasing and that eating alone is associated with poor diet. Furthermore, most food testing for companies is done with isolated subjects, that is, people eating alone.

Eating alone is a major factor in the normal context of daily life, but one that is not particularly constant in terms of person or day-part. Sobal[36] has pointed out that eating alone is devalued in most cultures. In research conducted in one U.S. region, Sobal and Nelson[44] conducted a cross-sectional mail survey. Most people reported that they ate alone at breakfast, alone or with coworkers at lunch, and with family members at dinner. For people in family settings, most commensal eating was with family. Unmarried individuals more frequently ate breakfast and dinner alone, and more frequently ate lunch with friends. Sobal and Nelson found that living alone did not determine whether a person would eat alone. Thus "lone diners" are not necessarily those living alone. Furthermore, living and eating alone do not necessarily characterize the elderly.

Further information on eating alone or commensally comes from the four country Nordic study on eating patterns.[45] The study was based on 1,200 surveys each in Norway, Sweden, Denmark, and Finland and confirms many of the conclusions of Sobal and Nelson, as listed below:

1. These data support the view that we all eat alone regularly. For example, about 2/3 of respondents reported that they ate alone at least once on the day prior to the survey.
2. People tended to eat more often with family in the evening, as Sobal and Nelson had noted.
3. Overall the proportion of people eating alone and with family members was about the same.
4. Furthermore, people who lived alone ate alone three times more often, and older people ate alone more frequently, than did younger people.
5. The study did not support the conjecture that eating alone leads to a poor diet. In actuality, the chance of eating a full (proper) lunch or dinner did not vary whether eating alone or with others, respectively.
6. Finally, eating with colleagues peaked at midday, during the typical lunch time. Eating with friends and others occurred on weekends and was very infrequent. The two latter findings confirmed what had been reported by Sobal and Nelson.

What happens when eating alone is relevant for product testing by companies who assess products for consumer acceptance. Much laboratory testing of food is done among consumers who are isolated in testing booths. This is thought to control other inputs and to help focus on the product being tested. Yet we ought to question whether isolated testing is good at predicting the real world. King et al.[46] suggested that isolated testing should not be a cause for worry. They compared isolated testing and social testing and observed that products were rated only very slightly higher in social testing. The researchers encouraged a true social environment by inviting people to bring colleagues and friends, ensuring that people knew other people at their table.

Another effect of social setting is the phenomenon of social modeling. People tend to "copy" the pattern of eating of those with whom they eat. Social modeling occurs quite often when the people involved are of opposite sex or enjoy different social status. Pliner and Rozin[37] summarize research pointing out that people model eating less in order to make a good impression with strangers and business coworkers (as opposed to family and friends), eat less with a person of superior rank (e.g., for soldiers and their superiors), and finally eat less with someone who eats very little. Social modeling also affects behavior of the obese; they react to social pressure differently. The obese eat less when eating with others (the opposite of social facilitation). Perhaps the most disturbing finding of all for those who conduct research on eating is that people eat less when under observation. Clearly, there is more need to understand the factors that promote eating alone and promote product acceptance when eating alone.

20.7 EFFECTS OF EFFORT, CHOICE, AND MEAL STRUCTURE

Although it has become clear that where food is consumed has an impact on how the food is perceived, research continues to explore the nature of the variables driving these effects of locations and the reason for their differences. The importance of consumer expectations has already been noted above. People have expectations of the quality of food based on the environment in which the food is served/consumed.

These expectations affect perception according to the regular expectation effects of assimilation and contrast.[29,30] With assimilation, product ratings move in the direction of the expectation; this is the basis of luxury branding, where it is hoped that the expectation of quality improves the product rating. With contrast, the opposite occurs, and the consumer reacts to expectation by moving in the direction opposite the expectation. Assimilation effects are more common than contrast effects.

Other variables affect location differences as well. The effort to acquire food is an important variable, affecting food perception in many environments. Meiselman et al.[47] reported that simple manipulations of effort in a natural eating environment, a university cafeteria, could reduce intake of popular foods to almost zero. Effort is one of the key components of consumer perceptions of convenience.[48,49]

When comparing laboratory testing to other natural eating environments, the freedom to choose foods appears to be important. In the laboratory, people usually have no choice regarding what they will sample. Once subjects volunteer for a test, they usually are presented with a series of food samples to evaluate. Subjects very rarely are offered choices and very rarely refuse a sample. This situation differs quite a bit from most natural eating situations where some degree of choice is possible. In two studies of product acceptability methods, King and colleagues have shown the importance of choice. In the first study,[46] King et al. reported that giving consumers a choice of products for testing generated higher scores for those selected products. In a follow-up study in a restaurant,[50] they showed that products ratings were higher only in the location with choice. The reason behind this effect might lie in the enhanced liking which has been demonstrated to occur when attention is focused on a product.[51] Conducting a product test in a real restaurant without choice and with a meal structure did not enhance product ratings. Therefore, the physical setting might not be as important as the procedures employed in product testing. Decorating the laboratory might not have beneficial effects. The working hypothesis is that choosing one product over another results in increased attention and increased liking. That is, if you choose one product over another, you pay more attention and you also "up-rate" it.

Meal structure was the other significant variable assessed in the aforementioned two studies. In the earlier study, acceptability increased when a product was served as part of a meal, in contrast to the components being tested alone.[46] When products were served in the laboratory or in the actual restaurant, but not in an actual meal format, the ratings were similar; when the same products were served at an actual meal, the ratings were enhanced.

We are just beginning to understand the importance of meal structure on overall meal acceptance and individual product acceptance. Pliner and Rozin[37] argue that the meal is the fundamental psychological unit of eating, and as such, the meal should influence most aspects of eating. More information on meals is available in several recent books.[51-54]

20.8   EFFECTS OF FOOD ON MOOD AND EMOTION

Assessing emotional responses to food products is the new frontier of the twenty-first century. Developing valid methods to identify and quantify emotional responses to foods in a commercial setting is one of the first challenges of this new research area.

1. Health*
2. Mood
3. Convenience*
4. Sensory appeal*
5. Natural content
6. Price*
7. Weight control
8. Familiarity
9. Ethical concern

FIGURE 20.12   Factors affecting food choices. (From Steptoe, A., et al., *Appetite*, 25, 267, 1995.)

Common sense tells us that foods affect our moods and emotions. This intuitively obvious notion is borne out by the results of the Food Choice Questionnaire.[55] When people were asked what factors are important in their food choices, mood was one of the factors listed. The major factors shown with stars in Figure 20.12 were health, convenience, sensory appeal, and price.

Defining mood and emotion: Psychologists distinguish the terms mood and emotion. Mood is thought to describe longer term affective behaviors that build up gradually, are more diffuse, and have no specific referents. Emotions, in contrast, are shorter term affective responses to specific referents. Emotions can be more intense than the longer term moods. In practice it is sometimes difficult to be clear whether one is measuring a mood or an emotion, especially since often the same term is used to describe both a mood and an emotion.

Measuring mood and emotion: Moods and emotions are measured using a variety of techniques including questionnaires and checklists, selection of a face on a facial scale, measurement of facial movement, and physiological measures.

The most common method for measuring moods and emotions are questionnaires and checklists. It is important to note that most of these were developed for psychiatric work and are used clinically and in academic research settings. Most categorizations of emotions follow this clinical orientation. No published questionnaire or checklist has been developed specifically for commercial use with food, although commercial interests certainly experiment with the measurement of moods and emotions. Coming from the marketing perspective, Laros and Steenkamp[56] emphasize the basic positive–negative distinction. They divide negative affect into anger, fear, sadness, and shame, and they divide positive affect into contentment, happiness, love, and pride.

Standardized mood questionnaires: One of the earliest mood questionnaires is the Profile of Mood States (POMS) which traces its roots to the American psychology of the 1940s and 1950s. The POMS survey can be specified toward a variety of time frames: feelings during the past week, today, right now, and the past 3 minutes. The POMS contains 65 mood terms, which cover 6 dimensions of mood: tension–anxiety, depression–dejection, anger–hostility, vigor–activity, fatigue–inertia, and confusion–bewilderment. The 65 mood terms are each rated on a 5-point rating scale. The POMS survey has been used extensively in research, and may well be the most widely used questionnaire for research in clinical and academic environments (e.g., Lieberman, 2005).[57]

Another widely used measurement form is the Multiple Affect Adjective Checklist or MAACL. The first version of MAACL was published in 1965,[58] and a revised version or MAACL-R published 20 years later.[59] A bibliography of 1,900 references was published in 1997.[60] The questionnaire is in an easy checklist format with instructions to "Check as many words as you need to describe your feelings." The words in the checklist are at a median reading level of sixth grade in the United States, with 90% of items below eighth grade reading level. The checklist takes about 3 min to complete.

The MAACL-R exists in two forms. The state form asks for moods to describe "how you feel now." The state form has high internal consistency and high and low test–retest reliability. The trait form asks "how you generally feel," and has high internal consistency and high and high test–retest reliability.

The MAACL-R is available in full length and shortened forms containing 132 items and 66 items, respectively. Both scales are divided into five scales. There are three negative scales (anxiety, depression, and hostility) and two positive scales. The first of these is positive affect which is more passive. The second is sensation seeking which is more active and energetic. The Sensation Seeking Scale correlates with other measures of the sensation seeking trait. Positive and negative affect scores are independent. Furthermore, positive affect and sensory seeking are modestly and positively correlated (0.5 on the state form), so both are relevant. The five subscales with sample terms on the checklist appear in Figure 20.13.

There are some commonalities between the mood scales on these two questionnaires as follows:

POMS   MAACL-R

Tension   ⟷   Anxiety
Depression   ⟷   Depression
Anger   ⟷   Hostility
Vigor   ⟷   Sensation seeking and positive affect

1. Anxiety: afraid, fearful, frightened, panicky, shaky, tense
2. Depression: alone, destroyed, forlorn, lonely, lost, miserable, rejected, suffering
3. Hostility: annoyed, complaining, critical, cross, cruel, disagreeable, disgusted, furious, hostile, incensed, mad, mean
4. Positive affect: free, friendly, good, happy, interested, joyful, loving, peaceful, pleasant, polite, satisfied, secure, tender, understanding, whole
5. Sensation seeking: active, aggressive, daring, enthusiastic, merry, wild, (keyed negatively) bored, mild, tame

FIGURE 20.13 Five major subscales for the Multiple Affect Adjective Checklist (MAACL-R). (From Zuckerman, M., et al., *J. Psychopathol. Behav. Assess.*, 5, 119, 1983.)

A shorter alternative to these longer scales is the Brief Mood Introspection Scale developed by Mayer and Gaschke.[61] Mood terms describing "your current mood" are rated on a 4-point scale: definitely do not feel, do not feel, slightly feel, and definitely feel. Terms include lively, happy, sad, tired, caring, content, gloomy, jittery, drowsy, grouchy, peppy, nervous, calm, loving, fed-up, and active.

Recently King and Meiselman[62] have reported development of a new questionnaire designed to test foods in a commercial context. This questionnaire emphasizes several main points:

1. Responses to commercial foods by healthy consumers are usually positive[63,64] and, hence, positive emotions should be emphasized. This is an important point because many existing questionnaires were developed in a psychiatric context and contain numerous negative emotions such as depression and anxiety.
2. A questionnaire designed for commercial use needs to be both practical in terms of length and comprehension, and also needs to be comfortable for consumers who are recruited for commercial tests.
3. A format in which emotions are scaled provides more information than checking all that apply.
4. A larger number of terms are needed to fully describe emotional responses to foods and to differentiate different foods from each other in emotional terms.

Another approach to measuring emotions and moods has been the recently developed facial recognition systems. These methods were developed for use in a commercial, rather than clinical setting. They can be viewed on their Web sites (see below).

A complete discussion of these methods is beyond the scope of this chapter and awaits scientific validation and marketplace performance. The systems are presented to see which emotions they contain and to emphasize the prevalence of negative emotions in analysis of emotion. The emotions are almost entirely negative, with the exception of happiness. Figure 20.14 shows that two of the facial recognition

Noldus	Emotionomics	PrEmo	
Happy	Happiness	Dissatisfaction	Admiration
Sad	Sadness	Boredom	Fascination
Angry	Anger	Disappointment	Satisfaction
Surprised	Surprise	Unpleasant surprise	Amusement
Scared	Fear	Indignation	Desire
Disgusted	Disgust	Disgust	Inspiration
Neutral	Contempt	Contempt	Pleasant surprise

Noldus: www.noldus.com

Emotionomics: www.sensorylogic.com

PrEmo: www.tustudiolab.nl/desmet/premo

FIGURE 20.14 Emotions in facial recognition systems.

methods use the identical sets of emotions, which include anger, fear, and sadness as proposed by Laros and Steenkamp,[56] but include only happiness and not contentment love and pride. The third system, PrEmo, uses animated cartoon characters representing 14 emotions, 7 positive and 7 negative. More can be found on the Web sites for these systems, along with relevant "white papers" that discuss the different approaches and provide rationales and reasons for the use of the technology.

20.9 SUMMARY AND FUTURE TRENDS

This chapter presents recent developments in consumer research. Such a review always reflects the bias of the author, and this chapter is no exception. Nevertheless, the chapter reflects important trends in consumer research on food. There are two key mega-trends with which to close the chapter.

The first trend is greater focus on the person and relatively less focus on the product. This is an important trend. Traditionally, the strong field of food technology, which drives new product development of food and beverage, focuses on the product. Consumer research has been relegated to a supporting technology, called upon when needed. New developments in consumer research, such as the use of psychographics and the measurement of emotional consequences, are shifting some of the focus from the product to the person or at least to the person and the product together. The question has changed to discovering that specific knowledge of the individual which can help to design better products. As we develop better techniques to better reflect the individual in the development process, this trend will continue and grow.

The second important trend is the venue of the research. As we see a better balance between product and person, so we see a better balance between laboratory-based research and research in other, more naturalistic settings. When the focus was the product, much consumer research took place in the laboratory settings where products were developed. The technologists developed products, and the consumer team tested those products all within the controlled environment of the laboratory. As this chapter points out, testing in the laboratory produces different results from testing in other more natural settings. And there is now a growing trend to use a range of testing environments, and in turn to discover why these different locations produce different results. This trend will also continue. The consequence is that consumer research will be seen to be much greater than what can be contributed to product development by the traditional within-laboratory context. Along with this shift in testing venues will be the impact of the Internet as a venue for presenting questionnaires or certain types of test stimuli (e.g., concepts, package designs, etc.), and where the respondent can key in answers privately, at great convenience, for the different types of consumers tests.

REFERENCES

1. E. Van Kleef, H.C.M. van Trijp, and P. Lunning, Consumer research in the early stages of new product development: A critical review of methods and techniques, *Food Quality and Preference* 16, 2005:181–202.
2. E.P. Koster and J. Mojet, Boredom and the reasons some new products fail, in *Consumer-led Food Product Development*, ed. H. MacFie (Cambridge, U.K.: Woodhead Publishing, 2007), pp. 262–280.

3. H. MacFie, ed., *Consumer-Led Product Development* (Cambridge, U.K.: Woodhead Publishing, 2007).
4. R. Schifferstein and P. Hekkert, eds., *Product Experience* (Oxford, U.K.: Elsevier, 2007).
5. L. Frewer and H.C.M. van Trijp, eds., *Understanding Consumers of Food Products* (Cambridge, U.K.: Woodhead Publishing, 2007).
6. B. Wansink, Environmental factors that increase the food intake and consumption volume of unknowing consumers, *Annual Reviews in Nutrition* 24, 2004:455–479.
7. H.C.M. van Trijp and J.-B.E.M. Steenkamp, Consumers' variety seeking tendency with respect to foods: Measurement and managerial implications, *European Review of Agricultural Economics* 19, 1992:181–195.
8. P. Pliner and K. Hobden, Development of a scale to measure the trait of food neophobia in humans, *Appetite* 19, 1992:105–120.
9. P.N. Ritchey, R.A. Frank, U.K. Hursti, and H. Tuorila, Validation and cross-national comparison of the food neophobia scale (FNS) using confirmatory factor analysis, *Appetite* 40, 2003:163–173.
10. S.C. King, A. Henriques, and H.L. Meiselman, The effect of choice and psychographics on the acceptability of novel flavors, *Food Quality and Preference* 19(8), 2008: 692–696.
11. A. Henriques, S. King, and H.L. Meiselman, Consumer segmentation based on food neophobia and application to product development, *Food Quality and Preference* 20(2), 2009:83–91.
12. H. Tuorila, A.V. Cardello, and L.L. Lesher, Antecedents and consequences of expectations related to fat-free and regular-fat foods, *Appetite* 23, 1994:247–263.
13. H. Tuorila, A. Andersson, A. Martikainen, and H. Salovarra, Effect of product formulation, information and consumer characteristics on the acceptance of a new snack food, *Food Quality and Preference* 9, 1998:313–320.
14. A. Arvola, L. Lahteenmaki, and H. Tuorila, Predicting the intent to purchase unfamiliar and familiar cheeses: The effects of attitudes, expected liking and food neophobia, *Appetite* 32(1), February 1999:113–126.
15. T. van Strien, C.P. Herman, R.C.M.E. Engels, J.K. Larsen, and J.F.J. van Leeuwe, Construct validation of the Restraint Scale in normal-weight and overweight females, *Appetite* 49, 2007:109–121.
16. C.P. Herman and J. Polivy, Restrained eating, in *Obesity*, ed. A.J. Stunkard (Philadelphia, PA: W. B. Saunders, 1980), pp. 208–225.
17. A.J. Stunkard and S. Messick, The three factor eating questionnaire to measure dietary restraint, disinhibition and hunger, *Journal of Psychosomatic Research* 29, 1985:71–83.
18. T. van Strien, J.E.R. Frijters, G.P.A. Bugers, and P.B. Defares, The Dutch Behavior Questionnaire (DEBQ) for assessment of restrained eating, emotional and external eating behavior, *International Journal of Eating Disorders* 5, 1986:293–315.
19. J. Westenhoefer, A.J. Stunkard, and V. Pudel, Validation of the flexible and rigid control dimensions of dietary restraint, *International Journal of Eating Disorders* 26(1), 1999:53–64.
20. R. Bell and D.W. Marshall, The construct of food involvement in behavioral research: Scale development and validation, *Appetite* 40, 2003:235–244.
21. W.O. Bearden and R.G. Netemyer, eds., *Handbook of Marketing Scales*, 2nd edition (London, U.K.: Sage, 1999).
22. H.L. Meiselman, The role of context in food choice, food acceptance and food consumption, in *The Psychology of Food Choice*, eds. R. Shepherd and M. van Raats (Wallingford, U.K.: CABI, 2006), pp. 179–200.
23. H.L. Meiselman, The impact of context and environment on consumer food choice, in *Understanding Consumers of Food Products*, eds. L. Frewer and H. van Trijp (Cambridge, U.K.: Woodhead Publishing, 2007), pp. 67–92.

24. H.L. Meiselman, Experiencing food products within a physical and social context, in *Product Experience*, eds. R. Schifferstein and P. Hekkert (Oxford, U.K.: Elsevier, 2007).

25. N. Stroebele and J.M. de Castro, Effect of ambience on food intake and food choice, *Nutrition* 20, 2004:821–838.

26. H.L. Meiselman, J.L. Johnson, W. Reeve, and J.E. Crouch, Demonstrations of the influence of the eating environment on food acceptance, *Appetite* 35, 2000:231–237.

27. J.S.A. Edwards, H.L. Meiselman, A. Edwards, and L. Lesher, The influence of eating location on the acceptability of identically prepared foods, *Food Quality and Preference* 14, 2003:647–652.

28. A.V. Cardello, R. Bell, and F.M. Kramer, Attitudes of consumers toward military and other institutional foods, *Food Quality and Preference* 7, 1996:7–20.

29. A.V. Cardello, Consumer expectations and their role in food acceptance, in *Measurement of Food Preferences*, eds. H.J. Mac Fie and D.M.H. Thomson (London, U.K.: Blackie Academic, 1994), pp. 253–297.

30. A.V. Cardello, Measuring consumer expectations to improve food product development, in *Consumer Led Product Development*, ed. H. MacFie (Cambridge, U.K.: Woodhead Publishing, 2007), pp. 223–261.

31. M. Hersleth, B.-H. Mevik, T. Naes, and J.-X. Guinard, Effect of contextual variables on liking for wine—Use of robust design methodology, *Food Quality and Preference* 14, 2003:615–622.

32. M. Herlseth, O. Ueland, H. Allain, and T. Naes, Consumer acceptance of cheese, influence of different testing conditions, *Food Quality and Preference* 16, 2005:103–110.

33. I. Boutrolle, D. Arranz, M. Rogeaux, and J. Delarue, Comparing central location test and home use test results: Application of a new criterion, *Food Quality and Preference* 16, 2005:704–713.

34. I. Boutrolle, J. Delarue, D. Arranz, M. Rogeaux, and E.P. Koster, Central location test vs. home use test: Contrasting results depending on product type, *Food Quality and Preference* 18, 2007:490–499.

35. H.L. Meiselman, S. King, and A.W. Hottenstein, *Laboratory Product Testing Produces an Underestimation of True Product Acceptability*, Paper presented in Florence, Italy: A Sense of Identity, (2004).

36. J. Sobal, Sociability and meals: Facilitation, commensality and interaction, in *Dimensions of the Meal*, ed. H.L. Meiselman (Gaithersburg, MD: Aspen, 2000), pp. 119–133.

37. P. Pliner and P. Rozin, The psychology of the meal, in *Dimensions of the Meal*, ed. H.L. Meiselman (Gaithersburg, MD: Aspen, 2000), pp. 19–46.

38. J.M. deCastro and E.S. deCastro, Spontaneous meal patterns of humans: Influence of the presence of other people, *American Journal of Clinical Nutrition* 50, 1987:237–247.

39. C.P. Herman, D. Roth, and J. Polivy, Effects of the presence of others on food intake: A normative interpretation, *Psychological Bulletin* 129, 2003:873–886.

40. J.M. deCastro, E.M. Brewer, D.K. Elmore, and S. Orozoco, Social facilitation of the spontaneous meal size of humans occurs regardless of time, place, alcohol and snacks, *Appetite* 15, 1990:89–101.

41. G.I. Feunekes, J.C. de Graaf, and W.A. Van Staveren, Social facilitation of food intake is mediated by meal duration, *Physiology and Behavior* 58(3), 1995:551–558.

42. R. Bell and P.L. Pliner, Time to eat: The relationship between the number of people eating and meal duration in three lunch settings, *Appetite* 41, 2003:215–218.

43. P. Pliner, R. Bell, E.S. Hirsch, and M. Kinchla, Meal duration mediates the effect of social facilitation on eating in humans, *Appetite* 46, 2006:189–198.

44. J. Sobal and M.K. Nelson, Commensal eating patterns: A community study, *Appetite* 41, 2003:181–190.

45. L. Holm, The social context of eating, in *Eating Patterns: A Day in the Lives of Nordic Peoples*, ed. U. Kjaernes (Lysaker, Norway: National Institute for Consumer Research, 2002), pp. 159–198.

46. S.C. King, A.J. Weber, H.L. Meiselman, and N. Lv, The effect of meal situation, social interaction, physical environment and choice on food acceptability, *Food Quality and Preference* 15, 2004:645–654.

47. H.L. Meiselman, D. Hedderley, S.L. Staddon, B.J. Pierson, and C.R. Symonds, Effect of effort on meal selection and meal acceptability in a student cafeteria, *Appetite* 23, 1994:43–55.

48. M.J.J.M. Candel, Consumers' convenience orientation towards meal preparation: Conceptualization and measurement, *Appetite* 36, 2001:15–28.

49. S.R. Jaeger and H.L. Meiselman, Perceptions of meal convenience: The case of at-home evening meals, *Appetite* 42(3), 2004:317–325.

50. S.C. King, H.L. Meiselman, A.W. Hottenstein, T.M. Work, and V. Cronk, The effect of contextual variables on food acceptability: A confirmatory study, *Food Quality and Preference* 18, 2007:58–65.

51. J. Prescott, Interactions between cognitive processes and hedonic states, Paper presented at the *6th Pangborn Sensory Science Symposium*, Harrogate, U.K., 2005.

52. U. Kjaernes, ed., *Eating Patterns: A Day in the Lives of Nordic Peoples* (Lysaker, Norway: National Institute, 2002).

53. H.L. Meiselman, ed., *Dimensions of the Meal* (Gaithersburg, MD: Aspen, 2000).

54. H. Walker, ed., *The Meal* (Totnes, U.K.: Prospect Books, 2002).

55. A. Steptoe, T.M. Pollard, and J. Wardel, Development of a measure of the motives underlying the selection of food: The food choice questionnaire, *Appetite* 25, 1995:267–284.

56. F.J.M. Laros and J.-B.E.M. Steenkamp, Emotions in consumer behavior: A hierarchical approach, *Journal of Business Research* 58(10), October 2005:1437–1445.

57. H.R. Lieberman, Human nutritional neuroscience: Fundamental issues, in *Nutritional Neuroscience*, eds. H.R. Lieberman, R.B. Kanarek, and C. Prasad (London, U.K.: Taylor & Francis, 2005), pp. 3–10.

58. M. Zuckerman and B. Lubin, *Manual for the Multiple Affect Adjective Check List* (San Diego, CA: Educational and Institutional Testing Service, 1965).

59. M. Zuckerman, B. Lubin, and C.M. Rinck, Construction of new scales for the multiple affect adjective check lisy, *Journal of Psychopathology and Behavioral Assessment* 5(22), 1983:119–129.

60. B. Lubin, S.E. Sweamgin, and M. Zuckerman, *Research with the Multiple Affect Check List (MCAACL & MCAACL-R): 1960–1996* (San Diego, CA: Educational and Industrial Testing Service, 1997).

61. J.D. Mayer and Y.N. Gaschke, The experience and meta-experience of mood, *Journal of Personality and Social Psychology* 55, 1988:102–111.

62. S.C. King and H.L. Meiselman, Development of a method to measure mood and emotion in a food product development context, Paper presented at the *Third European Conference on Sensory and Consumer Research a Sense of Innovation*, Hamburg, Germany, September 2008.

63. E.L. Gibson, Emotional influences on food choice: Sensory, physiological and psychological pathways, *Physiology & Behavior* 89, 2006:53–61.

64. P.M.A. Desmet and H.N.J. Schifferstein, Sources of positive and negative emotions in food experience, *Appetite* 50, 2008:290–301.

Part V

Positioned Correctly at the Shelf and in the Media

Success	=	Defining and meeting target consumer needs and expectations	×	The right food	×	Proper packaging and preparation	×	Positioned correctly at the shelf and in the media	×	Meet corporate logistics and financial imperatives

21 Getting the Package and Web Site Graphics Right with Consumer Co-creation

Alex Gofman

CONTENTS

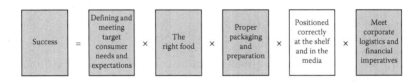

21.1 INTRODUCTION

This chapter shows how Rule Developing Experimentation (RDE) helps designers to use a disciplined approach which sifts through possible design features and their combinations and then discovers what works for a food package design and Web site design. The RDE output is a narrower set of design options, comprising the most feasible and acceptable (to consumers) ones, created out of the designer's talent, but generated by hard data early in the development cycle, in order to be more productive. Timing in the design process is critical. A pattern of productive guidance ought to occur before designers start spending lots of valuable time exploring

369

options that consumers could have told them "would not fly" anyway. Productive guidance to artists/designers ensures that they can concentrate on the more profitable direction.

21.2   OPTIMIZING PACKAGING

Food products and packaging are deeply intervened. People started using some form of packaging to store raw and prepared food about 15,000 years ago. Initial usage of primitive materials like tree leaves, bark, woven grass, etc. led to eventual utilization of animal intestines and skins and later to pottery, then glass, etc. For most of human history, the packaging for food was utilitarian. About 15,000 years ago, late Paleolithic settlers in Japan produced some types of pottery and it is quite conceivable that they or other early humans adorned their creations with the same fascinating images we find now on the walls of the caves they lived in.[1] Ancient civilizations witnessed possibly some of the early known usage of professional art and graphics on food-related packages in the form of artistic amphorae, etc., albeit limited to upper classes.

The main purposes of the packaging were to provide a safe and convenient storage for the food, protect it from spoilage and pests, and facilitate easy transportation. The esthetic side of wrapping the mainstream food came only in the last 200 years.[2]

Nowadays only a small fraction of products is not packaged in one form or another. Starting from a purely practical necessity, packaging has evolved into an industry by itself. In fact, between 10% and 50% of food cost goes to food packaging, amounting for more than $100 billion[3] and growing at a pace of 10%–15% each year.[4]

The extensive use of technology in packaging and distribution brings down the food wastage before it reaches the consumers to only about 2%–3% in developed countries, whereas in other parts of the world between 30% and 50% is wasted. In developed countries, less than 1% of packaged food goes waste compared to 10%–20% of unpackaged food.[5] These numbers indicate the waste attributed only to spoilage of unopened food packages—kudos for the industry. Another story is how much of prepared food we throw away which, according to FoodProductionDaily. com, could reach an incredible 50%![6]

While serving the four main functions of packaging—containment, protection, convenience, and communication[7]— the technological marvels that keep, for example, milk unspoiled for years were beautified by the top designers making it an art. Here is how Yuriko Saito describes the role of packaging in Japan: "The Japanese aesthetic tradition, exemplified in the art of packaging...demonstrates how aesthetic considerations can be thoroughly entrenched in, and integral of, our profound, yet everyday, concerns, such as moral virtues."[8] Unfortunately, this merited approach all too frequently leads to creations of art on the shelves of supermarkets without regard to consumer needs and tastes, sustainability and environmental issues, etc.

On the opposite end of the scale, there is no dearth of experts and readily available best practices that guide many producers in their efforts to create packages. Some suggest that the approaches that work well in other media, e.g., minimalist design, do not apply to package design graphics. The products must figuratively "jump out" at the consumer in order that the consumer picks it out from the shelf, where it competes with many other offerings. Experts advise that the graphics design should be as bold as package configuration, space, and stacking position allow, using lively, persuasive colors, striking typefaces and prominent, creative photography or illustration.[9]

Multiple stakeholders with very different views and goals are involved in packaging —marketers, designers, product developers, brand managers, etc. They all try to improve the creation resulting in many cases in the packages that are "too busy" with too much graphic and text information, some of which might overwhelm, other in contrast might be irrelevant. Although one would think that there is little or no harm in placing an extra message or a visual on the package, the desired consumer response, either comprehension or selection, might actually suffer as a consequence. There is an evidence that the irrelevant information weakens consumers' beliefs that the product will provide the benefit.[10] The authors argue that consumers search in a determined fashion for information on the package that supports the desire for the product to deliver the expected benefits. Consumers view any additional information, irrelevant or disconfirming, as not confirming and thus "diluting" the perception.

It is difficult to overstate the role of correctly choosing the right visual parameters for packaging. Recent studies showed that even when shoppers are open-minded and directly considering a category as opposed to picking up their usual brand, more than one-third of the brands displayed are completely ignored. However, a unique appearance consistently helps attracting shoppers' considerations and drives purchase.[11]

Perhaps the most dependable and effective way to satisfy the consumers is by involving them in the actual process of package creation. Focus groups and other forms of direct questioning on a post hoc basis, although still popular, do not usually produce actionable results. The groups or survey techniques ask the consumer to evaluate what has been created, and perhaps to identify aspects of the package or product that are liked versus those that are disliked. The problems associated with the "actionability" of simple post hoc evaluations have led to other approaches. For example, to increase the actionability of the consumer involvement, some researchers and practitioners have gone so far as to abandon completely attempts to *understand* user needs in detail in favor of *transferring* need-related aspects of product and service development to users through the usage of so-called toolkits.[12] This latter part is purely utilitarian, with the goal to create the product and service by an evolutionary approach that does not, however, produce knowledge of "rules" or reasons "why."

The truth, as usual, is somewhere in between direct questioning to understand but not create, and direct creation that does not necessarily seek understanding. One point is clear in either approach i.e., the consumers should be involved in the package creation. In fact, consumers should *co-create* the package in one form or another to ensure that they will eventually buy it. The full range of consumer involvement on every step of new package design creation is beyond the scope of this chapter (see, for example, [13]). We will concentrate on selecting the right package graphics here and the involvement of consumers in both rule development and co-creation.

RDE[14] widely used to optimize product composition and messaging has been successfully applied to packaging design as well. Although RDE does not replace the creative involvement, it allows the creation of new and attractive packages based on actionable consumer insights. The approach is based on the notion that *customers might not know or be able to articulate what they like if asked directly* (e.g., in focus groups). To find a right package design and graphics, one follows the same six RDE steps used in sensory and message optimization:

1. Create multiple permuted experimentally designed prototypes.
2. Expose them to a target group of consumers (e.g., via a Web-based tool).
3. Collect the ratings of their interest or purchase intent.
4. Find individual contributions of each graphics elements to the dependent variable assessed by a rating scale (e.g., purchase intent).
5. Search for latent segments in the population for better targeting.
6. Create rules that could be utilized by the designer to build new packages.

RDE applied to graphics design is very similar to RDE with just words or with text and pictures which has been widely used for the last 25 years or so.[14,15]

21.3   USING RDE TO IDENTIFY "WHAT WORKS" IN PACKAGE GRAPHICS

Let us explore this approach using the example of a package for pretzels. The task is to identify the design features for a pretzel package, both for the total panel and for latent segments that might exist. Rather than creating a single package for evaluation, the designer will create a "template" with various features (e.g., picture of the product) and a number of options for the features. The goal then is to identify "what works" to drive consumer interest and what does not.

The target package could be divided into six features (Figure 21.1): producer logo (upper left), health message (upper right), brand name (upper center), scenery

FIGURE 21.1   The background layer (center) and the features of the package.

(center belt), flavor (lower center belt), and product image (bottom). All the parts should match each other and the outlining package (in the center of Figure 21.1). RDE requires the creation of multiple packages based on an experimental design. The design shows the package "recipe," i.e., the combination of options for each test package. The design mixes and matches the four options of each feature to create what can essentially be considered a prototype design. The respondent participating in the study simply evaluates the combinations, without necessarily realizing that the different combinations have been created according to a structured approach.

Beyond the experimental design, there is the implementation of the study. Fortunately, today's computers are able to provide a fast and virtually error-free system that creates these combinations "on the fly," presents them to consumer respondents, obtains the ratings, and creates the RDE model. Let us see how.

The experimental design provides the different recipes for the package. The graphical elements (options) are combined/overlaid one with another in order to create some renderings of the prototypes that will be tested on the computer screen. The challenge is that, unlike a box of, say, cereal, a pretzels package is completely "un-flat" which makes it more difficult to mix and match the visuals into a prototype that looks realistic or at least coherent and seamless. The obvious question was how to work on a two-dimensional screen with a three-dimensional problem. The solution is surprisingly straightforward.

Each feature of the package can be thought of as a transparency layer. Adobe Photoshop users find this analogy very easy to understand. Think about layers in a Photoshop project, transparent everywhere except for the key object of the layer. For example, in Figure 21.2 the left frame represents such an option. The outlines of the package on this figure are just to show the option's positioning on the background package. Except for the rectangle in the middle with the picture of the field, the area is transparent.

The computer (browser) superimposes these transparencies according to the RDE design, thus creating different executions of the experimental packages. Each new combination defined by the RDE design corresponds to a different package. Over the course of the interview, the participant evaluates many different combinations of options. The participant never sees the individual transparencies, but only complete packages (Figure 21.3).

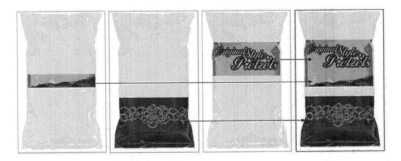

FIGURE 21.2 The process of "building" the experimental packages (the outlines of the package on the left three layers are for demonstration purposes only).

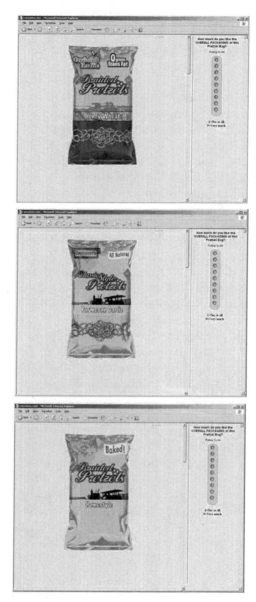

FIGURE 21.3   Sample experimental packages. The bottom package has a missing option.

The designers first created several executions of each of the six features of the pretzels package (see Figure 21.1). Each option was created on a transparent "layer" with the feature positioned properly to allow correct matching during overlaying. All the surface bending was consistently captured inside each option. The designer can do that quite easily based on a single template, shown in the middle of Figure 21.1. The result is a set of graphically realistic pictures, assembled by the principle of RDE.

Computer technology makes the presentation and packages fairly straightforward. RDE provides the design, i.e., the "recipe." The computer program needs only to combine the components for each specific package, present these to the respondents as a totality with transparencies superimposed, and then acquire the ratings. The individual transparencies are already in place at the participant's computer because the server uploaded them at the beginning of the interview (to make screen changes faster).[16]

When participants are exposed to these synthesized packages, they do not really know that the experimental design lies underneath the combinations, nor could they. The transparencies are combined so quickly that, to the participant, it looks like a single package. The participants evaluate package after package, one at a time. If the evaluations last about 12–18 min, most respondents have no problem and actually say that they enjoy the experience. The trick is that people assess visual stimuli much faster than it takes to read text. This speed of the response, almost a "gut-feel response," compensates for the increased number of concepts that graphical RDE uses.

RDE uses experimental designs that require the occasional absence of some components in some test concepts (prototypes). The in-depth analysis of the issue is beyond the scope of this chapter. Previous work by the author with Dr. Moskowitz confirms that despite designer worries that somehow participants will not be able to respond to anything but a complete package, reality is just the opposite. Participants have no problem evaluating both the complete and the partial packages (Figure 21.3, bottom screen). When asked after the interview through a so-called exit survey whether they felt uncomfortable, almost no participant reported feeling uncomfortable with the partial, incomplete packages.[17]

A very important difference from a text-based RDE is the need for a "filler" background image for the situations when the package on the screen occasionally must be without an option, as required by the experimental design. The best solution places a bare package in the back of each screen, behind all the layers. This way, a "zero condition" (the absence of an option in the design) does not create a disturbing image with "holes" in it. The background image of the bare package makes the test stimulus on the computer screen look like an acceptable package, meaningful to consumers even if it has an element missing (e.g., the bottom package in Figure 21.3).

The outcome of this exercise provides to the designer a merge of art, science, and consumer knowledge formalized by RDE. The consumer participants report on exit interviews that they have evaluated realistic-looking packages, rather than rough designs, ensuring a modicum of reality. The respondents, invited through panels, represent the target population of consumers rather than representing "experts," further leading to a sense that the study generates information about consumer reactions rather than expert opinions.

21.4   ANALYZING RDE RESULTS TO UNDERSTAND RESPONSES TO FOOD PACKAGES

Experimental design in RDE automatically creates unique balanced designs for each individual respondent. Each respondent evaluates different combinations, albeit with the same set of elements. RDE generates individual equations or models relating the presence/absence of each of the design feature to the respondent's rating. The model thus shows how every package element drives the response, whether the response be a rating (e.g., 1 = not interested → 9 = very interested) or the response be yes/no (0 → not interested, 100 → interested). The individual modeling, done by regression analysis, will generate understanding of the package features and help to design packages that are more impactful.

Figure 21.4 shows the type of package design data that emerge from the exercise. Recall that each respondent generated a set of impact numbers, or coefficients from the regression analysis. In Figure 21.4, we see these impact numbers, based upon a very simple analysis. The respondent ratings for the different package designs were originally assigned on a 9-point scale. However, following the conventions of market research, the ratings were re-coded. Ratings of 1–6, the lower part of the scale, were re-coded to 0 to denote that the respondent looking at the particular package design was not interested or at best marginally interested. In contrast, ratings of 7–9, the higher part of the scale, were re-coded to 100 to denote that the respondent was interested. Afterward, the RDE program ran the regression analysis, relating the presence/absence of the design features to the binary response of "not interested" or

	Total	Seg. 1 health oriented	Seg. 2 design oriented
Base size	**200**	**61**	**139**
A1 Golden sun	4	4	4
A2 Gregory's	6	6	6
A3 Wholesome foods	5	4	5
A4 Orchard farms	6	4	7
B1 Baked!	6	**9**	5
B2 All natural	6	6	6
B3 50% Less fat than potato chips	8	**11**	7
B4 0 grams Trans fat	8	**11**	7
C1 Homemade pretzels	**14**	5	**16**
C2 Original style pretzels	**15**	6	**17**
C3 Classic style pretzels	**13**	6	**15**
C4 Braided pretzels	**14**	4	**17**
D1 Sun Burst	**10**	*−1*	**13**
D2 Tractor (brown)	**10**	*−3*	**12**
D3 Tractor (black)	**10**	*0*	**12**
D4 Field	**11**	*−2*	**14**
E1 Homestyle	6	0	7
E2 Honey mustard	5	3	5
E3 Parmesan garlic	5	3	5
E4 Honey wheat	8	4	8
F1 Dark brown	**15**	*−14*	**21**
F2 Purple	**19**	*−13*	**27**
F3 Light brown	**14**	*−13*	**20**
F4 Gold	**11**	*−6*	**15**

FIGURE 21.4 Performance of options for the pretzels packaging study (total and two segments). Numbers in the body of the table are the impact values, after the ratings have been converted to a binary scale (ratings 1–6 → 0; ratings 7–9 → 100). The bold values add substantially to the liking; the bold italic detract from it.

"interested." Figure 21.4 shows the parameters of the regression for the total panel and for two emerging mind-set segments with different "points of view."

1. For the total panel, the most important features, i.e., those with the highest impact, are the product image and the brand name. The impact values are very high (from +11 for gold background for the product picture to +19 for the purple background; from +13 for "classic style pretzels" to +15 for "original style pretzels").
2. Scenery pictures, although very positive and important in increasing interest, are not different enough to cause variation in utilities (+10 to +11).
3. The same applies to brand logos (+4 to +6 which is considered neutral) and for the health messages (+6 to +8 which is moderately positive).
4. Flavors are slightly more differentiating with "honey wheat" as the most popular choice across the board, albeit with moderately positive utilities (total and segments).
5. The optimal package for the total panel appears in Figure 21.5 (left).

Consumers are not all created the same. People's preferences may differ substantially. Most traditional approaches divide people based on some demographic or purchase behavior criteria. A more effective approach for design and development divides people based on their mind-set and then specifies the development accordingly. People in the same mind-set segment like the same package design features or products. We can cluster respondents based on the patterns of their responses and try to optimize the package for each segment. This approach proved to be quite robust in many dozens of case histories.[17]

FIGURE 21.5 Optimized packages. The optimal package for total panel and for Segment 2 appears at the left (the same features). The optimal package for Segment 1 appears at the right.

The segmentation analysis revealed two substantially different mind-set of the consumers.

1. Segment 1—health oriented (roughly one-third of the total sample)
   a. The consumers from this segment are positive to the health messages. "50% less fat than potato chips" and "0 grams trans fat" produce utilities of (+11).
   b. This segment is positive to baking perhaps because baking is often associated with more healthy ways of cooking.
   c. The remaining design features and options do not have much impact on the segment with the exception of the background color. Segment 1 dislikes the different colors, except gold.
   d. All-in-all, the health-oriented segment cares only about how healthy the pretzels are and seems to prefer less colorful packages (this part needs further research).
   e. The optimized pretzel package for Segment 1 appears in Figure 21.5 (right-hand side).

2. Segment 2—design oriented (roughly two-thirds of the total)
   a. The majority of the consumers tested belong to this segment. They react much more positively to the bulk of features.
   b. The purple background making the package to stand out on the shelf drives their interest by extraordinary (+27) points meaning that an additional one in four consumers would like the package if this option were to be present. In fact, Segment 2 likes all the options in the product image feature.
   c. Segment 2 likes all brand names and scenery visuals that suggest that they are not picky and very positive to images and design elements.
   d. Flavors, health messages, and logos are neutral to moderately positive to them.
   e. Segment 2, the design-oriented group, thus responds strongly to the graphical elements of the package. They like something that stands out on the shelf, while being less influenced by the rest of the features.
   f. For a designer to create an appealing package, it is quite important to know both positively accepted (by the consumers) elements of the packaging as well as negative ones. The optimized package for Segment 2 appears on the left-hand side of Figure 21.5, and is similar to the optimized package for the total panel. This creates an excellent opportunity for marketing. In a typical case, a manufacturer would consider creating a separate SKU for each segment. In this case, the second segment is much bigger (more than twice) than the first one, and it happens to have an optimized package identical to the total panel. If a company wants to create a single item, this could be their best bet. To attract a smaller but nevertheless sizable segment of health-oriented consumers, the producer should consider a separate item for them as well.

21.5   OPTIMIZING WEB PAGES

It is a truism to say that creating a good product and packaging is not enough—one has to sell the product. After the dot com bust in the early years of this decade, online grocery stores faded away for a while. They are back now and growing very fast. This chapter introduces approaches relevant to optimizing food-related Web page design and graphics.

Business understand the importance of making their Web sites appealing and "sticky." For a long time, the only solution was to rely on gurus: Web designers who were just supposed to know the "right" answers. Yet what if the guru made a mistake or did not take into account all the variables? Indeed, people's perception changes, the target visitors have their own unique preferences, etc. Potential loss of not optimizing the landing pages may be staggering.[18] For example, many Web site designers do not consider payment pages important at all, from the appearance point of view. However, simple changes to those pages could bring a substantial improvement to revenue per visitor with some reporting boosting conversion rates as much as 300%.[19,20]

A recent study by researchers in Canada showed that the snap decisions that Internet users make about the quality of a Web page have a lasting impact on their opinions. They also reported that impressions were made in the first 50 ms of viewing.[21] The implication of these findings is that it is *mostly the main features and the general appearance of the landing page that make a difference, not necessarily the actual content.*[22,23]

Should we solely rely on the artistic taste of Web designers as the only solution? No one can replace a good designer—whether it is a package design or a Web page. Luckily, there are some new ways to help them to achieve much better results faster. In the last few years, an approach called landing page optimization (LPO) has become popular. The idea behind LPO is to create several prototypes and test them with consumers. In the simplest case, the A/B split test approach, there might be only two variations of a page.

On the other hand, the most advanced form of LPO, called multivariate landing page optimization (MVLPO), involves thousands and thousands of prototypes. Although MVLPO was developed in the late 1990s, it did not get the deserved attention until very recently, especially with the introduction of the Google Web site optimizer. A typical MVLPO creates multiple experimentally designed variations of a Web page and evaluates the difference in the reaction or behavior of the consumers who visit these pages. We will use RDE to guide us in the Web page creation process.

The operator of an online grocery store wanted to optimize the landing Web page to increase the conversion rate and revenue per visit. There were several options for the banner, feature picture, and different promotions that should be placed on the Web page (Figure 21.6). What is the best combination of these components?

The approach is similar to the foregoing case study with package design for a pretzel. A landing page can have different layouts. In this case, the site contains

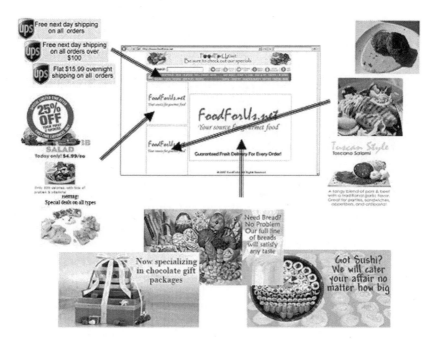

FIGURE 21.6 The template (in the middle, not to scale) and the tested elements of the Web site.

a feature picture, a banner, and three different types of promotions (see Figure 21.6). These placeholders are the *silos* or *categories* (banner, feature picture, etc.).

Each of the five silos on the page has three options, called *elements*. There are many more possible designs (combinations of categories and elements) readily available for different layouts.

The *template* is a schematic of the page; it places each element at a specific location. As we saw in the case of the pretzel package, for some combinations RDE requires that a silo be absent from the design. This absence enables the regression analysis to estimate the absolute values for the impact of the different design features.[14] For the landing page, the, the background of the template has some generic text and a neutral gray filling to be exposed instead of the missing options.

The RDE system dynamically created a relatively unique set of 27 landing page executions for each respondent. The number of executions depends on the design (number of silos and number of elements per silo).

In this case history, the respondents were invited from a Web panel. Fifteen hundred invitations were e-mailed to random panel members. There were 205 responses, and 172 of them completed the survey. Each respondent evaluated 27 unique pages out of hundreds created by the experimental design, as well as several demographic questions. Figure 21.7 shows two sample test screens. Respondents rated each of the screens on a 1–9 rating scale, answering this question: How interested are you in purchasing GROCERIES through this Web site? (1 = not at all interested → 9 = very interested).

FIGURE 21.7   Two sample screens of the interview. Each respondent has a unique set of landing page variations.

The analysis of the land pages follows the same regression modeling as the analysis of the pretzel graphics. The ratings were converted to the binary scale (1–6 → 0; 7–9 → 100). Following the conversion, the regression analysis showed the relation between the presence/absence of the 12 elements and the interest in the Web site. The regression, done on a person-by-person basis, shows the impact of each element and the baseline. The baseline, or additive constant, is the conditional probability of a person being interested in patronizing the store if no elements were present. This additive constant is a purely computed parameter. The 12 different elements each generate a single impact value, showing the conditional probability of the respondent saying he would patronize the store if the particular visual element were present on the Web site. The higher the impact value the more people will buy from the store. Negative impact values mean that fewer people will buy from the store if the element is put into the landing page.

Let us now look at the impact of the different elements for the Web site (Figure 21.8 shows the impacts for the total panel, as well as for two mind-set segments). Lest the reader thinks that only one extracts two mind-set segments from these types of experiments, we should keep in mind that we are trying to paint the simplest picture

		Total	Segment 1 value	Imaginary/ impulsive
	Base size	172	82	90
	Constant	18	20	16
A1	Free next day shipping on all orders over $100	−2	0	−4
A2	Free next day shipping on all orders	9	15	4
A3	Flat $15.00 overnight shipping on all orders	−3	−3	−4
B1	25% Off your first two orders	4	5	2
B2	Turkey cobb salad	0	2	−1
B3	Herring	−1	0	−2
C1	100% organic range chicken	1	2	0
C2	Tuscan style salami	0	3	−3
C3	Quick meals	−1	2	−3
D1	Sushi plate	−5	−12	2
D2	Breadbasket	6	−3	15
D3	Chocolate pyramid	4	−3	11

FIGURE 21.8 Performance of the Web site elements for the total panel and mind-set segments.

possible through the RDE experiment. Thus, if we find that two mind-set segments "tell the story coherently," then there is no reason to create more segments. In contrast, there are occasions when three or even four segments are needed, because the two-segment solution is simply not sufficiently coherent for one or even both segments:

1. The constant represents the basic interest of the respondents in this site if no elements are present. In this case, the constant is relatively low for the total panel (+18), which means that in general the respondents are not very interested in a grocery site (only 18% would be interested in shopping at the store in without seeing all the elements). What could change this perception?
2. For the total panel, only two visuals substantially increase interest. One is the "Free next day shipping on all orders" and the other is the picture of the breadbasket.

3. Other offers (middle-left panel) and food images (lower-left panel) have generally neutral to slightly positive impacts.

4. We can create the best performing landing page by choosing just the top-rated elements from each of the categories. We see this best performing Web site in Figure 21.9 (left panel).

5. The RDE impacts let us calculate the conditional probability of a person saying he will buy from this Web site. We need to only add the impacts of the elements to the additive constant. The value is the sum $P =$ additive constant + sum (utilities) $= 18 + 9 + 4 + 0 + 6 = 37$. This means that if we use the top elements from each category, we can increase the conditional probability of a visitor being inclined to purchase from the Web from 18% (the additive constant) to 37%.

6. If perchance we choose wrong elements (the lowest scores in each category), we would generate the Web page on the right of Figure 21.9. The sum of impacts is far lower: $P = 18 - 3 - 1 - 1 - 5 = 8$.

7. Therefore, by making a correct choice we can get the addition 106% of potential clients being interested in the site versus losing (–56%) of the visitors with a wrong choice. The difference between these two choices is a possibility of losing almost 80% of potential buyers! The result of a simple permutation of quite similarly looking elements may have a tremendous impact on the effectiveness of the Web site as a driver of interest to buy from the store.

FIGURE 21.9   The highest (left) and lowest (right) landing pages for the total panel, created by choosing the best (106% increase in interest) versus the worst (–56% decrease in interest) performing elements.

FIGURE 21.10   Optimized landing pages for the Segment 1 (left) and Segment 2 (right).

8. Even more insightful results emerge from segmenting the respondents, again on the basis of their impact values, just as we did for the pretzel packages. Two equally sized mind-set segments emerge. Segment 1 is the "value oriented" group. Segment 2 is the "imaginary/impulsive" group. The segment names come from the elements that score best.

9. Segment 1, the value oriented, shows a higher constant. We interpret that to mean that this segment is a bit more positive to the general idea of online grocery.

10. Respondents belong to Segment 1 love *free next day shipping* (+15) and positive to the offer of *25% off* offer.

11. The rest of the components leave them mostly indifferent except for the sushi picture. Segment 1 apparently believes that such perishable food items as sushi should not be even offered by an online shopper (−12).

12. Segment 2, the imaginary/impulsive, is generally neutral to everything except for the central mouthwatering images of a breadbasket and chocolate pyramid (+15 and +11).

13. Segment 2 is far less receptive to value offers, suggesting that they react to the items, not to cost savings.

14. We see the optimized landing pages for the two segments in Figure 21.10.

15. The size limitations of the chapter do not allow for the more in-depth RDE analysis of data, which usually includes detecting synergies and suppressions between the elements, etc. (see [15,24]). However, even the top-line results (automatically produced and available in real time) are powerful enough for immediate action.

21.6   CONCLUSION

In 2004, noted author Gladwell[25] dedicated his entire speech at the TED conference to explicating the notion that consumers might not know what they want deep inside, but they will easily react if given the options. This notion, originally promoted by Dr. Howard Moskowitz back in the 1970s and 1980s, resonated widely through the media and industry and has led to the resurgence of experimentation as a method for optimizing packaging and Web design.

To find a winner, one should experiment and test multiple prototypes that have been systematically varied. The goal is to identify patterns of stimuli that drive responses. The result of a simple permutation of quite similarly looking elements may have a tremendous impact on the purchase intent of consumers in a supermarket or the conversion rate on your Web site.

As mentioned in Section 21.1, the conventional methods of focus groups and simple concept tests often turn out to be simple "beauty contests." These research methods generally cannot create rules of the consumer mind. Furthermore, because they are limited to a few executions, they generally are an inefficient expenditure of research funds. They can find winners among the stimuli tested, but little else. In contrast, RDE, as presented here, tests thousands of packages and Web site combinations, develops rules, identifies segments, and generates winning propositions.

Web page design used to be an exclusive domain of designers and highly compensated consultants, at least for the companies with a sizable budget. For the rest, it was just the owners' best guess. With the introduction of new tools, MVLPO has made the field more democratic and available to virtually any Web site operator. For the professional designers, RDE tools create an opportunity to improve their design even further. The research efforts generate solid consumer data that reveal the anticipated effects of the design. Of course, computers and researchers as well as statistics and models do not work alone. For the best results in package design, the output of RDE should be taken as the input for the artistry of the designer, who adds the individual expression. Indeed, the combination of designer and RDE-based consumer insights might well create the best of all worlds—art and science combined for packaging and Web sites.

REFERENCES

1. K.K. Hirst, The invention of pottery, *Archeology* (About.com: Retrieved on January 19, 2008).
2. M.R. Klimchuk and S.A. Krasovec, *Packaging Design: Successful Product Branding from Concept to Shelf*, New York, NY: John Wiley & Sons, 2006.
3. Could be compared to the GDP of New Zealand or Egypt.
4. TheSite.org, Food packaging, http://www.thesite.org/healthandwellbeing/fitnessand-diet/food/foodpackaging (retrieved on January 18, 2008).
5. R. Coles, D. McDowell, and J. Kirwan, *Food Packaging Technology (Sheffield Packaging Technology)*, Ames, IA: Blackwell, 2003.
6. Half of US food goes to waste, FoodProductionDaily.com, November 25, 2004. http://www.foodproductiondaily.com/news/ng.asp?n=56340-half-of-us (retrieved on January 18, 2008).
7. G.L. Robertson, *Food Packaging: Principles and Practice*, 2nd Edition, Food Science and Technology Series, Volume 152, Boca Raton, FL: CRC Press, 2005.
8. Y. Saito, Japanese aesthetics of packaging, *The Journal of Aesthetics and Art Criticism* 57(2), 1999:257–265.
9. J.B. Jarman, Jr., How to take your package design to the next level, *Frozen Food Age* 48(4), November 1999:22.

10. T. Meyvis and C. Janiszewski, Consumers' beliefs about product benefits: The effect of obviously irrelevant product information, *The Journal of Consumer Research* 28(4), March, 2002:618–635.

11. S. Young, New and improved indeed: Documenting the business value of packaging innovation, *QUIRK'S Marketing Research Review* XXII(1), January 2008:46–50.

12. E. Hippel and R. Katz, Shifting innovation to users via toolkits, *Management Science* 48(7), July 2002:821–833.

13. J. Thomas, Begin at the beginning: Research should be involved at the outset of any (re)packaging process, *QUIRK'S Marketing Research Review* XXII(1), January 2008:52–56.

14. H. Moskowitz and A. Gofman, *Selling Blue Elephants: How to Make Great Products That People Want BEFORE They Even Know They Want Them*, Upper Saddle River, NJ: Wharton School Publishing, 2007.

15. A. Gofman and H. Moskowitz, Rule developing experiments (RDE) in package co-creation, in *Proceedings of 2006 IIR FUSE Brand Identity and Package Design Conference*, New York, NY: IIR FUSE, 2006.

16. H. Moskowitz, A. Gofman, M. Manchaiah, Z. Ma, and R. Katz, Dynamic package design and optimization in the Internet era, in *Technovate 2*, Barcelona, Spain: ESOMAR, 2004.

17. H.R. Moskowitz, S. Porretta, and M. Silcher, *Product Design and Development*, Ames, IA: Blackwell Publishing, 2005.

18. A. Theekson, Rocket conversion rates with multivariate testing, Ezinearticles, May 6, 2007. http://ezinearticles.com/?Rocket-Conversion-Rates-With-Multivariate-Testing&id = 554332 (retrieved on January 20, 2008).

19. J. Booth, Can multivariate tests reduce your shopping cart abandons? Real-life results.., Marketing Sherpa, October 03, 2006. https://www.marketingsherpa.com/barrier.html?ident = 29725 (retrieved on January 20, 2008).

20. Increase conversion rate by 300% and increase online sales with accelerated testing, The Search Engine Optimizer.com, 2007. http://www.the-search-engine-optimizer.com/boost-page-conversion-rates.htm (retrieved on January 20, 2008).

21. G. Lindgaard, G.J. Fernandes, C. Dudek, and J. Brown, Attention web designers: You have 50 milliseconds to make a good first impression! *Behaviour & Information Technology* 25(2), March 2006:115–126.

22. A. Gofman, Consumer driven multivariate landing page optimization: Overview, issues and outlook, *Transactions on Internet Research Journal* 3(2), 2008:7–9.

23. A. Gofman, Improving the 'stickiness' of your website, *Financial Times Press*, September 21, 2007.

24. A. Gofman, Emergent scenarios, synergies, and suppressions uncovered within conjoint analysis, *Journal of Sensory Studies* 21(4), August 2006:373–414.

25. M. Gladwell, What we can learn from spaghetti sauce, Speech at Ted 2004 Conference, Monterey, CA, 2004. http://www.ted.com/index.php/talks/view/id/20 (retrieved on January 20, 2008).

22 Getting the Positioning Right: Advertising Planning

Jeffrey Ewald

CONTENTS

$$\text{Success} = \begin{array}{|c|} \hline \text{Defining and} \\ \text{meeting} \\ \text{target} \\ \text{consumer} \\ \text{needs and} \\ \text{expectations} \\ \hline \end{array} \times \begin{array}{|c|} \hline \text{The} \\ \text{right food} \\ \hline \end{array} \times \begin{array}{|c|} \hline \text{Proper} \\ \text{packaging} \\ \text{and} \\ \text{preparation} \\ \hline \end{array} \times \begin{array}{|c|} \hline \text{Positioned} \\ \text{correctly} \\ \text{at the shelf} \\ \text{and in the} \\ \text{media} \\ \hline \end{array} \times \begin{array}{|c|} \hline \text{Meet} \\ \text{corporate} \\ \text{logistics and} \\ \text{financial} \\ \text{imperatives} \\ \hline \end{array}$$

22.1 OVERVIEW: THE CONCEPT OF "POSITIONING"

The "Positioning Era" can be traced to an article on the subject published by Jack Trout in 1969. In 1972, Al Ries and Jack Trout published a series of articles on the topic in *Advertising Age*. But it was Ries and Trout's 1981 bestselling book, *Positioning: The Battle for Your Mind,* which firmly established and popularized the concept on Madison Avenue (quickmba.com).

The breakthrough part of the Ries and Trout's conceptualization is that "a positioning" exists only in the mind of the customer. Ries and Trout felt that, in an era of information overload, which at the time was driven by continuous streams of advertising messages, the consumer would only be able to accept and absorb those messages consistent with prior knowledge or experience. "Positioning" would help the advertiser break through the message clutter. "Positioning" presents a simplified message consistent with what the consumer already believes by focusing on the perceptions of the consumer, rather than on the reality of the product.

The idea that consumer perceptions are critical to the success of a product changed the very basis on which new products could be developed. In their 1987 article entitled "Psychological meaning of products and product positioning," Friedmann and Lessig (1987) argued that products can engender important psychological meaning to customers, and that these psychological meanings can be both complementary and convergent from differentiation strategies based only on rational product attributes.

Today, the concept of positioning is embraced by the marketing mainstream, with the vast majority of marketers using the term as part of their professional lexicon. The term "positioning" has evolved (or devolved, depending on one's point of view) generally to describe any number of techniques by which marketers try to create an image or identity for a product, brand, or company in the mind of a target audience. Popular tools to assess positioning include graphical perceptual mapping, market surveys, and certain statistical techniques (valuebasedmanagement.net).

The marketing community seems to have lost the simple elegance of the original "positioning" idea. Marketing strategists have layered, expanded, and refined elements in the concept of "positioning." Researchers have created a plethora of techniques to measure "positioning." Yet, in the end what matters is how potential buyers *perceive* the product as it is expressed *relative* to the position of competition.

22.2   SEGMENTATION, TARGETING, AND POSITIONING

Although the concepts of segmentation, targeting, and positioning are all too frequently used interchangeably, they are distinctly different and should be considered as sequential relative to each other.

22.2.1   SEGMENTATION

Market segmentation divides the market into groups, called market segments, using characteristics, behaviors, and needs (functional or emotional) that are homogenous within each segment and heterogeneous across segments. Developing a useful segmentation is really a matter of discovering the "best" dimension(s) on which to divide the market into groups. Groupings of people inherently have many bases on which a useful segmentation can be formed. Common examples include demographics, geography, attitudes, need-states, and behaviors (media consumption, shopping behaviors, and product usage). These dimensions coexist. Any of them, individually or in combination, could be an appropriate and useful basis for dividing the market into groups.

In theory, benefit segmentation (i.e., dividing the marketing into groups of people who share the same desired benefit from the product) allows the marketer to develop different product offerings which optimize the delivery of that specific benefit to a targeted group.

In practice, marketers tend to delineate segments based on observable characteristics. Segments created in such a way are easier to operationalize; that is, to identify and address with specific marketing activities (products, services, channel offerings, messages, etc.). It is important to realize that a specific segmentation based on easily observable characteristics best works when the resultant groups are correlated with underlying (and often unobservable) benefit segments.

Whereas the segmentation process is conceptually straightforward, a useful segmentation requires quite a bit of experience and creativity (Sarvary and Elberse, 2006). The "best solution" typically involves compromises across competing marketing objectives and emerges from the insightful analyses of multiple data-streams from many sources. Examples of such data-streams include store scanner data, panel data, attitude and usage surveys, concept and positioning tests, new product tracking reports, consumer trends, etc.

22.2.2 TARGETING

Target market selection involves evaluating each of the segments that were previously identified in terms of their attractiveness to the marketer. This evaluation should consider

1. Overall market structure.
2. Nature of the competition and the competition's ability to serve the needs of each segment.
3. Company and the ability to marshal the resources needed to develop product which meets the segments' needs and its ability to launch and sustain the marketing and organizational support necessary for success.

Approaches for conducting a thorough competitive analysis are outside the scope of this chapter, but generally, the analysis should include assessment of each competitor's ability to

1. Conceive and design new products.
2. Produce products at the appropriate quality and quantity.
3. Launch a new product to both the trade and the consumer.
4. Finance the necessary activities.
5. Manage and execute all necessary activities.

The competitive analysis should not be just a snapshot in time, but should anticipate and predict the competitive reactions that might be faced when each specific segment is targeted and the opportunity exploited.

22.2.3 POSITIONING

Once a segmentation schema has been identified AND specific segments have been targeted, with the belief that a differentiated offering can be produced to deliver the benefits that are sought by the targeted segment(s), THEN the exercise of developing a desirable positioning can be initiated.

Our _____ *is*   _____

    (Product/Brand)   *(single most important claim)*

              among all   _____

              because   *(competitive frame)*

                                           _____

                                           (single most important support point)

FIGURE 22.1   An illustrative template for a positioning. (From Sarvary, M. and Elberse, A., Marketing segmentation, target market selection, and positioning, Harvard Business School Module Note 506-019, 2006. With permission.)

Developing a positioning is an activity. It involves defining a positioning goal—how the marketer would *like* the targeted segment(s) to think about its product relative to competing products. An ideal positioning will optimize the balance of distinctiveness and attractiveness to the target audience relative to competitive offerings. To illustrate why an appropriate balance between distinctiveness and attractiveness is required, consider the following new product concept: "Tuna Meringue Pie." Although certainly unique and distinctive, it is perhaps not very attractive to most.

A clearly stated positioning goal provides an important direction for the development and coordination of Kotler's classic "4 P's of Marketing": product, price, placement, and promotion. (Kotler, 1980)

In the author's experience, many marketing organizations, including both client companies and advertising agencies, have developed their own version of a standardized positioning statement template. The proper template provides a disciplined approach to the clear articulation of a desired positioning. An illustrative example cited by Sarvary and Elberse (2006) appears in Figure 22.1.

22.3 PSYCHOLOGY OF POSITIONING—A THEORETICAL FRAMEWORK

If the goal of a strong positioning is to create in the consumer's mind an understanding of how *one's* product is attractively differentiated from the competition, then logic suggests that a direct comparison between one's product and the relevant competitor would be the strongest demonstration. In fact, direct comparison is often used, particularly when the point of difference comes from a rational, functional product attribute itself. This approach can be labeled the "-*er*" approach. Positioning the product itself as fast*er*, bett*er*, or cheap*er* on an attribute that is meaningful and relevant to the targeted segment should result in preference or choice for one's product over competitor products.

Often a meaningful differentiation may not be based on differences between products on an attribute that allows for a direct, head-to-head comparison. When attributes are not comparable, how then do consumers make meaningful mental comparisons between competitive offerings, i.e., comparisons that influence purchase decisions? And, how do consumers make comparisons between offerings that may not even be in the same product category, but yet solve similar underlying needs, and might justifiably be considered as competitors?

It is well established that consumers do categorize products to enable them to identify and evaluate product-related information (Cohen and Basu, 1987). A useful and perhaps better way to understand the cognition of positioning might come from understanding the underlying psychological mechanism of how consumers process direct and indirect comparisons of potentially disparate product/service offerings (Punj and Moon, 2002).

In general, people categorize "agents" in two interdependent phases, association and differentiation, respectively.

During the *association* phase, the mind looks for patterns of similarity that identify the new agent as being a member of a category of like agents (i.e., the consideration set). These similarity judgments are frequently based on mental comparisons on those key attributes of a brand representative of the product category. This type of comparison is called an "exemplar-based positioning." Alternatively, the new agent can be analyzed on its similarity to an idealized composite of brands which are already in the mental consideration set, using attributes that are in common among the brands being associated (Punj and Moon, 2002). This type of comparison is called "abstraction-based positioning." In either case, the search is for similarity to something, hence the name association. As we will see just below, this type of comparison leads naturally to one type of positioning strategy.

Marketing management's decision to position a product using either exemplar- or abstraction-based comparison approach should vary, depending on the specific situational characteristics. Exemplar-based comparisons focus on a specific brand, leading consumers to define the category more narrowly. In addition, explicit mentioning of the exemplar can have the unintended consequence of highlighting the competitive offering. In contrast, abstraction-based comparisons avoid using a competitor as a reference point and allow consumers to define the category more broadly. However, an abstraction-based positioning requires more mental effort. First the consumer must first "mentally" develop a composite "mean brand" based on an assessment of the various attributes from different types of products which are accessible from their consideration set. After developing this "mean brand," the consumer is ready to compare the new offering to the abstract composite on attributes most meaningful.

Figure 22.2 summarizes Punj and Moon's discussion on the type of association comparison most appropriate for different types of positioning situations.

By way of example, consider two different new products launched under the Vlasic pickle brand name. When Vlasic launched a refrigerated dill pickle product, it was positioned directly against a leading competitor, Claussen, on the important consumer attribute of crunch. This exemplar-based comparison was appropriate due to the fact that the Vlasic offering was seeking to compete directly with an established leader within a target segment of the market, namely refrigerated pickle eaters willing to pay a premium price.

Type of brand comparision which should be emphasised	
Context factor	Positioning option
Product market definition	
	Generic need level Compare brand with abstract category prototype
Industry organization and history	
	When entering a market with a dominant leader Benchmark against dominant (exemplar) brand When entering a market with no dominant leader Compare brand to abstract prototype
Market share condition	
	For a high-share brand Compare brand to abstract prototype For a low-share brand Benchmark against dominant (exemplar) brand
Category life cycle	
	New category Compare brand to abstract prototype Mature category Benchmark against dominant (exemplar) brand
Consumer knowledge	
	More knowledgeable targets Compare brand to abstract prototype Less knowledgeable targets Benchmark against dominant (exemplar) brand

FIGURE 22.2 Association type comparison. (Adapted from Punj, G. and Moon, J., *J. Business Res.*, 55, 275, 2002.)

Contrast this exemplar-based comparison with another Vlasic strategy. The second product was the new Vlasic brand "Snacker's Pickles" which featured a unique cut and shape in a new, portable, single-serve and easy-to-open package. Snacker's Pickles was targeted at a segment of the market that enjoyed the taste and texture of pickles in nontraditional and away-from-home usage occasions. Although there was no direct category competitor in the market, Snacker's Pickles was really competing for consumption across a broad range of portable and savory finger snack foods. The benefits that this Snacker's Pickles delivered were a composite from other products which occupied the targeted away-from-home usage occasion space. The positioning developed for this product was of the abstract type.

22.4 EXTENDING CATEGORIZATION THEORY TO POSITION PRODUCTS ON EMOTIONAL DIMENSIONS

As mentioned earlier, the positioning concept has evolved to a mainstream marketing idea. Many authors and practitioners have developed their own frameworks which categorize the alternative dimensions on which a positioning can be developed. Many of these frameworks span from rationale product features at one extreme (consistent with exemplar-based positioning) to emotional or psychological attributes at the other (consistent with abstract-based positioning).

Mahajan and Wind (2002) provide a scheme by which to understand the different ways of positioning a product or service. In their system, the focus is not on exemplar or abstract, but on emotional response. They assert that "positioning can be based on cognition OR affect."

In their view, cognitive positionings depend on logical arguments which focus on how a product can solve a problem or provide a benefit that the consumer seeks. Comparative advertising executions, which compare the features of the product or service directly with competitors, offer an example of cognitive product positioning. In contrast to cognitive approaches, "affect goes straight to the heart by focusing on emotions, feelings, or drives associated with a product or service."

Mahajan and Wind (2002) discuss classic examples of advertising campaigns that depict an affective approach include Nike's "Just do it" and Budweiser's "Frogs and Lizards." Nike's campaign affectively communicates the inner drive and courage required to achieve the ultimate in sports, and by implication, in any chosen endeavor. And Bud's campaign associates the brand with fun by featuring the whimsical talking frogs and lizards. Logical? No, but effective, because these campaigns resonate at an emotional level and connect their respective brands to core human feelings and aspirations.

To integrate the cognitive and the affective approaches, Mahajan and Wind (2002) identify six positioning strategies that incorporate cognitive and/or affect elements in different proportions:

1. Based on product features
2. Based on benefits, solutions, or needs
3. For a specific usage occasion
4. For a specific user category (target segment)
5. Against another product
6. Disassociated from existing product class or category

They conclude by recommending that marketers not overlook the affective component of positioning alternatives, stating that affect "makes a more powerful impact" and can "forge deeper and more enduring relationships with your customers." Of course, the implication is that deep relationships will ultimately result in increased sales!

22.5 BRAND AS A CONSTRUCT TO FACILITATE ANCHORING OF A POSITIONING

Throughout this chapter, we have tended to use the terms "brand" and "product" somewhat interchangeably. At this point, we introduce the concept of "brand" as a construct by itself. The "brand" facilitates quick mental access to the set of information that, when attached to the brand name, forms the basis for creation of a unique positioning in the mind of the consumer.

The traditional definition of a brand has its roots in the Old West's practice of placing a particular ranch's brand on its cattle. For modern practice, the American Marketing Association has adopted the following definition for the term brand: "a name, term, sign, symbol, or design, or a combination of them intended to identify the goods and services of one seller or group of sellers and to differentiate them from those of competition" (Keller, 1998). This definition, "brand as stimulus," conceives of a brand as a set of visible signals that the brand manager can *physically* control names, logos, colors, packages, advertising, etc. Implicit in this notion is that a brand's components (i.e., its symbols), together with specifically targeted product attributes, can actively be manipulated in order to achieve the selected positioning goal.

An alternative definition of brand emerges from the consumer perspective. In this view, "a brand name represents a collection of concepts that consumers learn to associate with a particular product" (Nagle, 1979). Over time, and with repeated exposure to both the symbols and the product, this internally held collection of notions congeals into a deeply held set of beliefs or impressions that act as a time-stable reference. When accessed and compared to the set of beliefs held about competitive offerings, these deeply held beliefs become the raw materials underlying the creation of brand equity. Kapferer (1997) argues that these impressions create a cognitive filter. Through this filter, dissonant attributes are discounted, and reinforcing attributes are retained. The result is the illusion of permanence and coherence within a brand's perceptual space. Thus, over time a consistent approach to positioning can help to create the important perceptual asset of brand equity.

The foregoing is theory, but theory with a tangible outcome. The tangible outcome is that product developers need to understand how consumers view the perceptual brand-space. Consumers choose brands because they expect that brand to perform in accordance with the set of perceptual attributes they hold in their memory, i.e., the mind-print of the brand. Furthermore, today (2008), an era awash in product proliferation with enormous customer choice, consumers are increasingly demonstrating a cowardly preference for the familiar (Travis, 2000). That is, in the face of choice, consumers opt for brands that consistently deliver against expectations. The brands that consistently deliver are remembered. Knowing how to deliver the sensory characteristics which are consistent with a brand's core positioning is one of the keys to success. The successful product developer must therefore first develop a thorough understanding of the target audience's expectations for the product, not only as an absolute, monadic "preference," but also in the context of the unique positioning carved out among the category's competitors.

22.6   DEVELOPING A POSITIONING

Developing a positioning is an analytic problem-solving activity. The objective is to solve for two simultaneous, desired outcomes derived from the psychological theory of categorization already introduced. These two outcomes are association and differentiation. In practice, according to Kapferer (1997), the marketer has two distinct tasks:

1. Association: Indicate to which category the brand should be associated and compared
2. Differentiation: Indicate what the brand's essential difference and raison d'être are in comparison to the other products and brands of that category

Particularly in a well-established category where multiple product offerings are aimed at each well-recognized consumer segment, the process of choice often depends on both the perceived uniqueness of a brand's associations and on the strength of those unique associations to the brand. Keller (1998) describes these two tasks as establishing the correct point-of-difference and point-of-parity associations:

> "Points-of-difference are those associations that are unique to the brand that are also strongly held and favorably evaluated by consumers.... it is similar to the notion of 'unique selling proposition' (USP), a concept pioneered by Rosser reeves and the Ted Bates advertising agency in the 1950s."
>
> "Points-of-parity, on the other hand, are those associations that are not necessarily unique to the brand but may in fact be shared with other brands."

Keller points out that these common category associations can play one of two roles. Some attributes must be associated with a brand simply to be considered a legitimate and credible offering. One can think of these attributes as necessary, but not sufficient conditions to drive selection of the brand. The second role for a point-of-parity attribute is defensive—to negate a competitor's attempt at creating a point of difference. A competitor may introduce a new attribute. The new attributed may either be real through product development initiatives or perceptual through communication activities. The brand may make countermoves in order to blunt the new advantage claimed by the competitor, thus returning to the "status quo ante," i.e., the basis for choice most uniquely and strongly associated with the brand.

Determining the brand/product positioning which best establishes both an important point-of-difference balanced with appropriate points-of-parity consists of three steps:

1. Do the homework: Conduct in-depth consumer and competitive analyses to identify segments, and to target the appropriate ones.
2. Generate alternative positionings.
3. Evaluate and choose the "best" alternative.

Kapferer (1997) offers the 10 criteria below as a framework for evaluating alternative positionings. Although no single positioning can be expected to meet all these criteria perfectly, the "best" alternative should emerge through careful consideration using this framework as a starting point:

1. Compatibility: Are the product's current looks and ingredients compatible with this positioning?
2. Motivation: How strong is the assumed consumer motivation behind this positioning?

3. Opportunity: What size of market (segment) is targeted by this positioning?
4. Credibility: Is this positioning credible?
5. Competitive advantage: Does it capitalize on a competitor's actual or latent weakness?
6. Financials: What financial means are required by such a positioning?
7. Distinctive: Is this positioning specific and distinctive?
8. Protected: Is this a sustainable positioning which cannot be imitated by competitors?
9. Recovery: Does this positioning leave any possibility for an alternative solution in case of failure?
10. Pricing: Does this positioning justify a price premium?

Recall that earlier, we introduced an example of a standard "positioning statement" template (see Figure 22.1 above). The primary goal of the positioning statement is to codify the selected positioning. The position should provide unambiguous direction to the various activities (internal departments and external agencies/vendors). The output of these activities (i.e., the executional elements of the brand/product) will influence the sum total of the consumers' experiences and perceptions.

22.7 DEVELOPING EXECUTIONAL ELEMENTS TO ACHIEVE THE DESIRED POSITIONING

A review of six new product development strategies and theories (Rudder et al., 2001) revealed an interesting commonality. All of the approaches were conceptualized as a linear and sequential series of steps. Most started with "screening"—either of new business concepts or ideas—and progressed through development, production, trials, and then product introduction.

Based on the author's real-world experience, and in particular, when one formally introduces the step of positioning development, it seems clear that a linear model for developing new products underestimates the complexity of interrelations between the reality-based world of the product developer and the perceptual-based world of the marketer. Alternative approaches to better link marketing and product design through formal hierarchical processes have been proposed (Michalek et al., 2005). Central to these approaches is the key role of iteration: marketing and product design decisions must be updated iteratively as technical realizations of perceptually based targets uncover gaps between the ideal and the doable.

Michalek et al. (2005) also suggest that conjoint-based planning models could be used to include product characteristics, perceptual attributes, product positioning, and consumer heterogeneity (segmentation). Conjoint measurement refers to a class of research and analytic procedures whose objective is to estimate the contribution of components in a mixture (e.g., concept, advertisement, and actual physical product) from measuring reactions to the combination (Green and Srinivasan, 1990). Conjoint measurement becomes a critical tool for blending together the complexity of product, brand, positioning, messaging, and advertising. Ewald and Moskowitz (2007) describe a database of 20 conjoint studies where respondents evaluated concepts comprising product features, brand names, emotional statements, positioning statements, and the like. Analysis of the database demonstrated that marketing efforts (e.g., brand names and positionings) and product attributes do indeed have strategic interactions. Different brands and positionings support different types of product characteristics, and therefore the alignment between the two is what is really critical.

Beyond the iterative effort required to align the brand, positioning, and key product characteristics, there are many other executional elements that must be "in sync" in order to effect a clear and consistent positioning. Examples of these elements include logos and symbols, characters, slogans, packaging, and pricing. Alternative executions for each of these elements can also be included in a conjoint-based study to help identify optimal combinations.

22.8 ANTICIPATING COMPETITIVE EFFECTS AND REACTIONS

To reiterate, the concept of positioning has as its foundation a simple idea. What matters is how potential buyers *perceive* the product as the product is expressed *relative* to the position of competition. Either consciously or unconsciously, competitors may blunt the points-of-difference that the marketer attempts to establish with his own new products and/or communications. Consideration of how competitive positionings might evolve and the potential ripple-effect on consumer perceptions of one's brand are important when developing the optimal position for one's product or brand.

As a longer-term strategic variable, positioning should be relatively stable unless the structural characteristics of the market change. Forward-looking "what-if" analyses should be used to develop positioning strategies with the potential to maintain differentiation under a number of competitive scenarios. However, the very notion of a single sustainable positioning suitable over the life of a product ignores the dynamic nature of markets. Sustainability, while absolute as a goal, is a relative term: a given positioning is only more or less sustainable than the other alternatives under consideration.

In dynamic markets, shifts in the structure of consumer preferences and/or shifts in the competitive structure can necessitate consideration of an adjustment in our product's positioning. One example of such a situation might be a new competitive product entry aimed at one's target segment. Another example might be the positioning effort initiated by an existing competitor that encroaches on the unique positioning of one's offering. This shift or adjustment in the face of competition or changing preferences by consumers is commonly referred to as "repositioning."

One approach is defending the current turf. In theoretical work aimed at better understanding the generalized optimal strategies for products defending against a new competitor, Hauser and Shugan (1981) demonstrated that, at the margin, repositioning efforts should be increased in the direction of the defending brand's strengths. They also determined that any investments in product improvement should be focused at increasing the ability to defend those differentiating attributes that lie at the core of one's own brand. That is, maintain all the market "territory" that has been already conquered by the brand.

Formal approaches, such as game theory, have also been used to analyze the effectiveness of positioning moves and countermoves (Marks and Albers, 2001).

For example, one study assessed whether in their reaction through pricing and positioning, competitors follow a profit-maximizing Nash equilibrium. A Nash equilibrium exists when, in a competitive, noncooperative "game" (i.e., a competitive marketplace), no "player" (i.e., marketer) can improve his outcome by changing strategy (*ISCID Encyclopedia of Science and Philosophy*). In the study previously mentioned, a marketing simulation game was run with 240 advanced marketing students. The results showed that pricing and budgeting decisions were very well described by the Nash equilibria (assuming that all players attempt to maximize their profit). However, product positioning decisions were not explained by the Nash equilibria. It appears that, when it comes to positioning, decision makers have a more difficult time staying focused on delivering the maximum value to an identified target segment and instead, are prone to position their product too close to the competition. The consequence is that the marketers dilute the profit potential of their own positioning. (Marks and Albers, 2001)

Similar results have been observed in the "real world." While consulting for a Fortune 100 company, the author developed a formal game theory analysis to predict a key competitor's reaction should the client introduce a new product which would be positioned very close to the competitor's existing profit-generating product. The analysis indicated that introducing the new product would be a mistake—it would result in a significant reduction in overall profitability for the client firm. Pressure from the distribution channel to have a product "to match competition" forced management to ignore the game theory prediction, with subsequent results that far underperformed expectations.

22.9 APPLICATION OF THE POSITIONING CONCEPT TO THE FOOD INDUSTRY

In a market such as the food industry, with strong competitive dynamics that include both large well-resourced conglomerates and smaller, more nimble and innovative firms, one might assume that a product-focused positioning could be easily copied. This is especially true because the technology does not change quickly. In light of the dynamics of technology, the relative slow-movement of the food industry, and the ease of copying, perhaps the better, more sustainable approach for differentiation in positioning might be the psychological one. Many potential dimensions exist for creating an important perceptual difference versus competitors. Systematic exploration of specifics by food category and by consumer type is required to identify the best alternative.

REFERENCES

Cohen, J. B. and Basu, K. Alternative models of categorization: Toward a contingent processing framework. *Journal of Consumer Research*, 9, 1987, 225–235.

Creating the perception of a product/brand/company identity (http://www.valuebasedmanagement.net).

Ewald, J. and Moskowitz, H. R. Market forces: The push-pull of marketing and advertising in the new product business. *Accelerating New Food Product Design and Development*, Beckley, J. H., Foley, M., Topp, E. J., Huang, J. C., and Prinyawiwatkul, W. (Eds.). Blackwell Publishing, Ames, IA, 2007.

Friedmann, R. and Lessig, V. P. Psychological meaning of products and product positioning, *Journal of Product Innovation Management*, 4(4), 1987, 265.

Green, P. E. and Srinivasan, V. Conjoint analysis in marketing: New developments with implications for research and practice, *Journal of Marketing* 54(4), 1990, 3–19.

Hauser, J. R. and Shugan, S. M. Defensive Marketing Strategies, University of Chicago Working Paper #1243-81, May 1981, p. 27.

International Society for Complexity, Information and Design (ISCID). Nash Equilibrium (http://www.iscid.org/encylopedia).

Kapferer, J.-N. *Strategic Brand Management: Creating and Sustaining Brand Equity Long Term*, 2nd ed. Kogan Page, London, 1997.

Keller, K. L. *Strategic Brand Management: Building, Measuring, and Managing Brand Equity*. Prentice-Hall, Englewood Cliffs, NJ, 1998.

Kotler, P. *Marketing Management: Analysis, Planning, and Control*. Prentice-Hall, Englewood Cliffs, NJ, 1980.

Mahajan, V. and Wind, Y. (Jerry). Got emotional product positioning? *Marketing Management*, 11(3), May/June 2002, 36.

Marks, Ulf G. and Albers, S. Experiments in competitive product positioning: Actual behavior compared to Nash solutions. *Schmalenbach Business Review*, 53, July 2001, 150–174.

Michalek, J. J., Feinberg, F., and Papalambros, P. Linking marketing and engineering product design decisions via analytical target cascading. *The Journal of Product Innovation Management*, 22, 2005, 42–62.

Nagle, T. T. A Theory of Brands and Brand Management, University of Chicago Working Paper, *Journal of Sensory Studies*, 22(2), 1979, 126–175.

Positioning as popularized by Al Ries and Jack Trout (http://www.quickmba.com/marketing/ries-trout/positioning).

Punj, G. and Moon, J. Positioning options for achieving brand association: A psychological categorization framework. *Journal of Business Research*, 55, 2002, 275–283.

Rudder, A., Ainsworth, P., and Holgate, D. New food product development: Strategies for success? *British Food Journal*, 103(9), 2001, 657–670.

Sarvary, M. and Elberse, A. Marketing segmentation, target market selection, and positioning, Harvard Business School Module Note, 506-019, 2006.

Travis, D. *Emotional Branding*. Prima Publishing, Roseville, CA, 2000.

Trout, J. Positioning is a game people play in today's me-too market place, *Industrial Marketing*, 54(6), June 1969, 51–55.

Part VI

Meet Corporate Logistics and Financial Imperatives

Success	=	Defining and meeting target consumer needs and expectations	×	The right food	×	Proper packaging and preparation	×	Positioned correctly at the shelf and in the media	×	Meet corporate logistics and financial imperatives

23 The Importance of Product Innovation

Phillip S. Perkins

CONTENTS

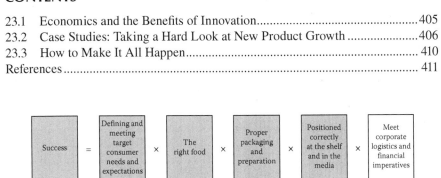

Recently, I was challenged to answer the question "what is the value of innovation to your company?" My scientific training naturally led me to start with a survey of literature published in the last decade. After scouring hundreds of references, reading dozens of articles, then culling it all down to a little more than 60 pages, I can confidently say that "the simple answer to the question is that there is no simple answer!"

Having said that, the fact that innovation is important to the success of a company has never been in dispute. Indeed, innovation has been a key topic for many consultants, academics, and practitioners who have written on the subject during the past decade. Analyses shows that one of the greatest difficulties lies in measuring how important innovation is to a company. One brand asset management study (Kuczmarski, 2002) showed that 44% of managers felt that innovation success must be assessed if a company is to grow—yet the study declared that none of them knew how to do it. Several methods of measuring innovation have been proposed; some break innovation into 3 or 4 categories (i.e., process, portfolio, performance, overall return on innovation, etc.), whereas others propose 16 separate metrics (Juttelstad, 1997). One article on the metrics of evaluation cites a scorecard approach using 33 tracking parameters that would keep a small army of scientists and accountants gainfully employed for months (Geisler, 2002).

One of the main reasons why innovation is so difficult to measure is that there is no clear definition of what innovation is. Without doubt, it means different things to different organizations. Professor Teresa M. Amabile of the Harvard Business School provides us with a broad definition namely "Innovation is the successful implementation of creative ideas within an organization" (APQC, 2002) ... but what does "successful implementation" mean? I suggest that a better definition might be "Innovation is the process of turning creative ideas into commercially successful products to achieve sustainable growth and profitability." Let us agree on this as a starting point. Moving forward, then, just how much sustainable growth and profitability does innovation lead to?

An analysis of 195 new product cases from 125 industrial product firms (Cooper, 2001) shows that highly innovative products do especially well in terms of measures such as

1. Success rate
2. Overall success as gauged by profitability
3. Domestic and foreign market shares
4. Opening new windows of opportunity
5. Meeting sales and profit objectives

Clearly, these five factors can make-or-break a product's success in the marketplace, and later on in this chapter, we will see case studies which illustrate where innovation has had this kind of impact in varying degrees.

Conversely, lack of innovation can cause products to fail, erosion of profits, lost opportunity, and severe loss of market share. For example, let us take a high-level look at the computer industry. In 1988, IBM's share of total market capitalization within the computing and office equipment domain amounted to 45.9%. Since then, IBM has lost market share to more innovative companies. Indeed, some analysts estimate that if IBM had maintained its share, then it would be worth approximately $140 billion more than its current value. So, one answer to the question is that "lack of innovation" can literally "cost" a company "hundreds of billions of dollars."

Let us also pause for a moment and think about Apple. For a company that looked doomed in the 1990s, Apple is literally an iconic company with a market capitalization that almost tripled in 18 months (think iPod and iPhone) from $40 billion to $114 billion, passing both Dell and Oracle in the process and looking like it may pass Intel and IBM whose market caps are $145 billion and $161 billion, respectively (as of July 2007). Clearly, using this example great product innovation can be "worth hundreds of billions of dollars."

Looking at these numbers, it is not surprising that shareholders love companies with innovation built into their strategy. For almost all companies, the goal is to improve shareholder returns. Many companies plan to achieve this from revenue growth, not from improvement in operating margins. A 2006 review of the Strategic Plans for 22 leading food and beverages companies in the United States showed that a third of them specifically call out innovation as part of their strategy. Specifically, these companies are Anheuser-Busch, Coca Cola, General Mills, Hershey, Hormel, Kraft, McCormick, Nestle, Proctor & Gamble, and Wrigley (Food Industry Report, 2006).

23.1   ECONOMICS AND THE BENEFITS OF INNOVATION

One of the reasons for a focus on innovation is that companies are simply running out of headroom on margin improvement. In one study of Fortune 1000 companies (Hamel, 1998; The Economist, June 2007), the top 40 performers averaged a compounded annual growth in revenues of 25.3% over a 10-year period. During the same period, their operating margins improved at a rate of just 6.7% per year. So the top 4% of companies grew their revenue by 25% but only grew their margin contribution by 7%, that is, a ratio of 3.5 to 1. This can be broadly translated by stating that if a company aspires to be in the top 4% of its industry then it ought to invest at least 3.5 times the amount of effort into revenue growth (a.k.a. innovation) that it invests in margin improvement (cost reductions). That is, not to state or even imply the absence of innovative ways to reduce costs, but it does reinforce the old adage "you can't cost save your way to prosperity."

Let us now go one step further in this analysis. The foregoing assumes that every revenue growth opportunity is a success, whereas typical numbers for the success rate of new products range from 30% to 60% (Brand Strategy, 1999; Business Insights, 2004; Narendra, 2007). When we factor in those numbers, the data shows that companies need to invest somewhere in the region of 7 or even 10 times more effort into revenue growth than they invest in improving margins. The metrics vary from company to company but the facts remain that in order to be successful, companies must grow revenue and that requires innovative new products, processes, and packaging along with innovative sales and marketing plans.

Let us move to some specifics and focus on new product innovation in food and beverage industries. A study by Information Resources International (IRI) investigated the success or failure of 2,250 new launches in the United Kingdom which accounted for 7% of the 32,000 brands within 260 categories on supermarket shelves (Brand Strategy, 1999). Fewer than one in five of these new launches succeeded, according to IRIs criteria for success.

The picture in the United States is similar although on a larger scale. The rate of new product introductions in the United States has grown dramatically since the turn of the millennium (Figure 23.1) yet out of the 22,000 New Food Products launched in 2002, fewer than 4,000 of them still could be found in 2007. Why is this dismal survival rate the rule, rather than the exception? We might speculate that it is due to the fact that 86% of new product launches are "me-too" or incremental improvements to existing products. Furthermore, and additionally discouraging, these "line extensions" account for only 39% of new profits (Dru, 2002). In contrast (and it is not difficult to do the math here), the remaining 14% of launches—those that are truly innovative—generate 61% of profits.

Companies with a successful innovation program can achieve new product success rates in excess of 66% (unpublished data), so it is reasonable to expect that successful innovation strategies will at least double the new product success rates that we see today. If improved innovation were to double the new product success rate, then this incremental volume would be worth a minimum $2 billion for the U.K. supermarket industry alone. Similar calculations for industries in the United States show that innovation could be worth in excess of $12 billion to the food industry, $65 billion to the

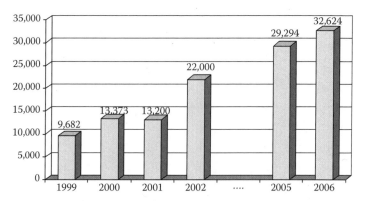

FIGURE 23.1 Food product introductions in the United States from 1999 to 2006. (From USDA Economic Research Service, Food Institute Report and Stagnito's New Product Magazine.)

chemical industry, $40 billion to the paper industry, $20 billion to the drug industry, and $16 billion to the automotive industry. Without question, successful innovation is worth hundreds of billions of dollars.

In order to survive, products must have a recognizable point of difference and fulfill a real consumer need—how to achieve that is covered elsewhere in this book and in the book *Selling Blue Elephants* (Moskowitz and Gofman, 2007). This applies equally to the success of new products and the survival of existing products. Needless to say, companies must continually innovate to develop successful new products and to maintain core business vitality.

To sum up, whether one considers the value of innovation from the perspective of market capitalization or the impact to the bottom line of successful new product launches, the "importance" of innovation is considerable and the "value" of innovation can be counted in billions of dollars, regardless of the industry in which one might operate.

23.2   CASE STUDIES: TAKING A HARD LOOK AT NEW PRODUCT GROWTH

Having made the point regarding the importance—and value—of innovation, now let us take a look at some case studies from the business units of food companies all of whom have active new product development (NPD) programs that invest in Consumer Research, Innovation, R&D, and Marketing. Each of these studies shows that we need to look beyond sales growth to truly understand the return on our investment in innovation. Our first example looks at sales data for a division of a food company that appears to be growing at a healthy rate, with sales growth over a 6-year period from $350 million to $480 million, a compound annual growth rate (CAGR) of 6.5%. These sales are shown graphically in the bar chart in Figure 23.2.

Conversations with the management team uncovered that this growth was predominantly achieved by the division launching one new product every year and that

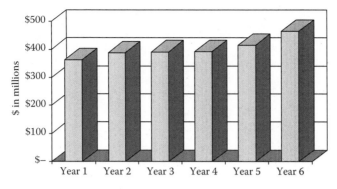

FIGURE 23.2   Six years of sales growth … good or bad?

the management team appeared to be happy with the growth of this division. The strategy and the business goals for this division were set knowing that NPD is expensive and that new product launches carry the burden of millions of dollars of slotting fees, promotional dollars, advertising costs, trade dollars, etc. The business team also knew that new product introductions are important to the growth of the company. Unfortunately, few parameters were set regarding the profit objectives for the new products, their marketplace sustainability, or their likelihood of cannibalizing existing products.

In order to get a true picture of the growth this group has "enjoyed", we need to understand what drove the growth. This understanding requires us to take a closer look at the sales data for both the new product launches and the core business during this time period.

Figure 23.3 suggests quite a different story. The bottom (largest) bar illustrates what is happening to the core business over this 6-year period. Starved of attention, possibly due to all the activity on NPD, the core business is declining. The core business is losing both volume and market share, some of it no doubt, due to cannibalization by the new products whose contribution is shown by the shaded bars on top of the

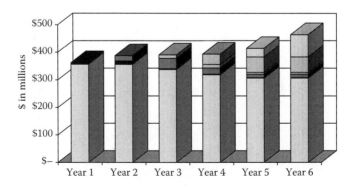

FIGURE 23.3   Core and new products in decline despite overall growth.

core business. Yet total sales are still growing with a CAGR of 6.5%. Should the management team be concerned? Absolutely! Despite the growth!

Let us take a closer look at this so-called NPD success. Each new product appears to do well following its launch with even more sales in year 2. Of course, this apparently strong performance might well be due to the so-called pipeline fill—sales to the trade that are filling the shelves—rather than replenishing stock that is due to consumer take-ways (i.e., actual demand being driven by consumers taking the product off the shelves). Yet the overall numbers still look good. Now move ahead to year 3 after launch. The sales of the new product decline, and to make the situation worse, the core business has also declined. Furthermore, the lack of sales for the new products carries with it the risk that it will be delisted and taken off the shelves by the retailers. "Fear not" the business team leader explains, "for we have more new products to launch" and so the game goes on. Year after year, the costly new product launches are able to mask the poor performance of last year's launch and the decline of the core business. To the outside observer (and may be to the Board of Directors) this business unit looks healthy and growing, but the growth is very expensive and it is unsustainable. I refer to this as "inappropriate innovation" on top of an "undernourished" core business. Sadly this scenario is all too common in the United States food industry today and it is played out in company after company, all dedicated to innovation at any cost.

Case Study 2

In this example (Figure 23.4), the business unit is struggling to maintain sales. There is some increase in sales in the first 3 years, but then there is a decline resulting in an almost flat period of growth over the 6-year period when adjusted for inflation.

A closer look at the data—by breaking it out into core products and the contribution of new products—shows that the core products are declining in sales and it is the new product launches which are solely responsible for maintaining total sales (and they are struggling to do so). Indeed, the product launched in year 3 has dwindled to the point where it is barely discernable on the chart by year 6. It is likely that unwarranted—and probably unnecessary—line extensions (or flavor changes) are the cause here. Whereas the line extensions may bring "new news" to the category (e.g., Double Dutch

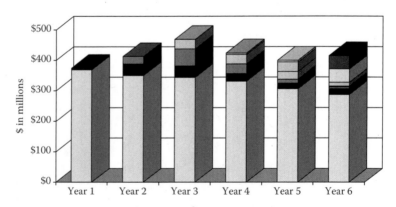

FIGURE 23.4 No growth despite significant NPD activity.

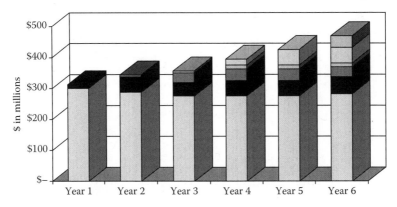

FIGURE 23.5   Growth attributed to NPD.

Chocolate in place of Belgian Chocolate!), such innovation does very little to truly grow the business and is very expensive to maintain. We see here an example of "wasted innovation"—the resources spent on the line extension could have been deployed on products that could really drive consumer demand. What might be necessary here is a deeper understanding of consumer needs which would focus the innovation program on meaningful new products rather than the knee-jerk reaction to the reflex-hammer of upper-management who are demanding new news.

Case Study 3
In the third case study shown in Figure 23.5, the performance of the business unit starts looking somewhat better.

The core business is largely stable, perhaps suffering a little from cannibalization, and the new products appear to have some staying power. This NPD team appears to have its act together and is doing a good job of launching sustainable new products, while keeping the core business on track. Moreover, they are not launching new products for new-products-sake, and we see this in year 5 when no new products were launched. Either the management team was happy with a CAGR of 7.1% or they decided that they did not have a product good enough to launch in year 5. Clearly, not launching a bad product is a good business decision. Any business manager with this kind of sales result should be very pleased with the success of their NPD team, but they could set the bar even higher—they could aspire to grow the core business, while launching profitable and sustainable new products. This is shown in our final case study.

Case Study 4
The fourth and final case study shows a business where sales are growing with a CAGR of 10% as a result of successful and sustainable new products in addition to core business growth (Figure 23.6). This growth model is remarkable in that it happens rarely, although I am pleased to say that I have experienced it personally.

The reality is that this growth pattern is difficult to achieve and an analysis of the food industry shows it happens only about 15% of the time. Yet this is frequently reflected in business plans. In one strategic business review of a multibillion dollar

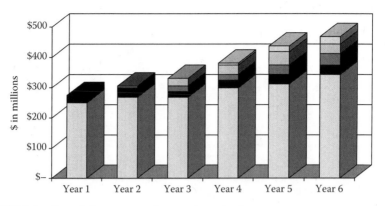

FIGURE 23.6 Growth from new products and core business.

international food company, all 13 business teams submitted plans that aspired to this kind of growth even though the likelihood is that only two or three of the divisions could achieve results like this. If senior management recognized this limitation, then they would allocate their resources strategically and their overall business plans may be more realistic and less disappointing to investors. If this view sounds cynical, remember that a cynic is what an idealist calls a realist. This type of growth—driven by successful new products launched on top of a growing core business—is what most NPD managers aspire to achieve, and the information shared in this book will help readers achieve this kind of growth.

23.3 HOW TO MAKE IT ALL HAPPEN

Clearly developing and launching products that consumers want is key to growing a successful business via new product innovation, but I need to offer up a caution here: Do not take everything consumers tell you on face value — "If I had asked customers what they wanted, they would have asked for a faster horse" (Henry Ford).

Clearly innovation is important and the right innovation is critical. The wrong innovation can be very expensive. So, the question is really quite simple: How should companies go about making profitable innovation happen? In reality, you cannot make innovation happen, but there are several things an organization can do in order to nurture an innovative environment.

It is important to have clarity and focus around what the vision for the company (or business unit) is. Once the parameters or "side-boards" are clearly understood, a team can proceed to the creative job of coming up with new and innovative ideas. Details regarding the processes used are covered elsewhere in this book but a few key lessons are worth sharing here:

1. Innovation is not the domain of one group or one department. Good ideas come from anywhere. Indeed 90% of food companies in the United States claim that they deploy cross-functional teams for product development. Moreover, good ideas can, and should, be solicited both from inside and

outside the company. This is becoming more prevalent recently with companies learning to leverage "open innovation" and to leverage Internet technology to solicit ideas from consumers, customers, and academia.

2. It is important to design new products around consumer wants and needs, not around process capabilities or technology limitations. An excellent tool for doing this is conjoint analyses, which should be used in conjunction with focused market research and well-designed protocepts (prototype concepts) as part of a stage-gate process for developing new products.

3. Smart companies should sometimes ignore what the market says it wants today. An example of this is the mouse, which we all use along with our computer's keyboard. Early in the development of computer systems, consumer research was conducted with computer users who were asked to test a mouse device. The results showed clearly that there was no interest in such an interface with the computer. Why? At the time, all computer users were well-versed with the keyboard and thought it would be a major disadvantage to remove one hand from the keyboard in order to manipulate a mouse.

In closing this chapter, let us reflect for a moment on some quotable quotes throughout history that all relate to innovations that have shaped the way we live today. Thomas Edison stated that "A phonograph is not of any commercial value"; Lord Kelvin said that "Heavier than air flying machines are impossible"; Harry Warner (of Warner Brothers) said "who in the hell wants to hear actors talk"; and T.J. Watson (chairman of IBM in 1943) stated that "There is a world market for about five computers." And one of my favorites from the U.S. Patent Office in 1899 "Everything that can be invented already has been."

The lesson with all of these examples is that it should not be left to an individual—no matter how smart, or successful, or important he or she might be—to decide whether or not a new product is a good idea and has the potential to become a commercial success. Good ideas can come from anywhere, but then again, so do bad ideas. Once formulated, an idea must be tested with consumers using all the tools at our disposal to ensure that what they tell us can be leveraged in developing successful new products. But, be warned, even with all the tools at our disposal, the best and most successful new product companies still fail 30% of the time.

I would like to finish this chapter with one final and very relevant quote from Mike Vaughn, Vice President of Innovation for Ball Corporation: "Innovation is expensive, but good innovation creates value."

REFERENCES

American Productivity & Quality Center (APQC), Using knowledge management to drive innovation. *Consortium Benchmarking Study*, 2002.

Cooper, R. *Winning at New Products: Accelerating the Process from Idea to Launch*. DaCapo Press, New York, 2001, p. 19.

Dru, J.-M. *Beyond Disruption: Changing the Rules in the Marketplace*. Wiley Press, New York, April 12, 2002.

Food Industry Report 18(6), Redstone Publishing, Bensalem, PA, 2006.

Geisler, E. The metrics of technology evaluation: Where we stand and where we should go from here. *International Journal of Technology Management* 24 (4), 2002, 341–374.

Hamel, G. Strategy innovation and the quest for value. *Sloan Management Review* 39 (2), Winter 1998, 7–8.

Juttelstad, A. Measuring your new product success rate. *Food Formulating* October 1997.

Kuczmarski, T. D. Measuring your return on innovation. *Journal of Product Innovation Management* 19 (3), May 2002, 246–252.

Moore, L. Future Innovations in Food and Drinks to 2006: Forward-focused NPD and consumer trends. *Business Insights*, September 2004, 116.

Moskowitz, H. and Gofman, A. *Selling Blue Elephants*. Wharton School Publishing, Upper Saddle River, NJ, 2007.

Narendra, R. The keys to new product success (Part 1)—Collecting unarticulated & invisible customer-needs. *Product Management & Strategy*, June 19, 2007.

Simple steps to maximizing new product launch success rates. *Brand Strategy*, March 19, 1999.

24 Alternative Processing Methods for Functional Foods

Dietrich Knorr, Ana Balasa, Doerte Boll,
Henry Jäger, Alexander Mathys,
Esma Oba, and Marcus Volkert

CONTENTS

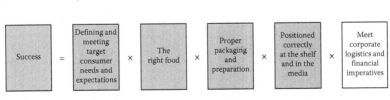

24.1 INTRODUCTION

Among today's most important trends in food processing is the development and perfection of gentle, energy efficient, and waste-free technologies. With the marketer's goal to design fresh-like yet shelf stable functional products comes the technologist's need to develop highly sophisticated systems. These systems should be adapted to the specific characteristics of the raw materials to be processed. The systems should be designed with certain requirements and situations in mind, such as subsequent or combined processes for optimum product quality, safety, and function. In addition, the processing of functional foods and food constituents should integrate the entire food chain in order to produce maximum function in the raw materials and to maintain or even increase it during processing.[1]

Today's social demand for sustainable food processes,[2] the consumer demand for "bio-guided processing,"[3] and the third demand to create products of high acceptance that satisfy consumer needs leads to the concept of "reverse engineering." The opportunity offers the potential to tailor-make foods via mastering of food process–structure–function relationships.[1]

This chapter shows how two alternative processing methods, high hydrostatic pressure (HP) and pulsed electric field (PEF) treatment,[4] contribute to the production of new highly functional foods or food constituents.

24.2 HIGH-HYDROSTATIC PRESSURE PROCESSING

24.2.1 BASIC PRINCIPLES

Two different concepts of high-pressure (HP) processing have been developed for different kinds of foods (Figure 24.1). One uses an internal intensifier (see Figure 24.1a).

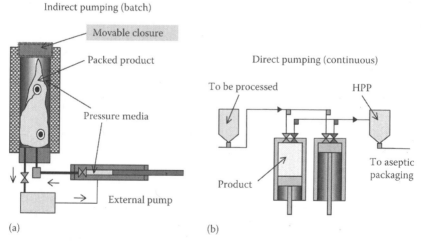

FIGURE 24.1 Two concepts of HP processing: (a) indirect (batch) and (b) direct (continuous) pumping (From Rovere, P., *Ultra High Pressure Treatments of Foods*. Kluwer Academic/Plenum Publishers. New York, 2002. With Permission.)

With this process, the maximum size of the particulates is limited by the rating of valves and pumps. However, this solution is more suitable for HP processes, because the cold inflow of pressure medium in the batch system (see Figure 24.1b) during pressurization may produce unwanted inhomogeneities in temperature. In any case with the appropriate treatment parameters, both batch and continuous systems have a high potential for HP processes.

Both concepts of HP processing shown in Figure 24.1 include an intensifier, where the simplest practical system is a single-acting, hydraulically driven pump. The two main parts of an intensifier are the low-pressure and the high-pressure cylinder. A double-acting arrangement enables a continuous, uniform flow. While one of the double-acting pistons is delivering, the other cylinder is being charged during its intake stroke. The pressure level is the same in each volume element. Thus the heat of compression is homogeneously distributed and improves microbial inactivation. As a result, the process efficiency can be maximized. Instantaneous adiabatic heating allows the system to reach the final temperature quickly and can result in a new approach to food sterilization with a significant improvement in food quality.

It is no wonder that the use of high-pressure technology in food processing has increased steadily during the past 10 years (Figure 24.2a), and in 2009, 113 industrial installations exist worldwide, with volumes from 35 to 420 L and an annual production volume of more than 120,000 tons.[5] Most of the vessel volume is used for meat and vegetable products (Figure 24.2b).

Currently, most of the industrial applications of high pressure, which were introduced in Japan in 1990 and in Europe and United States in 1996, are used for pasteurization purposes with some applications also geared toward product modification such as gelatinization of proteins and starch.[5-8]

24.3   HIGH-PRESSURE–LOW-TEMPERATURE PROCESSING

The fact that not only the shelf life but also the quality of food is important to consumers led to the concept of preserving foods using nonthermal methods. Therefore, alternative or novel food processing technologies are being explored and implemented. These methods provide safe, fresher-tasting, nutritive foods without the use of heat or chemical preservatives, foods that have fresh-like characteristics and satisfy new consumer demand.[9]

The application of HP processing has shown considerable potential as an alternative technology to heat treatments, in terms of assuring safety and quality attributes in minimally processed food products.[8] Extending the field of HP processing into the subzero domain introduces the novel technology of high-pressure–low-temperature (HPLT) processing which covers a wide range of processes and interesting alternatives to conventional freezing.[10]

24.3.1   PROCESS OPTIONS IN THE HPLT DOMAIN

Analogous to the standard HP processing, the field of HPLT processing embraces all HP processes in the subzero temperature domain. According to Benet, Schlüter, and Knorr[11] and Knorr, Schlüter, and Heinz,[12] the relevant processes including phase

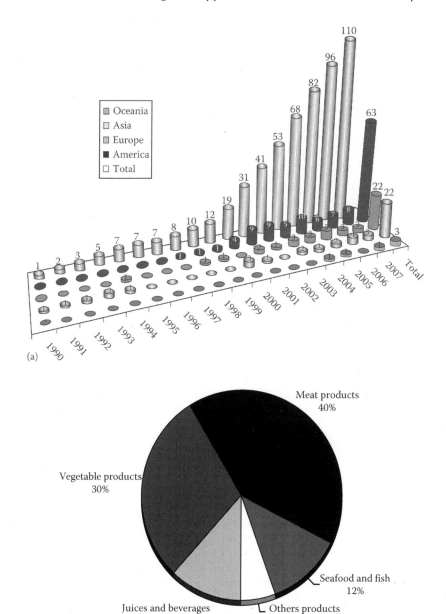

FIGURE 24.2 (a, b) HP machines in the world and total vessel volume versus food industries. (From Saiz A.H. and Samson, C.T. in *NC Hyperbaric Innovation Through Pressure* (Burgos, Spain: NC Hyperbaric, 2007, Personal communication). With permission.)

transition of water from liquid to solid under high pressure can be summarized as high-pressure shift freezing (HPSF), pressure-assisted freezing (PAF), and pressure-induced freezing (PIF).

During HPSF an aqueous matrix is pressurized in the liquid state and cooled under pressure to a temperature well below its freezing point at atmospheric pressure without ice formation. Subsequently, freezing is triggered by an instant pressure release, which causes sudden supercooling of the water with respect to the atmospheric freezing point of the matrix. As a result of the supercooling, homogeneous nucleation occurs instantly throughout the whole matrix leading to a uniform distribution and a high number of small ice crystals.[11,13] During the pressure release water is instantly frozen to ice.

The amount of instantly formed ice depends on the pressure–temperature conditions at the time of pressure release: the higher the pressure and the lower the temperature at which expansion takes place, the higher the percentage of instantaneously formed ice. Otero and Sanz[14] pointed out that the amount of ice instantaneously formed in the process can be increased by increasing the pressure before expansion. However, not all water is transformed to ice during pressure release, because the latent heat of crystallization of the product has to be removed.[12] Another positive aspect of the instantaneously produced ice is that ice crystals formed during pressure release improve the heat transfer. They also serve as nucleators during subsequent atmospheric freezing. As other freezing methods, the HPSF treatment may reduce the vitality of microorganisms; however, due to high freezing rates the lethal effect on viable cells is rather negligible.[15]

PAF is a freezing process at pressures above the atmospheric pressure, following the laws of atmospheric freezing. Depending on the pressure level, the pressurized water freezes to ice I, ice III, or ice V at temperatures below the atmospheric freezing point. Different from the HPSF the nucleation during PAF is triggered at the product surface and ice crystal growth proceeds to the product center. Thus, the number of ice crystals formed during PAF is considerably low and the crystals are larger.[16] Figure 24.3 schematically shows the mechanism of nucleation and ice crystal growth during HPSF and PAF and atmospheric freezing.

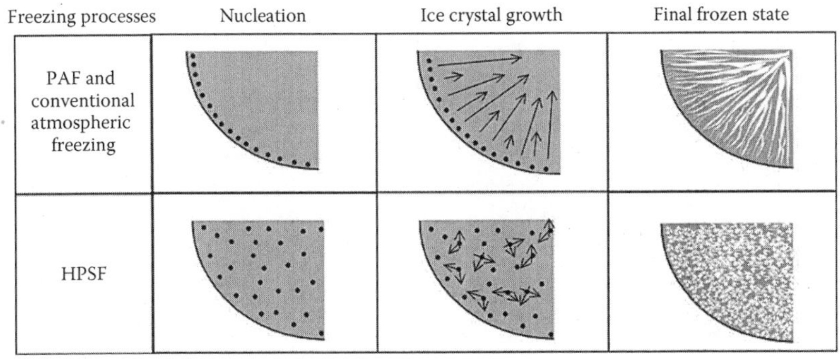

FIGURE 24.3 Hypotheses for nucleation and ice crystal growth by the different freezing processes. (Redrawn from Lévy, J., et al., *Lebensm. Wiss. u. Technol.*, 32, 396, 1999.)

PIF involves the induction of a phase change from liquid to solid by pressure increase. Ice I cannot be obtained through this process, as the phase transition temperature decreases with pressure and, therefore, pressurization cannot induce ice I crystallization. This is a process applicable only for higher ice modifications.

24.3.2 EFFECT OF HIGH PRESSURE AND LOW TEMPERATURE ON BIOLOGICAL MATTER

Complex reaction effects like protein denaturation, microorganism inactivation, or the loss of color are altered in their reaction rates by pressure as well as by temperature. Proteins are particularly affected by pressure treatments. They may unfold and denature, reversibly or irreversibly, depending on the kind of protein and the intensity of the treatment.

The majority of relevant processes in the HPLT domain is based on the formation of ice I, III, and V and operate in a pressure range from 0.1 to 360 MPa. Consequently less effect on biological structures occurs, in comparison to conventional HP treatments ranging up to 800 MPa. Nevertheless, the stability of proteins is highly specific. High pressure at high temperatures does not necessarily cause a higher degree of denaturation as a comparable pressure treatment at lower temperature. The coherencies of pressure, temperature, and the effect on proteins are schematically drawn as a phase diagram in Figure 24.4.

During HPLT treatment biological matter is affected by a number of changes. Biochemical changes are only one. Mechanical damage also occurs. Since freezing is involved in most of the HPLT processes, mechanical damage occurs with both

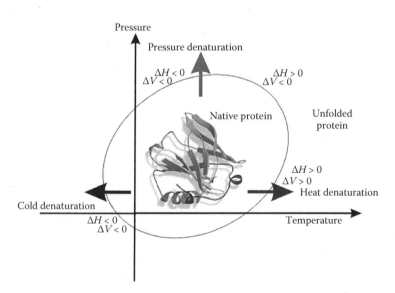

FIGURE 24.4 Typical phase transition curve of proteins in the pT-diagram. The relation between heat, cold, and pressure denaturation of proteins is presented by the sign of enthalpy changes (ΔH) and volume changes (ΔV).

plant and animal tissues. However, the damage may be less than in other methods. With respect to physical damage that comes along with the conventional freezing and is mainly due to ice crystal growth, HPLT treatment offers attractive processes that may decisively limit such quality losses.

When the texture of food after thawing becomes an important quality criterion, HPLT freezing processes have the potential to produce products with superior quality compared to the quality of conventionally frozen products. One effect to be exploited in this context is the formation of small ice crystals during HPSF. These ice crystals are triggered by homogeneous nucleation throughout the product. Textural damages are limited, however, because of a high number of small ice crystals in contrast to large, needle-shaped crystals that grow at slow freezing rates during atmospheric freezing.

A second effect that potentially improves texture of frozen foods is based on the density of water. When water is frozen at atmospheric conditions, ice I is formed, and the volume of water is increased during the phase transition. Therefore, histological damage to tissue and excessive textural damage occurs. Conversely, when water is frozen under high pressure the volume of water is decreased. Quality improvement by freezing under high pressure of different foods, such as fish, potatoes, strawberries, or tofu was the focus of numerous studies.[17–19]

24.3.3 Process Induced Variability of Probiotics

Probiotic bacteria differ from the lactic acid bacteria that are commonly used for the fermentation of yogurt (e.g., *Lactobacillus bulgaricus* and *Streptococcus thermophilus*). Probiotic bacteria have a positive effect on human health, going beyond the nutritional ones commonly known. The two main mechanisms of action have been suggested and are summarized as follows:[20]

1. *Nutritional effect*, characterized by the reduction of metabolic reactions that produce toxic substances, stimulation of indigenous enzymes, and production of vitamins.
2. Antimicrobial substances and *health or sanitary effect* which is distinguished by increase in colonization resistance, competition for gut surface adhesion, and stimulation of the immune response.

Several beneficial functions have been suggested for probiotic bacteria.[21] These functions span a wide range of effects including nutritional benefits, vitamin production, availability of minerals and trace elements, production of important digestive enzymes (e.g., β-galactosidase), barrier/restoration effects, cholesterol-lowering effects, stimulation of the immune system, enhancement of bowel motility/relief from constipation, adherence and colonization resistance, and maintenance of mucosal integrity.

Probiotic products represent a strong growth area within the functional foods group. Intense research efforts are under way to develop dairy and nondairy products into which probiotic bacteria such as *Lactobacillus* and *Bifidobacterium* species are incorporated.[22] It has been suggested that their health-promoting effect is only achieved when the person consumes living bacteria.[23,24] A daily dose of at least 10^8 living cells has been suggested to assure health-relevant effects following the consumption of probiotic products.[25] To maintain this proposed number of living bacteria, the survivability of probiotic bacteria during processing as well as in food has to be guaranteed up to the end of the shelf life.

24.3.4 TECHNOLOGY ASPECTS

For the development of dairy-based functional foods containing high numbers of viable probiotic cells, the starter culture should have the ability to grow in milk-based media, remain viable, and retain probiotic properties during production and storage (shelf life) of the probiotic food product.[26]

Drying and freezing are the two basic unit operations that can lower the water activity to a level that preserves living cells and provides sufficient shelf life of the bacteria. Among the conventional drying processes, spray drying (4000–6000 kJ/kg H_2O removed) is a promising process option that for some strains shows survival rates comparable to the survival rates achieved by freeze drying (100,000 kJ/kg H_2O removed).[27] In addition spray drying is claimed to be more cost effective and less time consuming.[26] Besides the media composition, the critical process parameter affecting the survival rate during spray drying is the outlet temperature of the dryer. The outlet temperature describes the temperature at the dryer exit and at the same time the maximum temperature of the product during the process. Figure 24.5 shows the survival rate of *L. rhamnosus* GG after spray drying as a function of the outlet temperature and pressure.

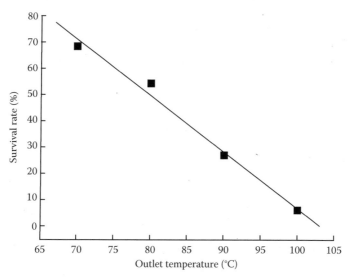

FIGURE 24.5 Effect of outlet temperature during spray drying on the survival of *L. rhamnosus* GG.27.

Another strategy freezes at high-freezing rates, creating a beneficial surface to volume ratio is the spray freezing.[15] During the spray–freezing process, small droplets are sprayed into a subzero atmosphere and rapidly cooled down below the freezing point. The frozen products can easily be separated from the cooling gas resulting in a snow-like powder.

24.4 PULSED ELECTRIC FIELD PROCESSING

PEF treatment is a nonthermal technology with a possible application in a wide range of food industries. PEF is based on the permeabilization of biological membranes. PEF processing involves the application of short pulses (in the range of microsecond to millisecond) of high electric fields. The result of PEF is a disintegration of the cell membrane consisting of a liquid-like bilayer of phospholipids. Depending on treatment intensity, the generated membrane pores can be permanent or temporary.

Electroporation (poration of the cell membrane caused by PEF treatment) remains irreversible if under the action of an electric field the pore radius grows beyond a critical value. The semipermeable character of the membrane becomes permanently destroyed and results in cell death. This property makes it possible to use PEF for microbial inactivation as nonthermal pasteurization method for liquid food. Furthermore, mass transfer processes like pressing, extraction, or drying can be improved through PEF.

PEF treatment combines gentle food preservation with short treatment times, continuous operation, and easy implementation into existing product flow. The result is an alternative to traditional thermal processes since PEF is capable of destroying microorganisms while maintaining fresh-like physical and chemical characteristics of food products.[28,29]

For application as nonthermal pasteurization, PEF comprises treatment with very short electric pulses (1–100 μs) at electric field intensities in the range of 20–50 kV/cm and an input of electrical energy of around 100 kJ/kg treated product. PEF takes place at moderate temperatures and affects the integrity of cell membranes by electroporation.

24.4.1 PRODUCTION AND RECOVERY OF SECONDARY PLANT METABOLITES

Secondary metabolites offer a unique opportunity to produce tailor-made beneficial products with far reaching health and nutrition ramifications. Detailed information about the relevance of secondary plant metabolites is still limited. Secondary metabolism is involved in the relation of the organism with its environment, e.g., in resistance against microorganic pests and diseases, as attractant of pollinators or as signal compound.[30,31]

There has been an increasing interest in the possibility of using low intensity, sublethal PEF treatment for stress induction in plant systems. The electrically insulating properties of the membrane can be recovered within a short time, restoring vitality and metabolic activity of the cells.[32] Similar to a plant defense reaction to pathogenic infection, this temporary permeabilization of the membrane may stimulate plant defense mechanism associated with a production of secondary metabolites.

In studies with soybeans, an increase in isoflavonoid daidzain and genistein content of 20% and 21% compared to the reference sample in the oil were found.[33] In both cases, the maximum amount was reached at low treatment conditions, where membrane permeabilization was reversible. We see a similar case for the impact of PEF treatment on polyphenolic content in grape products.[34] Total polyphenolic content was increased in both juice and grape pomace extracts obtained from grapes treated with different electric field intensities.

PEF treatment at conditions when reversible membrane poration occurs could be a promising tool for the manipulation of secondary metabolism for the production of special compounds and to produce high quality food with high yield of functional ingredients.

Current methods for the extraction of intracellular compounds are freeze–thawing and mechanical expression. Other methods use solid–liquid or liquid–liquid extraction with strong organic solvents, which penetrate in the cells establishing intensive contact with targeted compounds, and withdraw them out of the cells. Solid–liquid extraction of food matrices strongly depends on diffusion through the cells, and disintegration of material. To achieve high disintegration of solid tissue, thermal, mechanical, or chemical methods are used. Grinding to increase the surface area and separation of the product components after size reduction causes loss of nutritional valuable components due to oxidation of fatty acids and carotenoids. Losses of vitamin C in sliced and chopped fruits and vegetable are substantial.[35] To avoid such losses of nutritionally valuable components, PEF technology could be applied as a new method for cell disintegration. The occurring increase in membrane permeabilization by exposure of biological cells to an external electric field positively affects the mass transfer rate. The consequence is that the diffusion of intracellular components in extracellular liquid is increased, while leaving the product matrix relatively unchanged.[36–38]

The results from PEF have been shown to be promising from other studies as well. For example, the extractability of carotenoids and vitamin C was studied by Ade-Omowaye.[39] PEF treatment was able to enhance extractability. More than 60% of the β-carotene present in the fruit was extracted into the juice as compared to about 43% extracted by the untreated or about 44% extracted by enzyme treatment. This indicates that an increase of about 20% was possible when compared to the untreated.

PEF treatment was implemented in apple juice production as pretreatment of the mash to increase permeability of the cells.[40] In contrast to the expectations, no increased yield of polyphenolic compounds in the juice could be achieved as a result of improved disintegration of the plant cells as it was shown for total polyphenolic content and anthocyanins in grapes.[41] PEF treatment of red beetroot tissue was shown to be an effective method of permeabilization and red pigment (betalain) extraction.[38] Sensoy and Sastry[42] showed an enhanced leaching of solutes from tea leaves after PEF extraction.

All of the foregoing results suggest that PEF could replace a number of "operations" in food technology and return with enhanced quality. These operations include pressing, extraction, drying, or conventional disintegration techniques. The replacement by PEF could offer sustainable degree of tissue permeabilization and therefore, improvement in the efficiency of recovering valuable components.

24.4.2   SAFETY AND QUALITY ASPECTS

A number of factors affect microbial inactivation during PEF treatment. These are variables such as process factors (electric field intensity, pulse width and shape, treatment time, and temperature), microbial factors (type, concentration, and growth stage of microorganism), and media factors like pH, antimicrobials and ionic compounds, conductivity, and medium ionic strength.

Effective inactivation for most of the spoilage and pathogenic microorganisms has been shown, but in comparison to the treatment of plant or animal tissue the treatment intensity needs to be much higher. The potential to achieve sufficient reduction of microbes in various food products like fruit or vegetable juices,[43–51] model beer,[52] or milk[53–55] has been investigated.

Reports on the effects of PEF on enzymes are limited, and different experimental setups and processing parameters make them difficult to compare.[56–58]

Milk was the first product commercially pasteurized by the direct application of electrical energy using the "Electropure" process established in the 1920s. Hence, it is not surprising that also applicability of pulsed electric energy within a treatment for milk preservation has been studied extensively.[53,54,59–62]

24.4.3   PULSED ELECTRIC FIELD TREATMENT OF OILSEEDS

Oils can be obtained from oilseeds by different methods such as solvent extraction or pressing. To achieve high yields of oil with a high-nutritive value, one can use PEF before the oil separation phase. The yield of functional oil ingredients like phytosterols and tocopherols (vitamin E) can also be increased with the PEF treatment.[63] Phytosterols are beneficial due to their cholesterol-lowering effect. Furthermore, they may also decrease the risk of certain types of cancers. The application of PEF has also led to an increase of the phytosterol content in the maize germ—hulled and nonhulled rapeseed—and in the sunflower seed oil.

Plant oils are also rich sources of antioxidants like tocopherols, which inhibit lipid oxidation in foods and biological systems. They retard the oxidation of unsaturated fatty acids and hence avoid the production of rancidity. Vitamin E has also a nutritional high-value and plays an important role in the prevention of chronic diseases. With PEF treatment, higher amounts of tocopherol were up to 20% obtained in the hulled rapeseed oil in comparison to the untreated reference.

In summary, with the application of PEF the content of functional food ingredients can be increased under gentle conditions. Consequently, PEF can be used as an effective pretreatment for oil recovery. PEF increases the concentration of functional constituents as well as provides the basis for increased shelf life of plant oils.

24.5   SUMMING UP

Alternative processing methods such as HP and PEF offer unique and gentle ways to recover, modify, and preserve functional foods or food constituents. These methods furnish new technologies that can be utilized for new product development. The methods offer unique quality, functionality, and health benefits.

ACKNOWLEDGMENTS

Parts of the work presented have been funded by grants from the German Research Foundation (DFG), the German Industrial Research Foundation (AIF-FEI), the German Ministry for Education and Research (BMBF), and the European Commission (Projects: SAFE ICE, PROTECH, NOVEL Q).

REFERENCES

1. European Technology Platforms, Food for Life, http://www.etp.ciaa.be (downloaded July 24, 2008).
2. S. Toepfl, A. Mathys, V. Heinz, and D. Knorr, Review: Potential of emerging technologies for energy efficient and environmentally friendly food processing, *Food Reviews International* 22(4), 2006:405–423.
3. R.E. Ward, H.J. Watzke, R. Jimenez-Flores, and J.B. German, Bioguided processing: A paradigm change in food production, *Food Technology* 58(5), 2004:44–48.
4. M. Hendrickx and D. Knorr. *Ultra High Pressure Treatments of Foods* (New York: Kluwer Academic/Plenum Publishers, 2002).
5. A.H. Saiz and C.T. Samson, Aplicaciones de las altas presiones en la Industria Agroalimentaria, in *NC Hyperbaric Innovation through Pressure* (Burgos, Spain: NC Hyperbaric, 2007, Personal communication).
6. J.C. Cheftel, Review: High pressure, microbial inactivation and food preservation, *Food Science and Technology International* 1, 1995:75–90.
7. B. Tauscher, Pasteurization of food by hydrostatic high pressure: Chemical aspects, *Zeitschrift für Lebensmittel-Untersuchung und Forschung* 200, 1995:3–13.
8. E. Palou, A. López-Malo, G.V. Barbosa-Cánovas, and B.G. swanson, High-pressure treatment in food preservation, in *Handbook of Food Preservation*, ed. M.S. Rahman (New York: Marcel Dekker, Inc., 1999), pp. 533–576.
9. L. Goularte, C.G. Martins, I.C. Morales-Aizpurúra, et al., Combination of minimal processing and irradiation to improve the microbiological safety of lettuce (*Lactuca sativa, L.*), *Radiation Physics and Chemistry* 71, 2004:155–159.
10. G.U. Benet, High pressure low temperature processing of foods: Impact of metastable phases on process and quality parameters, PhD dissertation (Berlin University of Technology, Berlin, Germany, 2005), pp. 196.
11. G.U. Benet, O. Schlüter, and D. Knorr, High pressure–low temperature processing. Suggested definitions and terminology, *Innovative Food Science & Emerging Technologies* 5(4), 2004:413–427.
12. D. Knorr, O. Schlüter, and V. Heinz, Impact of high hydrostatic pressure on phase transitions of foods, *Food Technology* 52(9), 1998:42–45.
13. J. Lévy, E. Dumay, E., Kolodziejczyk, and J.C. Cheftel, Freezing kinetics of a model oil-in-water emulsion under high pressure or by pressure release. Impact on ice crystals and oil droplets, *Lebensmittel-Wissenschaft und-Technologie* 32, 1999:396–405.
14. L. Otero and P.D. Sanz, High-pressure shift freezing. Part 1. Amount of ice instantaneously formed in the process, *Biotechnology Progress* 16, 2000:1030–1036.
15. M. Volkert, E. Ananta, C.M. Luscher, and D. Knorr, Effect of air freezing, spray freezing, and pressure shift freezing on membrane integrity and viability of *Lactobacillus rhamnosus* GG, *Journal of Food Engineering* 87, 2008:532–540.
16. P.P. Fernández, L. Otero, B. Guignon, and P.D. Sanz, High-pressure shift freezing versus high-pressure assisted freezing: Effects on the microstructure of a food model, *Food Hydrocolloids* 20, 2006:510–522.

17. M. Fuchigami, N. Kato, and A. Teramoto, Effect of pressure-shift freezing on texture, pectic composition and histological structure of carrots, in *High Pressure Bioscience and Biotechnology*, eds. R. Hayashi and Balny (Amsterdam: Elsevier Science B.V., 1996).

18. O. Schlüter, Impact of high pressure-low temperature processes on cellular materials related to foods, PhD dissertation (Berlin University of Technology, Berlin, Germany, 2004), pp. 172.

19. C.M. Luscher, Effect of high pressure-low temperature phase transitions on model systems, foods and microorganisms, PhD dissertation (Berlin University of Technology, Berlin, Germany, 2008), pp. 158.

20. A. Anadón, M.R. Martínez-Larranaga, and M. Aranzazu, Probiotics for animal nutrition in the European Union: Regulaton and safety assessment, *Regulatory Toxicology and Pharmacology* 45, 2006:91–95.

21. W.H. Holzapfel and U. Schillinger, Introduction to pre- and probiotics, *Food Reviews International* 35, 2002:109–116.

22. E. Ananta, Impact of environmental factors on vitality and stability and high pressure pretreatment on stress tolerance of *Lactobacillus rhamnosus* GG (ATCC 53103) during spray drying, PhD dissertation (Berlin University of Technology, Berlin, Germany, 2005), pp. 218.

23. A.C. Ouwehand and S.J. Salminen, The health effects of cultured milk products with viable and non-viable Bacteria, *International Dairy Journal* 8, 1998:749–758.

24. Joint FAO/WHO Working Group, Joint FAO/WHO Working Group Report on Drafting Guidelines for the Evaluation of Probiotics in Food (London, Ontario, Canada: Food and Agriculture Organization of the United Nations/World Health Organization, April 30–May 1, 2002).

25. A. Lourens-Hattingh and B.C. Viljoen, Yogurt as probiotic carrier food, *International Dairy Journal* 11, 2001:1–17.

26. D. Knorr, Technology aspects related to microorganisms in functional foods, *Trends in Food Science and Technology* 9, 1998:295–306.

27. E. Ananta, M. Volkert, D. Gloyna, and D. Knorr, Cellular injuries and storage stability of spray dried probiotic bacterium *Lactobacillus rhamnosus* GG, *International Dairy Journal* 15, 2005:399–409.

28. A.J. Castro, G.V. Barbosa-Canovas, and B.G. Swanson, Microbial inactivation of foods by pulsed electric fields, *Journal of Food Processing and Preservation* 17(1), 1993:47–73.

29. G.V. Barbosa-Cánovas, M.M. Góngora-Nieto, U.R. Pothakamury, and B.G. Swanson, *Preservation of Foods with Pulsed Electric Fields* (San Diego, CA: Academic Press, 1999).

30. L. Taiz and E. Zeiger, *Physiologie der Pflanzen* (Heidelberg-Berlin, Germany: Spektrum, 1999).

31. R. Verpoorte, Secondary metabolism, in *Metabolic Engineering of Plant Secondary Metabolism*, eds. R. Verpoorte and A.W. Alfermann (New York: Springer, 2000): pp. 1–29.

32. H. Angersbach, V. Heinz, and D. Knorr, Effects of pulsed electric fields on cell membranes in real food systems, *Innovative Food Science & Emerging Technologies* 1, 2000:135–149.

33. M. Guderjan, S. Toepfl, A. Angersbach, and D. Knorr, Impact of pulsed electric field treatment on the recovery and quality of plant oils, *Journal of Food Engineering* 67(3), 2005:281–287.

34. A. Balasa and D. Knorr, Extraction of total phenolics from grapes in correlation with degree of membrane poration, in *Proceedings of the COST Meeting 928-300606* (Reykjavik, Iceland, 2006).

35. Y. Chalermchat, M. Fincan, and P. Dejmek, Pulsed electric field treatment for solid–liquid extraction of red beetroot pigment: Mathematical modelling of mass transfer, *Journal of Food Engineering* 64, 2004:229–236.

36. D. Knorr, M. Geulen, T. Grahl, and W. Sitzmann, Food application of high electric field pulses, *Trends in Food Science & Technology* 5, 1994:71–75.

37. D. Knorr and A. Angersbach, Impact of high-intensity electric field pulses on plant membrane permeabilization, *Trends in Food Science & Technology* 9(5), 1998:185–191.

38. M. Fincan, F. DeVito, and P. Dejmek, Pulsed electric field treatment for solid–liquid extraction of red beetroot pigment, *Journal of Food Engineering* 64, 2004:381–388.

39. B.I.O. Ade-Omowaye, Application of pulsed electric fields as a pre-treatment step in the processing of plant based foods, PhD dissertation (Berlin University of Technology, Berlin, Germany, 2002).

40. S. Schilling, T. Alber, S. Toepfl, et al., Effects of pulsed electric field treatment of apple mash on juice yield and quality attributes of apple juices, *Innovative Food Science & Emerging Technologies* 8, 2007:127–134.

41. W. Tedjo, M.N. Eshtiaghi, and D. Knorr, Einsatz nicht thermischer Verfahren zur Zell-Permeabilisierung von Weintrauben und Gewinnung von Inhaltsstoffen, *Flüssiges Obst* 9, 2002:578–583.

42. I. Sensoy and S.K. Sastry, Extraction using moderate electric fields, *Journal of Food Science* 69(1), 2004:7–13.

43. Q. Zhang, A. Monsalve-González, B.L. Qin, G.V. Barbosa-Cánovas, and B.G. Swanson, Inactivation of *Saccharomyces cerevisiae* in apple juice by square-wave and exponential-decay pulsed electric fields, *Journal of Food Process Engineering* 17, 1994:469–478.

44. G.A. Evrendilek, Z.T. Jin, K.T. Ruhlman, X. Qiu, Q.H. Zhang, and E.R. Richter, Microbial safety and shelf-life of apple juice and cider processed by bench and pilot scale PEF systems, *Innovative Food Science & Emerging Technologies* 1, 2000:77–86.

45. C.J. McDonald, S.W. Lloyd, M.A. Vitale, K. Petersson, and F. Innings, Effect of pulsed electric fields on microorganisms in orange juice using electric field strengths of 30 and 50 kV/cm, *Journal of Food Science* 65(6), 2000:984–989.

46. Z. Ayhan, Q.H. Zhang, and D.B. Min, Effects of pulsed electric field processing and storage on the quality and stability of single-strength orange juice, *Journal of Food Protection* 65(10), 2002:1623–1627.

47. A.M. Hodgins, G.S. Mittal, and M.W. Griffiths, Pasteurization of fresh orange juice using low-energy pulsed electrical field, *Journal of Food Science* 67(6), 2002:2294–2299.

48. V. Heinz, S. Toepfl, and D. Knorr, Impact of temperature on lethality and energy efficiency of apple juice pasteurization by pulsed electric fields treatment, *Innovative Food Science & Emerging Technologies* 4(2), 2003:167–175.

49. S. Min, Z.T. Jin, and Q.H. Zhang, Commercial scale pulsed electric field processing of tomato juice, *Journal of Agricultural Food Chemistry* 51, 2003:3338–3344.

50. N. Lechner and Z. Cserhalmi, Pulsed electric field (PEF) processing effects on physical and chemical properties of vegetable juices, in *Proceedings of the Safe Consortium Seminar: Novel Preservation Technologies in Relation to Food Safety* (Brussels, Belgium, 2004).

51. P. Molinari, A.M.R. Pilosof, and R.J. Jagus, Effect of growth phase and inoculum size on the inactivation of S. cerevisiae in fruit juices by pulsed electric fields, *Food Research International* 37(8), 2004:793–798.

52. H.M. Ulmer, V. Heinz, M.G. Gaenzle, D. Knorr, and R.F. Vogel, Effects of pulsed electric fields on inactivation and metabolic activity of Lactobacillus plantarum in model beer, *Journal of Applied Microbiology* 93(2), 2002:326–335.

53. S. Bendicho, G.V. Barbosa-Cánovas, and O. Martin, Milk processing by high intensity pulsed electric fields, *Trends in Food Science & Technology* 13, 2002:195–204.

54. D.D. Sepulveda, M.M. Góngora-Nieto, J.A. Guerrero, and G.V. Barbosa-Cánovas, Production of extended shelf-life milk by processing pasteurized milk with pulsed electric fields, *Journal of Food Engineering* 67, 2005:81–86.

55. S. Toepfl, H. Jaeger, V. Heinz, and D. Knorr, Milk preservation by pulsed electric fields-utilization of native antimicrobial activity, in *Proceedings of the IUFOST* (Nantes, France, 2006).

56. A. Van Loey, B. Verachtert, and M. Hendrickx, Effects of high electric field pulses on enzymes, *Trends in Food Science & Technology* 12, 2001:94–102.

57. H. Schuten, K. Gulfo-van Beusekom, I. Pol, H. Mastwijk, and P. Bartels, Enzymatic stability of PEF processed orange juice, in *Proceedings of the Safe Consortium Seminar: Novel Preservation Technologies in Relation to Food Safety* (Brussels, Belgium, 2004).

58. R.J. Yang, S.Q. Li, and Q.H. Zhang, Effects of pulsed electric fields on the activity of enzymes in aqueous solution, *Journal of Food Science* 69(4), 2004:241–248.

59. K. Smith, G.S. Mittal, and M.W. Griffiths, Pasteurization of milk using pulsed electrical field and antimicrobials, *Journal of Food Science* 67(6), 2002:2304–2308.

60. F. Sampedro, M. Rodrigo, A. Martínez, D. Rodrigo, and G.V. Barbosa-Cánovas, Quality and safety aspects of PEF application in milk and milk products, *Critical Reviews in Food Science and Nutrition* 45, 2005:25–47.

61. S. Toepfl, H. Jaeger, V. Heinz, and D. Knorr, Neues Verfahren zur Haltbarmachung von Milch, *Deutsche Molkerei Zeitung* 2, 2006:24–28.

62. H. Jaeger and D. Knorr, Einsatzmöglichkeiten gepulster elektrischer Felder zur Haltbarmachung hitzeempfindlicher Produkte in der Milchindustrie, *Deutsche Molkerei Zeitung* 24, 2007:22–25.

63. M. Guderjan, Influence of pulsed electric fields on the production of vegetable oils, PhD dissertation (Berlin University of Technology, Berlin, Germany, 2006).

25 Accelerated and Parallel Storage in Shelf Life Studies

I. Sam Saguy and Micha Peleg

CONTENTS

$$\text{Success} = \begin{array}{c}\text{Defining and}\\\text{meeting}\\\text{target}\\\text{consumer}\\\text{needs and}\\\text{expectations}\end{array} \times \begin{array}{c}\text{The}\\\text{right food}\end{array} \times \begin{array}{c}\text{Proper}\\\text{packaging}\\\text{and}\\\text{preparation}\end{array} \times \begin{array}{c}\text{Positioned}\\\text{correctly}\\\text{at the shelf}\\\text{and in the}\\\text{media}\end{array} \times \begin{array}{c}\text{Meet}\\\text{corporate}\\\text{logistics and}\\\text{financial}\\\text{imperatives}\end{array}$$

25.1 INTRODUCTION

We take for granted that all foods, regardless of their kind, must be always attractive and safe to eat. This is so obvious that we hardly give it a second thought. Yet, defining "attractiveness" is not a simple matter. This is especially true because the time can vary substantially between when a food is bought and when it is actually consumed. The same can be said about pet foods and animal feed. The establishment of quality and safety criteria, therefore, is one of the major concerns in food products development. Moreover, all foods, even when properly packaged, undergo biochemical, physical, and other changes that can affect their quality and safety. The rate at which these changes occur is influenced by the product's characteristics and the conditions under which it is produced, handled, and stored. In many cases, however, most conditions are beyond the manufacturer's control, which raises a responsibility issue and the need to define what "normal handling" and "abuse" are. The term "food" itself covers a wide range of agricultural and industrial products. Except for being eaten, directly or by being admixed with other edible products, they can have little in common. A fresh bell pepper, a ripening banana, a chocolate bar, canned tuna, bottled apple juice, frozen shrimps, ice cream, yogurt, breakfast cereal, powdered garlic, and hard candy differ in almost every aspect: from composition, chemistry, physical state, and packaging requirements to how long they can be stored and under what conditions. For making each a successful product, it is imperative to know and understand its chemistry, biology, and physics—there is no way around it. Intimate familiarity with the product, and how it is manufactured, packaged, transported, stored, and consumed, is the first requirement for its development or improvement. Yet, there are general guidelines that, when followed by the developer, will greatly increase the probability of a high-quality product, and its success in the marketplace. Recently, the concept of total food quality has been introduced in order to account for the multitude of properties that determine a food product's sensory attributes, safety, nutritional value and wholesomeness, functionality, and stability.[1] It deals with the physical, chemical, biological, and physiological aspects of food consumption as well as the psychological aspects that affect it.

In this chapter, we will not deal with specific food products and the factors that affect their acceptability and durability. Our focus will be on the strategies that the new product developer can adopt in order to guarantee a product's shelf life and the principles on which they are based. To facilitate the discussion, we will address the general principles in the chapter's main body and leave the more elaborate mathematical and kinetics models to Appendix A. Although frequently governed by the same principles, safety issues concerning stored foods will not be discussed. This is because health-related aspects of food production, transportation, and storage are regulated by specific legal requirements, such as those of good manufacturing practice (GMP), hazard analysis and critical control point (HACCP), etc., which are outside the scope of this chapter.

25.2 QUALITY CRITERIA

We can all recognize a spoiled food; no need for science here. But waiting for a food to "go bad" is not an option—it must be consumed or withdrawn long before

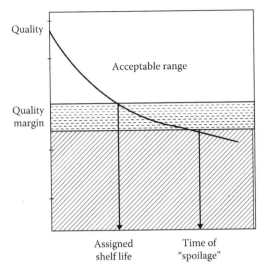

FIGURE 25.1   Schematic view of how a product's shelf life can be assigned.

this happens. For the food manufacturer, the important question is when. Early withdrawal of the product, or assigning it with a too early expiration date, will obviously be costly. But allowing the product to deteriorate on the shelf, thus endangering or upsetting the consumers, would be even costlier. Therefore, the product's assigned shelf life must be shorter than the time to reach noticeable quality loss (see Figure 25.1). However, whereas the principle itself is clear, it will be more difficult to determine the critical quality loss level that will require the product's withdrawal, as well as how to estimate the time at which this level will be reached.

Quality, as already mentioned, is multifaceted. In foods, the quality can be judged by the microbial load (as in dairy products, milk, refrigerated fruit juices, meats, or fish), loss of the desirable flavor, the development of off flavor (as rancidity in cereals and nuts), texture deterioration (as in bread staling), unpleasant changes in appearance (such as "blooming" in chocolate, browning of a marmalade), and any of their numerous possible combinations. The developer's first tasks are to identify the product's most pertinent quality attribute or attributes, search for the methods of their evaluation, and establish the level of deterioration deemed unacceptable. (The special case where a product's quality is improved through aging, such as in wines and certain cheeses, will not be discussed in this chapter.)

Common quality criteria that determine a food product's shelf life are:

1. Sensory—e.g., overall assessment of taste, smell, appearance, texture, and detection of off flavor or odor
2. Chemical—e.g., vitamin loss, peroxides formation, accumulation of a Maillard reaction product, and pH change

3. Physical—e.g., phase change or separation, sedimentation, color loss, turbidity, gelling, undesirable hardening through drying or softening by moisture sorption, and caking
4. Microbial—e.g., rise in the total bacterial count and the growth of a specific organism or type of organisms.

Frequently, at least two of the four are interrelated. In such a case, one can follow the changes in the food by monitoring a chemical or physical marker or markers that are easy to determine. This will simplify the analysis and reduce its cost considerably. Once the methods to monitor quality have been identified, the issue of sampling will arise. It has three issues that need to be considered:

1. How many samples should be inspected or tested and at what frequency?
2. Where should the samples be taken from?
3. How can we tell that the samples are representative, especially if taken when the product is already in the market?

There are no simple answers to these questions. However, there are several statistical sampling plans, with incorporated decision criteria, that can serve as guidelines. A notable example is the Military Standard 105E. It is now available as freeware on the Internet and can be used online. This standard allows the user to choose a sampling plan and examine the statistical implications of the choice. (The tables in the standard give the number of specimens to be tested in single and multiple sampling plans depending on the lot size and the "level of inspection," i.e., the plan's rigor. They also specify the number of faulty units that will be required for the lot's rejection or for the testing of additional specimens. The standard also provides "operational characteristics curves" that depict the probabilities of accepting a "bad" sample depending on the sampling plan and the actual number of faulty units or faults.) One should always remember, however, that increasing the "level of inspection" and its reliability by increasing the number of tested samples always comes at a cost—a factor that must be taken into account.

25.3 SHELF LIFE

Every food properly prepared, packaged, and stored are wholesome only for a certain time, which we call shelf life. The term itself, however, has no generally accepted definition. Consequently, what is considered a desirable or acceptable shelf life might depend on the manufacturer or the regulating agency that requires its specification on the package or the label. Some known shelf life definitions adapted from Taoukis et al.[2] are:

1. High-quality life—the time from preparation to the first noticeable change in a sensory attribute or attributes
2. Practical storage life—the time a properly stored product remains suitable for consumption
3. Time of minimum durability—defined by an EEC directive as the time during which the food retains its declared specific properties when properly stored

25.3.1   Methods of Shelf Life Assessment

25.3.1.1   Consumer Consideration in Shelf Life Determination

Consumers' acceptance data are vital to quality standards development. They tell the developer how the product's history, from preparation to consumption, in its totality is manifested in the food's quality perception. Such data also reveal whether or not the product matches the consumer's expectations. "Meeting the expectations" has a profound impact on whether the particular individual and most probably many other consumers will purchase the product again. The complaints rate, n, expressed as the number of complaints per a selected specific volume of units sold, N_s (e.g., 1000 cases, million units) should be continuously monitored and evaluated. The documented consumers' responses form a valuable database and should be constantly updated. In most cases, the consumer's acceptance will be the ultimate criterion and will determine the product's actual shelf life.

Figure 25.2 demonstrates the possible differences between shelf lives established on the basis of consumer complaints data and a physiochemical quality index. A plot of the consumer complaints data versus time resembles the characteristic "bathtub curve" as depicted in the figure. It shows that some individual units will fail fairly early. This is known as "infant failure," which is frequently caused by faulty packages—a problem that can be rapidly solved. Some units will last longer until "wear-out" sets in, whereas a few others will fail sometime during the relatively long period dubbed the product's "normal life." It can also be called a "consumer-driven shelf life," after which there will be a noticeable rise in the number of consumer complaints. Actual data about consumer complaints indicate a sudden increase of dissatisfaction with the product. As also shown in the figure, this abrupt change may not correspond with the "more objective" physical or chemical quality index, which continues to fall monotonically. The emergence of such a scenario may require reassessment of whether the use of the particular objective index should continue or a new quality measure be sought.

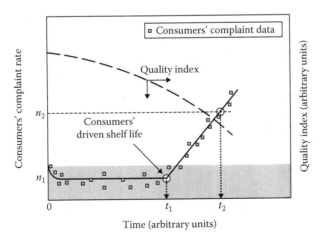

FIGURE 25.2   Schematic view of a typical relationship between physiochemical deterioration of a food product on the shelf and its manifestation in the number of customer complaints (for explanations of terms, see text).

One should also always keep in mind that for every logged complaint (e.g., a call to 800 number center, e-mail, and letter), there are additional unsatisfied consumers who simply did not bother to respond. Each complaint, therefore, may represent a large number of "quiet" or "silent" unhappy consumers who will not purchase the product again. Worse still, some of the disappointed consumers may boycott the brand name or even the whole range of the company's products in their future purchases. Some may even go further and spread the word among acquaintances, or through social networks, thus deterring others from buying the product. The actual complaint rate is known to be a strong function of the product price—the higher the expectation, the deeper is the dissatisfaction when it occurs. The estimated ratio between "responsive" and "silent" consumers, C_F, is usually unknown and therefore must be estimated. A common assumption is that C_F is between 65 and 100. To get an idea of the actual number of disappointed consumers, one should multiply the number of received complaints by the aforementioned ratio estimate.

Barring special situations, mostly related to package defects, a product leaving the manufacturing plant is in prime condition. Thus as already mentioned, the initial complaints, arriving soon after the product's introduction, have little to do with its durability. These complaints are typically related to faulty packages (e.g., pinholes, and seals integrity) that can be easily fixed. Usually, the number of such complaints declines rapidly as the problem is being solved (Figure 25.2). During the relatively long "normal" or "useful" shelf life, the complaints rate is low and the occasional product failure that results in a complaint is due to random circumstances. This low complaints rate period ends at about a time marked as t_1 on Figure 25.2. From that point on, the number of complaints progressively rises as the product starts to deteriorate. The time t_1, therefore, marks the "end of the product's life" or the beginning of its "wear-out period." The rise in the complaints' number indicates that the product's quality loss becomes increasingly noticeable by the consumer, especially in places where its storage conditions are far from optimal. The rise may also be a manifestation of the increased probability of the product's misuse or abuse by the consumer, known as "hazard" in the field, as time goes by. At a certain point, marked t_2 in Figure 25.2, the number of received complaints, and, more importantly, what they imply in terms of the actual number of dissatisfied consumers, reaches an unacceptable level, despite of that other quality indices are still within their permitted ranges. Thus, although laboratory-based shelf life indices might indicate that the product is still saleable, the consumer complaints data would dictate that they should be overruled and the product withdrawn.

An estimate of the lost opportunity of repeat sales due to consumers' dissatisfaction as a result of product deterioration on the shelf can be estimated by the loss function, Loss($):

$$\text{Loss}(\$) = N\left[N_s(n_2 - n_1)\right]C_F P_u \qquad (25.1)$$

where

N is number of units sold

n_1 and n_2 are the number of complaints per number of units sold, N_s, at the beginning of the "normal life" at time t_1 (e.g., days), and when it reaches an unacceptable "wear-out" level at time t_2 (e.g., days), respectively

C_F is the ratio between "responsive" and "silent" consumers

P_u is the unit's price ($)

The potential monetary loss due to consumer dissatisfaction can have a profound effect on future sales and profitability. Therefore, food companies try to find ways to extend their products' shelf lives, i.e., to decrease n_2 in terms of Equation 25.1, and get it as close as possible to n_1. In other words, they try to ensure that the increase in the number of complaints will remain low until a time close to t_2. Accomplishing this can be a costly endeavor and may require the product's reformulation and/or modification of the process by which it is made. Examples of the needed changes are the introduction of nitrogen flushing or vacuum sealing, the use of a less permeable package, refrigeration, etc. The alternative is to post an earlier expiry date on the package. This will shorten the time interval between t_2 and t_1 that will also reduce the difference between n_2 and n_1; see Figure 25.2. This will not only reduce the losses caused by consumers' dissatisfaction but also improve the product's perceived freshness and quality, thus increasing its sales. Also, moving from a "push" to "pull" production mode may result in a significant reduction of the required inventory. Deliberate shortening of the assigned shelf life is not a hypothetical option. Its implementation by a national U.S. distributor of a frozen food product has resulted in considerable savings to the company.

The decision regarding whether or not to adopt a shorter shelf life strategy should be based on the product characteristics, the quality–time relationship, the rate at which the consumer complaints rise when the product approaches the end of its shelf life, the unit price, the volume sold, the availability and reliability of the consumers' responses, the company's marketing philosophy, and the experience with the similar products already in the market.

The staggered sampling design is a method of shelf life assessment based on consumers' responses. Originally developed by Gacula,[3] it progressively increases the number of samples evaluated as the product approaches the end of its shelf life. Another analysis used in shelf life assessment is based on a hazard versus time plot, with the Weibull distribution used to estimate the storage time at which 50% of the consumers will find the product unacceptable. This approach is known as the "Weibull hazard method."[4] Staggered testing combined with a hazard analysis using the Weibull distribution makes it possible to determine a product's shelf life based on consumer quality perception. The Weibull hazard method has been employed in a wide range of food products, as reported by Cardelli and Labuza.[5] The staggered sampling design is part of a more general testing approach known as "reliability studies,"[6] where accumulated follow-up data are used to determine the probability that the product will not fail during its planned shelf life.

25.3.1.2 PARALLEL STORAGE

Once a product is shipped out from the plant or the distribution center, usually the manufacturer loses control over it's fate. Thus, in case of a problem, the manufacturer will become aware of its existence only after the first complaints arrive. Such a situation would be avoided and the manufacturer could respond proactively, if samples of the product were kept in the plant and inspected periodically. An early indication of an emerging durability problem will allow the manufacturer to manage the product based on its actual quality, circumventing the usual FIFO (first in, first out) scheme. Having lead time, the manufacturer could also identify the specific problematic lots and may initiate selective withdrawals. Flexible inventory management based on actual quality rather than on predetermined scheduling can be assisted by time–temperature indicators. The interested reader can find information on their applications in several publications.[7] At least some of the samples kept in the plant can be subjected to accelerated or even abusive storage conditions, i.e., high temperature, relative humidity, and/or oxygen tension (see Section 25.4). This will provide an early indication of a problem emergence and give the manufacturer more time for remedial action. Accelerated storage can also help to avoid emergency situations where the product must be recalled immediately and from everywhere. Storage of "parallel samples" and their periodic testing can also indicate that the product is exceeding its planned shelf life. Such information can be implemented immediately or in future shipments. Moreover, a comparison of the storage results with those obtained by testing samples collected from retail stores and distribution centers can help the manufacturer generate the correlation between the product's shelf life in the laboratory and in the market.

Parallel storage would be particularly useful to new products on the market, where previous experience with similar products might not be sufficient to establish their durability. In principle, the more samples are stored and tested, the more reliable the results are. This testing versus cost is demonstrated in Figure 25.3. However, increasing the "level of inspection" comes at a cost, as also depicted in the figure. In many cases, the benefits of intensive testing might not increase as steeply as the testing cost.[8] This is manifested in the relatively flatter "cost of the risk curve" shown in the figure. In products made on many dates and/or in numerous lots, adequate testing might become cost prohibitive. Besides the direct cost of labor and reagents, for example, the operation would require considerable organization, bookkeeping,

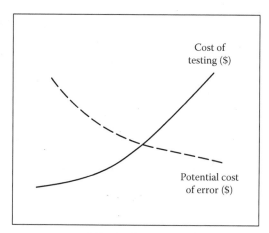

Cost of testing ($)

Potential cost of error ($)

Testing frequency

FIGURE 25.3 Schematic view of the inverse relation between the cost of testing and the risk of unanticipated quality deterioration or spoilage.

and storage space that might be difficult to manage, especially for a small company. A way to mitigate the problem but still retain some of the benefits of parallel storage is to follow only selected samples at critical times. These samples would be taken from the plant or wholesaler's warehouse and tested only toward the end of the product's planned shelf life or, if needed, after a problem with a particular lot is suspected or has been discovered.

It is worth mentioning that, unlike in pharmaceuticals, where shelf life is defined as the time at which at least 90% of the specified dose is retained, the shelf life of foods has never had a universally accepted and legally binding definition. Barring actual spoilage, what constitutes "shelf life" in foods depends on the product, the manufacturer or buyer's own standards, the consumer's expectations, and other factors. Inevitably, shelf life estimated from laboratory stability tests has a certain degree of uncertainty due to experimental errors. To reduce the potential consequences of such errors, a common practice in the pharmaceutical industry is to define and report the shelf life as the earliest time at which the lower 95% confidence limit of the mean is reached. The assumption is that the measurement scatter is random and that the confidence limits, or percentiles, can be calculated with the normal or student distribution function.[8] No such statistical criteria of shelf life have yet been implemented in foods.

25.4 PRINCIPLES OF ACCELERATED STORAGE

The temperature effect on biochemical and biological processes is a common everyday experience. We all keep foods in the refrigerator in order to slow down their deterioration, thus extending their useful "life." We also know that milk left at room temperature turns sour long before refrigerated milk, because of the accelerated growth of the *Lactobacilli* in it. Temperature manipulation is an effective means to control the rate of biological and biochemical processes, which is widely exploited in industrial and domestic situations. Generally, increasing the temperature accelerates biological and chemical processes and therefore their outcome becomes detectible or evident sooner. Accelerated storage is based on this idea. Instead of waiting for a year, say, in order to notice an off flavor, discoloration, or the onset of corrosion at ambient temperatures, we can shorten the time to several weeks when the food is stored at an elevated temperature. However, in order to interpret the information obtained from elevated temperature experiments and use it to predict changes that would occur under normal storage conditions, we need a kinetic theory and mathematical models. These not only explain the results but also enable their extrapolation by quantifying the temperature's role. A comprehensive and critical review of the kinetic models used in foods' quality assessment and how their parameters are affected by temperature can be found in [9].

A schematic presentation of the principle governing accelerated shelf life studies (ASLT) is given in Figure 25.4.

The nature of kinetic models and how they can be used to predict shelf life will be discussed in more detail in the next section and Appendix A. Regardless of the kinetic model chosen, we must always keep in mind that there is a limit to how high a product's temperature can be raised during its accelerated storage. This is

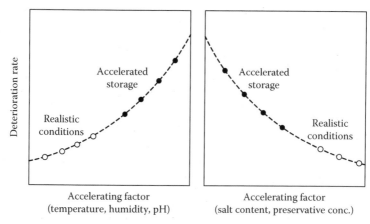

FIGURE 25.4 Schematic view of the principle underlying the use of accelerated storage to predict a process's deterioration rate. Notice that accelerated deterioration can be accomplished by either intensifying a promoting factor like temperature, humidity, and oxygen tension (left) or through removing an inhibitor, by lowering the salt or chemical preservative concentration, for example (right). (Courtesy of Dr. Maria G. Corradini.)

because exposure of a food to a too high temperature can cause qualitative changes that will never occur under ambient conditions. A classic example is fertilized eggs incubation, where overheating will kill the embryos inside. (Obviously, no chick has ever hatched from a hard boiled egg—a dramatic qualitative change that elevated temperature can produce in a protein.) Chocolate, for example, would simply melt if stored at a temperature above 37°C, imposing an upper limit to the temperature elevation. Similarly, the physical stability of ice creams cannot be studied at temperatures that will allow even partial melting, and the same applies to other frozen foods. Carbonated beverages are also a case in point because the gas solubility is dramatically affected by temperature.

Temperature is not only the most prominent factor that affects foods deterioration rate, but also the easiest to control under laboratory conditions. There are also well-developed theories of its effect on chemical and biochemical reactions[10,11] and on microbial growth. No wonder, therefore, that "accelerated storage" is almost synonymous with storage at elevated temperature(s). Recommended temperatures for accelerated storage of foods can be found in several publications, e.g., Labuza and Schmidtl[12] and Institute of Food Science and Technology.[13] They are summarized in Figure 25.5.

Ideally, at least four constant storage temperatures would be recommended. It is also advisable to store the product under variable temperature conditions, too. The results of such storage can be used to validate the kinetic model and test its ability to predict changes that occur in the market, where the temperature almost always fluctuates.

Acceleration of foods deterioration can be achieved by other means, too. Storage at high relative humidity and under high oxygen tension are two of the most notable examples. One can also accelerate the process by combining elevated temperature

Product Type	Storage Temperatures (°C)	Test Control (°C)
Dry, intermediate moisture	30, 35, 40, or 45	40
Canned	4, 35, 40	−18
Frozen	−5, −10, −15	<−18
Chilled	5, 10, 15, 20	0
Aseptically filled	30, 35, 40	0

FIGURE 25.5 Recommended accelerated shelf life temperatures for foods according to Ref. [12].

and high relative humidity, with or without oxygen-enriched atmosphere.[10] Like in the case accelerating the deterioration through temperature elevation alone, the relative humidity level also should be selected with great care. This is in order not to induce kinds of changes that will never occur under low relative humidity. For example, at excessively high relative humidity, the food can spoil by microbial growth well before the accelerated biochemical reactions would have any visible effect. (Microbial growth can be suppressed by adding a chemical preservative, but this might change the food in other ways.) Similarly, at a certain point, excessive high oxygen tension, especially at an elevated temperature or humidity, can produce oxidation and deterioration patterns that might not occur in the normal life of the product.

As already mentioned (Figure 25.4), accelerated deterioration can also be accomplished by removing an inhibitor or protective agent. Salt in cured meats and cheeses, for example, plays such a role by suppressing microbial growth. Thus, at least in principle, preparing and testing such products with low salt contents can be used to accelerate their deterioration at the low temperatures at which they are usually held. But here again, the range of low salt concentrations should be carefully chosen because otherwise the microorganisms that will grow in the experiment might not be of the same kind as those that spoil the original product. The same could happen in pH elevation to accelerate the spoilage by microbial growth. Here, too, the habitat's properties modification can affect the composition of the growing microbial population. Successful accelerated storage experiments and meaningful interpretation of their results frequently require knowledge of the role of heat and mass transfers between the product and its environment and within the product itself. For example, cases inside a shipped container, and sometimes even individual packages in the same box, can experience different thermal histories. This should be taken into account in the translation of accelerated storage results into predicted spoilage patterns in the market.

25.4.1 ACCELERATED STORAGE DATA INTERPRETATION

The changes in most stored foods, at both ambient and elevated temperatures, are caused by simultaneous and in many cases interactive physical processes, chemical reactions, and/or microbial growth. In an accelerated storage experiment, one usually monitors the changes in several attributes such as visual, organoleptic, chemical, and/or textural. A food product ends its useful shelf life when even one of these has become unacceptable. Sometimes, it is possible to identify a chemical or physical marker of the product's deteriorative state. When such a marker is found, nascent changes that are yet to be manifested in the product's appearance or flavor can be monitored and quantified. Once this has been done, the information can be used to predict the food's remaining shelf life. What follows is a discussion of different modeling approaches to the kinetics of such changes, with special emphasis on the temperature's role. An excellent summary of the state-of-the-art in accelerated storage kinetics can be found in [10]. Much of the development in this field has been done by a group of investigators associated with Professor Karel's laboratory at MIT. It includes Professor Labuza at the University of Minnesota, Shimon Mizrahi at the Technion Israel Institute of Technology in Haifa, the first author of this chapter (Sam Saguy) at the Hebrew University of Jerusalem, and many of their students and peers.

25.4.2 CHOOSING A QUALITY MARKER

The choice of one or more quality markers is based on the following considerations:

1. If not the limiting factor itself, the marker must be relevant to the product's state. This means that its value can be correlated with that of those attributes that do determine the product's shelf life.
2. It can be determined by a simple chemical, physical, or biological assay, i.e., its evaluation should not require an elaborate preparation procedure and/or long time.
3. Its determination does not require a highly trained technician, costly reagents, sophisticated instrumentation, and/or elaborate data processing and calculations.
4. The results of its determination are clear, unambiguous, and require little interpretation.

For example, microstructure determined by electron microscopy will not be a good choice while pH, Hunter color, or a directly titrated compound or group of compounds will.

Sensory analysis is the most direct way to detect changes of immediate implications, and sensory evaluation frequently serves as the ultimate criterion of a product's acceptability. However, it is not always easy to interpret the results of a sensory analysis, especially in marginal cases and when there is a substantial discrepancy between the assessments of an expert and consumer panels. Although taste is of utmost importance to consumers, their discrimination ability varies considerably. This factor needs to be taken into account, especially in marketing research.[14]

25.4.3 SHELF LIFE MODELING

25.4.3.1 Linear Deterioration Kinetics

Consider the simple scenario of an isothermal deteriorative process or reaction that can be monitored through an effective chemical marker whose concentration can be easily determined. Let us call its normalized changing concentration $Y(t)$, defined as

$$Y(t) = \frac{C(t)}{C_0} \qquad (25.2)$$

where $C(t)$ and C_0 are the momentary and initial concentrations, respectively.

$Y(t)$ versus time can be a decay function (e.g., vitamin loss) or a growth function (e.g., accumulation of an oxidation product and a brown color intensification). For simplicity, we will assume that the decay or rise is monotonic at least on the pertinent timescale, i.e., that $C(t)$, and consequently $Y(t)$, does not reach a peak.

Perhaps, the most common decay pattern is the one that follows the first-order kinetics, i.e.,

$$\frac{dY(t)}{dt} = -k(T)Y(t) \qquad (25.3)$$

where $k(T)$ is the exponential decay rate constant at a given absolute temperature, T.

On integration, this model yields

$$Y(t) = \exp\left[-k(T)t\right] \qquad (25.4)$$

Theoretically, there can be a simpler decay pattern. It follows what is known as zero-order kinetics, where the concentration of the monitored compound diminishes at a constant rate, i.e., at a rate that is independent of the momentary concentration. There are also more complicated patterns where the decay rate is proportional to the mth power of the momentary concentration ($m \neq 1$, $m \neq 0$). These are known as following the mth-order kinetics, where m need not be an integer. One can also construct "hybrid" kinetic models if needed. For example, only the initial part of the decay would be described by the first-order kinetics model, allowing for a finite asymptotic residual concentration instead of the complete elimination that the original model, Equation 25.4, implies.

Equations 25.3 and 25.4, or their equivalents in the zero- and mth-order kinetics, can be converted into growth or accumulation models by removing the minus sign, i.e., by replacing $-k(T)$ by $k(T)$. Either way, and regardless of the actual equation and whether it describes loss or gain, the model enables its user to calculate the value of $Y(t)$, or $C(t)$, at any given time within the range of the model's applicability. When the rate constant, $k(T)$, or the corresponding $k(T)$ and m of the more elaborate models, at a given temperature are known or can be estimated, one can generate a $Y(t)$ or $C(t)$ versus time curve at that temperature. With $k(T)$, or $k(T)$, and m known, and by solving the corresponding equation for time, one can also calculate the time at which $Y(t)$ or $C(t)$ would reach an alarming or unacceptable level at that temperature, which we can call Y_{crit} or C_{crit}.

25.4.3.2 Effect of Temperature on the Rate Constant

Consider a case of a zero-, first-, or any other mth-order kinetics. Traditionally, it has been assumed that the temperature dependence of such processes' rate constant follows the Arrhenius equation,[15] i.e.,

$$\log_e\left[\frac{k(T)}{k_{T_{ref}}}\right] = -\frac{E_a}{R}\left(\frac{1}{T} - \frac{1}{T_{ref}}\right) \tag{25.5}$$

where

$k_{T_{ref}}$ is the rate at a reference temperature T_{ref},

E_a an "energy of activation",

R the universal gas constant.

According to this model, a plot of $\log_e[k(T)]$ versus $1/T$ (in degree Kelvin reciprocals) is a straight line whose slope is E_a/R. Once E_a/R is known, one can choose any convenient reference temperature and rate in order to calculate the value of $k(T)$ at any given temperature T. Notice that E_a could at best be used as a comparative measure of a process rate's temperature sensitivity. It should not be confused with the energy barrier to a reaction between gases as originally formulated by Arrhenius. Also, this model will only hold if the "energy of activation" is itself temperature independent, which might not always be the case in complex reactions with interacting paths. This issue need not be of concern when dealing with the relatively small temperature gap between "normal" and accelerated storage. (The need for the logarithmic transformation of the rate constant and the temperature conversion into absolute temperature reciprocals is more controversial. The same can be said about the place of the universal gas constant in reactions that bear no similarity to those for which the Arrhenius equation had been originally developed.[16])

Once the temperature dependence of the rate constant has been determined, through Equation 25.5 or an alternative model, it can be combined with the original rate model to predict the decay or growth pattern under nonisothermal conditions, too. For example, if one could safely assume that a vitamin loss follows a first-order kinetics at all pertinent temperatures, then one could also generate a $Y(t)$ or $C(t)$ versus time curve for any temperature history, $T(t)$. If indeed the exponential rate constant $k(T)$ varies with temperature according to the Arrhenius equation (Equation 25.5), then the $Y(t)$ versus time relation would be

$$Y(t) = \int_0^t k_{T_{ref}}\exp\left\{-\frac{E_a}{R}\left[\frac{1}{T(t)} - \frac{1}{T_{ref}}\right]\right\}dt \tag{25.6}$$

The right-hand side of the equation can be integrated numerically for almost any conceivable continuous algebraic expression or discrete record of the temperature history, $T(t)$. The numerical integration itself can be done using any number of commercial mathematical programs or even with general-purpose software like MS Excel© using the Runge–Kutta algorithm, for example. Once a $Y(t)$, or concentration, $C(t)$, versus time relationship is derived, it can be plotted or used to extract numerical values for any chosen time.

After the relationship has been generated, its plot (or digital file) can be used to identify the critical time, t_{crit}, at which the product will reach an unacceptable level of deterioration, expressed in term of Y_{crit} or C_{crit}. This time can then be treated as the product's shelf life. To be on the safe side, one can define the actual shelf life as the time to reach $1.1Y_{crit}$ or $1.5C_{crit}$, say, leaving a margin for error and uncertainties. (In cases where the marker's value rises during storage, the safety margin will be achieved by setting the shelf life to $0.9Y_{crit}$ or $0.5C_{crit}$, for example.) Once the parameters $k_{T_{ref}}$ and E_a of the first-order Arrhenius equation combination have been determined experimentally or estimated from published data, they can be used to examine the consequences of any number of contemplated scenarios by generating the corresponding $Y(t)$ or $C(t)$ versus time curves for the corresponding temperature histories, $T(t)$'s. Their consequences in terms of the level of deterioration reached can then be examined and used as a guideline for the product's shelf life assessment. A similar procedure can be used to estimate the probabilities of failure under hypothetical but realistic conditions based on random temperature fluctuations that mimic typical conditions in the marketplace; see Figure 25.6.

The criterion of the product's useful shelf life can be expressed either in terms of a ratio or the chosen quality index's absolute magnitude. For example, one can

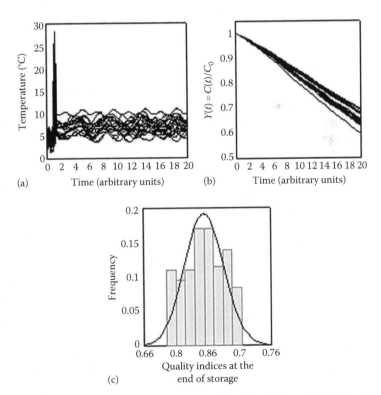

(a)

(b)

(c)

FIGURE 25.6 (a) Simulated random temperature histories of a hypothetical food product, (b) the corresponding degradation curves of its quality, and (c) the distribution of the quality index at the end of the storage period. (After M.G., Corradini, et al., in *Food Engineering: Integrated Approaches*, Springer, New York: 2008, pp. 47–71; courtesy of Dr. Maria G. Corradini.)

define 50% loss of a vitamin, say, as indicating the end of the product's useful life. But this vitamin's concentration drop to a level specified in parts per million (ppm) could also be used as an alternative criterion. In general, the "absolute option" will be more effective when using growth- or accumulation-based criteria. Reaching a noticeable level of off flavor or discoloration is, of course, more noticeable than whether a vitamin concentration has been reduced by 40% or 50%, for example. The same is applicable to microbial spoilage, where the actual count, and not the growth ratio, determines whether the product is still edible and safe to eat. Microbial growth, however, rarely follows a linear kinetics and therefore it will be discussed separately in the next section.

25.4.3.3   Nonlinear Deterioration Kinetics

The assumption that food deterioration can be characterized by a single rate constant need not be always justified. Microbial growth is a case in point. Even under isothermal conditions, the plot of the number of cells versus time is a sigmoid curve of a kind that requires at least three growth parameters for its mathematical characterization. Moreover, these parameters have their idiosyncratic temperature dependencies whose mathematical characterization requires additional parameters. As shown earlier, when the first-order Arrhenius model holds, knowing the rate constant, the reference temperature, and E_a is sufficient to generate a $Y(t)$ or $C(t)$ versus time curve under almost any conceivable temperature history. In most cases, to do the same for microbial growth will require the numerical solution of a rate model (i.e., a differential equation) having four to six coefficients, which need to be determined experimentally (instead of the two in the linear case).

Nonlinear kinetics can also be found in certain vitamin degradation processes (e.g., Corradini and Peleg[18] and van Boekel[9]), and lipid oxidation. In all such systems, the concept of a rate constant, linear or exponential, becomes blurred. This is because, even under isothermal conditions, the rate of change is not only a function of temperature but also a function of time as well. Conventional temperature dependence models, like the Arrhenius equation, lose much of their appeal in such cases because the temperature influence on the underlying process or processes is manifested in the magnitude of several different kinetic parameters instead of by a single rate constant. There are mathematical methods to handle such situations, two examples of which given in Appendix A. As shown schematically in Figure 25.7, such methods can be used to estimate the full deterioration curve under "normal conditions," by extrapolating not one rate constant but all the relevant kinetic parameters.

In principle, the choice of models to describe the isothermal deterioration curve and the temperature dependence of its parameters should have little effect on the prediction, as long as the gap between the accelerated and normal conditions is not too large.[16,18,19] This is demonstrated in Appendix A, where whole predicted microbial growth curves, obtained with two very different mathematical models, are shown together with the actual experimental data.

The reliability of the predictions would primarily depend on the quality of the accelerated storage data, the frequency at which they had been collected, and the number of elevated temperatures (or humidity levels, etc.) at which they had been determined. The predictions' accuracies will also depend on how far the extrapolation goes. The smaller

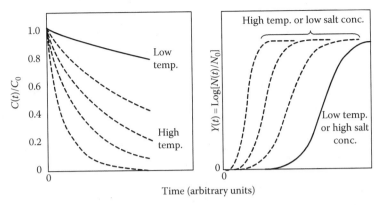

FIGURE 25.7 A schematic view of the construction of the whole deterioration curve from accelerated storage data. (Left) chemical degradation and (right) microbial growth. For details see Appendix A. (Courtesy of Dr. Maria G. Corradini.)

the gap between the accelerated and "normal" conditions, the better will be the agreement between the various models' predictions and between these and the actual values.

Today, application of nonlinear kinetics in food research and technology is fairly rare, except perhaps in issues related to microbial growth. But with a better grip on the kinetics of complex systems, the increased power of computers, and the rapidly growing availability of user-friendly mathematical programs, the situation may change in the coming years.

25.4.3.4 Spoilage Risk Assessment

Once a rate model has been constructed, as already mentioned, it can be run repeatedly for a variety of plausible scenarios. These may include random temperature oscillations within a specified range of amplitudes and frequencies. If a spoilage criterion has been established, e.g., an unacceptable low level of a nutrient loss or an unacceptable high level of oxidation, discoloration, or microbial growth, then computer simulations could provide the most probable times at which these levels would be reached. The deterioration curves produced by what is known as the Monte Carlo method can be used to construct a temporal distribution of the expected spoilage. An example of how simulations of this kind can be used to determine a product's shelf life was given above in Figure 25.6.

A distribution so generated can then be used to estimate the fraction or the percentage of the product reaching an unacceptable quality level as a function of time. This information in turn can be used to determine when an unacceptable fraction or percentage of the units would reach the end of their acceptability, and thus when the product should be withdrawn from the shelf. Or, conversely, one can repeat the simulations with various thermal histories and seek optimal storage and handling conditions that will ensure tolerable levels of deterioration while the product is on the shelf and in the consumer's possession.

Since food spoilage can be considered a "failure phenomenon," it would be anticipated that the times at which individual units perish would have the aforementioned

Weibull distribution. The Weibull distribution had been originally developed for degradation that stems from mechanical breakage but has since been found to be an excellent model of a variety of unrelated and very different kinds of "failure phenomena." This has indeed been observed in stored foods and has led to the use of the Weibull distribution function to estimate the risk of their spoilage on the shelf.[20]

Although most of the discussion has referred to temperature as the sole means to accelerate food deterioration, the principles and methods also govern systems where moisture, oxygen tension, or any other factor plays the decisive role in the product's shelf life determination. Conceivably, as has already been indicated, there are many situations where the deterioration pace is determined by the combined effects of several factors, temperature and humidity being the most prominent. The effects of several simultaneously changing factors on a food product's shelf life are harder to analyze and predict.[10] The main reason is that it is difficult to account for both the roles of the individual factors and at the same time their interaction effects. In principle, when the range of the temperature, humidity, and other factors' variations is relatively small, the response surface methodology could be used to identify the optimal storage conditions. But since it is quite possible that no set of optimal conditions exists in the sought range, the benefits of this method might be limited.

25.4.3.5 Single versus Multiple Quality Parameters

Unlike in pharmaceuticals, for example, where activity is the sole criterion, a food product's acceptability is rarely determined by a single quality attribute. During handling, transportation, and storage, the various deteriorative processes need not follow the same kinetics. The same can be said about their interactions. Multiple operating processes generate significant consequences. For example, it is theoretically possible that the changes in Property A, say, (see Figure 25.8) have only mild temperature dependence relative to that of the changes that Property B undergoes. Thus, at a low storage temperature, the product will most probably end its shelf life when Property A reaches an unacceptable level. In contrast, at a higher storage temperature, the changes in Property B can overtake those in Property A and will render the food unfit for consumption while Property A is still at an acceptable level (see Figure 25.8). Theoretically,

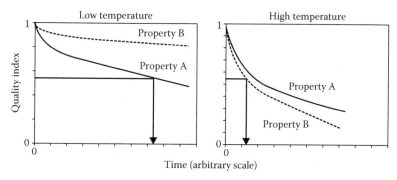

FIGURE 25.8 Schematic (exaggerated) view of the potential differential effect of temperature on a product's assigned shelf life when monitored through a process whose rate has weak or strong temperature dependence (Property A and Property B, respectively).

there can be numerous such combinations, which will result in the product's failure through different mechanisms at different temperatures. One possible response to a scenario of this kind would be to switch from monitoring changes in one property at the beginning of the storage to another property at the end. The actual time of the switch will depend on the particular kinetics of the processes in question.

Although the foregoing example merely demonstrates the theoretical possibility that such situations would occur, it highlights the need to be aware of their potential impact. This refers to the planning of accelerated storage experiments, interpretation of their results, and their implementation. Moreover, the decay curves shown in Figure 25.8 depict hypothetical scenarios where the kinetics of the two deterioration mechanisms are independent of each other. This might not be always true, and it is quite possible that the interactions between the various spoilage processes would also depend on the product's temperature history. The same can be said about the effects of humidity and oxygen tension on the product's stability; their synergism is most likely to be strongly influenced by temperature, too. And again, when this is the case, it would be advisable to intensify the testing toward the end of the expected product's shelf life by including additional assays and increasing the sampling frequency.

25.4.3.6 Confidence in Storage Results

No doubt, increasing the number of temperature (or humidity or oxygen tension) levels and expanding their range will produce more reliable accelerated storage data. The same can be said about the number of specimens in each sample and the frequency at which they are tested. The problem, however, as already mentioned, is that the testing has a direct cost to which one must add the costs of the samples themselves, their handling and storage, and the tests administration. Feasibility considerations dictate that a compromise must be struck between the attempt to improve the results reliability and the need to keep the cost at an affordable level. The available options as to how to resolve the conflict will be discussed below. It is sufficient to mention here that the subject has been studied intensively, albeit mostly in relation to quality assurance.

Research of this topic has produced a large body of knowledge that is valuable and pertinent. It includes the already mentioned statistics of sampling and decision, which can be very helpful in the design of accelerated storage experiments. Another aspect, more relevant to the data analysis stage, is how to deal with odd results, which in food testing are inevitable. Simple methods to identify outliers can be easily found on the Internet or statistical textbooks. Certain data points are outliers that should be recognized before a mean value is computed, but their presence should still be considered in the decision on how to proceed. Advice from a professional statistician will always be helpful, even to people with a good mathematics background. It will be essential to those having only basic training in the quantitative aspects of food science. The statistician's help will be by far more effective if received at the planning stage that occurs before the experiments even start. The professional statistician can help the developer to extract maximum information from a limited number of results and avoid gathering data that could not be effectively used later. Keep in mind, though, that most statisticians or industrial engineers have only limited knowledge of food science and technology. Therefore, a lot of explanation would be needed to develop a fruitful cooperation.

25.5　SHELF LIFE ASSESSMENT STRATEGIES IN AN INDUSTRIAL ENVIRONMENT

In principle, and as already repeatedly stated, the more extensive the testing and follow-up are, the more reliable the results become. Updated information on product state and quality can be used to extend its life and avoid mishaps. This information, however, comes at a cost that may or may not be recovered. On the other hand, a lower level of inspection, which reduces the testing cost, may force the manufacturer to assign a too early expiration date in order to err on the safe side. This preventive strategy also entails a cost, of course, which needs to be balanced against the direct expenditure on research, potential impact of failure on the product marketing, and other considerations. The compromise between these two conflicting demands would be determined by the product's value to the company in immediate and long-range terms, and the available resources at the time of the product's introduction.

Here are some options available to the product developer and the manufacturer.

Minimal testing and short planned shelf life. This option can be attractive to small-scale operations (like a deli or bakery) or market trials. The product is deliberately assigned a shorter shelf life than would be needed on the basis of experience or the results from a previous study. While the product is on the shelf, samples can be periodically inspected, thereby using the market as a laboratory and part of the experiment. Once enough information has been gathered, the assigned shelf life can be extended to be commensurate with the product's actual durability. For a totally new kind of a product, it will be prudent for the new product developer to initially limit its shelf life. Then, as more information and data are collected, the shelf life can be gradually expanded. This "dynamic option" enables the manufacturer to launch a new product fairly rapidly with only a small investment beyond what is required by law. For a very small company, especially with a localized market, this option can be adopted as an ongoing permanent strategy. For a large or small company planning expansion, the experience so gathered can serve as a stepping stone for a more elaborate strategy aimed at capturing a bigger market. At that point, however, larger variations in storage conditions and consumption patterns will have to be dealt with and a longer shelf life would have to be guaranteed. Nowadays, temperature, humidity, and other relevant factors can be monitored with inexpensive computerized data loggers. These data can be combined with consumers' responses, positive or complaints, and considered together in the decisions on how to proceed with the product.

Intensive testing and long planned shelf life. This option is almost mandatory for companies aiming for a large regional or national market where the product will inevitably encounter diverse temperature and humidity histories and possibly some abuse as well. In such an environment, the ability to respond effectively to an emerging problem or complaints is minimal, and therefore the product must be designed to withstand the market vagaries before it is launched. This may require comprehensive studies of the product's deterioration patterns to assure its shelf life even under a "worst case scenario." Accelerated storage experiments can shorten the time to obtain the relevant information. Their results, however, would be of little value unless the deterioration kinetics has been either established or could be inferred from the literature or data obtained on similar products. Again, one of the study's main objectives would be to identify the most effective quality marker or markers for the particular product. Once found, they could be monitored in stored samples at the laboratory and, more importantly, in specimens collected from the market. After the methodology has been developed or adapted, the product can be assigned an expiration date that would match its actual shelf life. How exactly this information is communicated to the consumer is another matter. Most marketers prefer that the expiration date on the package could be exceeded without any sign of deterioration or noticeable quality loss, especially if the product is stored under favorable conditions.

Combined laboratory, market, and consumers monitoring. It is worth repeating that a follow-up on a product has significant merits. The most notable is that it enables the detection of an emerging problem in time to avoid consumer dissatisfaction and a forced recall. The monitoring can be done by placing samples under accelerated storage conditions in the laboratory and checking them periodically for early signs of deterioration or quality loss. The follow-up in the laboratory can be supplemented by testing samples retrieved periodically from the market. The results of both can then be compared with consumer complaints data to generate an empirical correlation between changes in the product at the laboratory and the market, as well as how either or both translate to the consumer response. Such a correlation would be most useful to the company both in decisions concerning the particular product and in considerations about future operations. When recorded temperature histories of specimens gathered from retailers' shelfs and/or a distribution center are also available, they can be used to test the applicability and predictive ability of the kinetic models used to assess the product's quality loss from accelerated storage data. Analysis of the combined information would indicate whether the kinetic models need adjustment, correction, or replacement. Again, temperature and humidity monitoring can be done inexpensively with data loggers, and hence gathering the information need not be a serious hurdle. The decision on whether the accelerated and actual storage data should be compared routinely or as part of a special study would be based on logistic and strategic considerations. These will also determine the study's scope and the allocated resources to carry it out. Some products undergo reformulation even after their introduction to the market, primarily as a result of changes in the raw materials availability and cost. In such cases, special care is required to make sure that the changes do not adversely affect the product's shelf life.

Feedback and update. Data collected during the product quality monitoring, in the laboratory, marketplace, or both, should be interpreted as quickly as possible. "Postmortem" analysis, although not without some merits, may not be the most effective manner to handle a study's results. For example, temperature records can indicate smaller or larger fluctuations than initially thought or that the product's deterioration rate and/or its temperature dependence differ from those originally assumed. The implications of such discrepancies, once revealed, should be assessed while the product is still being monitored. When significant, the discovered differences between what has been anticipated and reality can be used to revise the product's estimated shelf life and improve its accuracy. This improvement could be achieved by rerunning the computer simulations in a manner discussed in Section 25.4.3.4, using an updated kinetic model and/or with more realistic limits on the temperature range, for example. Handling the results should always be a dynamic process with a focus on the changing circumstances. Special attention should be paid to these and their possible manifestations whenever the product has been reformulated after its introduction to the market. Naturally, the "level of scrutiny" should be as high as possible at least initially. It could be subsequently lowered as the accumulated information on the reformulated product's durability enables the manufacturer to establish its shelf life within acceptable and safe margins.

APPENDIX A

SHELF LIFE ESTIMATION BASED ON NONLINEAR DETERIORATION KINETICS

Consider a nutrient loss as an index of deterioration and that its diminishing concentration can be described by the two-parameter Weibullian model[16,21]:

$$Y(t) = \exp\left[-b(T)t^{n(T)}\right] \tag{25.A.1}$$

where $Y(t) = C(t)/C_0$, as before, and $b(T)$ and $n(T)$ are temperature-dependent parameters.

A decay pattern following Equation 25.A.1 emerges when the disintegration or expiration events (in our example the nutrient's reduced concentration) follow a Weibull temporal distribution, hence the name. According to this model, the "survival curve," when plotted as $\log Y(t)$ versus time, will have downward concavity if $n(T) > 1$ and upward concavity if $n(T) < 1$. The log-linear case, i.e., the "first-order decay kinetics," is a special case of Equation 25.A.1, where $n(T) = 1$. Thus, everything said in this section will apply to deterioration patterns that follow the first-order kinetics, in which case $b(T)$ would be equal to $k(T)$ in Equations 25.5 and 25.6, but not vice versa. The parameter $b(T)$ in Equation 25.A.1 is a "rate parameter," albeit a nonlinear one. This is because the actual exponential rate in a Weibull decay is a function of both temperature and time, even under isothermal conditions.

The Weibull model is by no means unique. In many cases, the same decay pattern can be described by an alternative mathematical expression with the same degree of fit. For example, when $n(T) < 1$, an alternative two-parameter model can be

$$Y(t) = 1 - \frac{t}{k_1(T) + k_2(T)t} \tag{25.A.2}$$

where $k_1(T)$ and $k_2(T)$ are the temperature-dependent parameters.

According to this particular model, the asymptotic residual value of $Y(t)$, i.e., when $t \to \infty$, is $1 - 1/k_2(T)$, and $Y(t)$ will reach 50% at $t = k_1(T)/[2 - k_2(T)]$. This is true, of course, only for a limited time range where the two models overlap. Eventually, the decay curves depicted by the two models (described by Equations 25.A.1 and 25.A.2) must diverge as shown in Figure 25.A.1.

In principle, one can use accelerated storage data to plot or describe mathematically the temperature dependence of $b(T)$ and $n(T)$, $k_1(T)$ and $k_2(T)$, or the equivalent

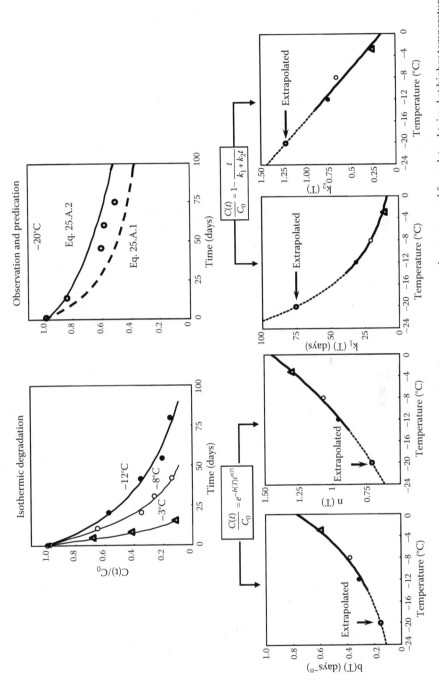

FIGURE 25.A.1 An example of how a vitamin's degradation curve at a low temperature can be constructed from data obtained at higher temperatures using two different models. (After M.G. Corradini and M. Peleg, *Trends Food Sci. Technol.*, 18, 37, 2007; courtesy of Dr. Maria G. Corradini.)

parameters of any other model deemed appropriate. The mathematical description of these dependencies can be an *ad hoc* empirical expression as shown in Figure 25.A.1 or derived from theoretical considerations. Once determined, extrapolation of the plots or corresponding mathematical relationships will provide an estimate of the parameters' values at any chosen "normal storage temperature." These extrapolated values can then be used to construct the whole decay curve at that temperature, as shown in the figure. The procedure can be repeated for other chosen temperatures in its neighborhood or to predict the outcome of nonisothermal storage.

Figure 25.A.1 demonstrates that the two models (Equations 25.A.1 and 25.A.2), which had the same and almost perfect fit to the experimental "accelerated storage data," rendered predicted decay curves at the lower temperature, which were far from being identical. Still, and despite the gap, the predictions reasonably agree with the actual values that were determined experimentally. This suggests that it would be worthwhile to use more than one model for mutual verification of the predictions. And, in case of a substantial discrepancy between the predictions, it might be prudent to use either an averaged curve or the one that indicates the fastest deterioration.

All the above also pertain to processes that result in growth or accumulation. The rise of the Maillard reaction's products in nonenzymatic browning or that of the other oxidation products is a typical example. The kinetics of some such deteriorative processes, at least initially, could probably follow the models based on inverted versions of Equations 25.A.1 and 25.A.2, for example:

$$Y(t) = \exp\left[b(T)t^{n(T)}\right] \qquad (25.A.3)$$

or

$$Y(t) = 1 + \frac{t}{k_1(T) \pm k_2(T)t} \qquad (25.A.4)$$

The resulting curves in such a case will be a mirror image of those shown in Figure 25.A.1 above. For example, when, in Equation 25.A.3, $n(T) > 1$ or there is a minus sign before $k_2(T)$ in Equation 25.A.4, the curve will look like it is exponentially climbing. But when $n(T) < 1$ and there is a plus sign before $k_2(T)$ in Equation 25.A.4, the monotonic rise of $Y(t)$ will proceed at a progressively diminishing rate.

Another example is microbial growth. Microbial growth can be accelerated by raising the temperature to a level that will not destroy the organism(s) or by removing an inhibitor like salt, as shown in Figure 25.A.2.

Notice that before mortality sets in, the microbial growth curve has a characteristic sigmoid shape. A mathematical description of such curves, with few exceptions, requires at least three parameters. They have to account for the exponential growth onset ("lag time"), exponential rises sleepness, and the population level at the "stationary phase". At least the first two model parameters must have strong temperature dependence, whereas the third to a lesser extent, if at all. This is because the growth curve's asymptote is thought to represent the habitat's carrying capacity. Consequently, it primarily depends on the amount of internal resources

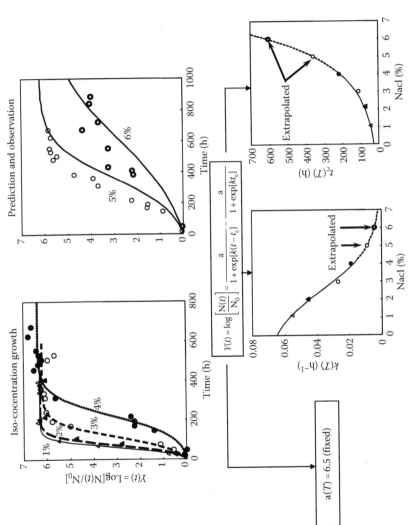

FIGURE 25.A.2 An example of how a microbial growth curve at high salt concentrations can be constructed from data obtained at low salt concentrations using Equation 25.A.5 as a model. (After M.G. Corradini and M. Peleg, *Trends Food Sci. Technol.* 18, 37, 2007; courtesy of Dr. Maria G. Corradini.)

rather than on external conditions like temperature. Again, various mathematical models can describe the same isothermal growth data. Two such models are the classic Gompertz model[22] and its alternative, the shifted logistic function[18]:

$$Y(t) = \frac{a(T)}{1 + \exp\{k(T)[t_c(T) - t]\}} - \frac{a(T)}{1 + \exp[k(T)t_c(T)]} \tag{25.A.5}$$

where $a(T)$, $k(T)$, and $t_c(T)$ are temperature-dependent parameters.

In this model, $Y(t)$ can be defined as the net growth ratio, $[N(t) - N_0]/N_0$, or the logarithmic growth ratio, $\log[N(t)/N_0]$. Both definitions entail that at $t = 0$, $Y(0) = 0$, and therefore that the growth curve described by this model always starts at zero. The asymptotic growth levels according to this model are $a(T)(1 - 1/\{1 + \exp[k(T) t_c(T)]\})$. Since, in most cases, $1/\{1 + \exp[k(T)t_c(T)]\}$ is much smaller than 1, $a(T)$ is a good approximation of the growth ratio's asymptotic level. The growth rate at the exponential regime and the inflection point's location is determined by $k(T)$ and $t_c(T)$, respectively. Once the temperature dependence of the growth model's coefficients has been determined under accelerated storage conditions, they can be extrapolated to the chosen "normal temperature" and used to generate the whole growth curve. The same procedure can be employed to accelerated storage caused by an inhibitor removal. An example is shown in Figure 25.A.2, where the inhibitor was salt.

Deterioration under nonisothermal conditions. Once the deterioration parameters' temperature dependencies have been determined from accelerated storage data, they can be used to generate "deterioration curves" under almost any conceivable temperature history, at least in principle. This is regardless of whether the underlying isothermal model is of decay (e.g., Equation 25.A.1 or 25.A.2) or growth (e.g., Equation 25.A.5). The underlying assumption is that the momentary rate of change is the isothermal rate at the momentary temperature, at a time that corresponds to the momentary state of the system as expressed by the momentary value of $Y(t)$.[20,22] The assumption is implemented in the construction of a rate model in the form of a differential equation that ought to be solved numerically.[19] For all the isothermal model equations presented in this appendix, the corresponding nonisothermal rate models are ordinary differential equations that can be solved with almost any commercial mathematical software and even general-purpose software like MS Excel©. They have to account for exponential growth onset ("lag time"), exponential rise steepness, and the population level at the "stationary phase." For details see [22,23].

REFERENCES

1. A.M. Giusti, E. Bignetti, and C. Cannella, Exploring new frontiers in total food quality definition and assessment: From chemical to neurochemical properties, *Food Bioprocessing Technology* 1, 2007: 130–142; DOI 10.1007/s11947-007-0043-9.
2. P.S. Taoukis, T.P. Labuza, and I.S. Saguy, Prediction of shelf life from accelerated storage studies, in *Handbook of Food Engineering Practice*, eds., K.J. Valentas, E. Rotstein, and R.P. Singh, Boca Raton, FL: CRC Press, 1997, pp. 363–405.
3. M.C. Gacula, The design of experiments for shelf life study, *Journal of Food Science* 40, 1975: 399–404.

4. B. Fu and T.P. Labuza, Shelf life prediction: Theory and application, *Food-Control* 4, 1993: 125–133.
5. C. Cardelli and T.P. Labuza, Application of Weibull hazard analysis to the determination of the shelf life of roasted and ground coffee, *Lebensmittel-Wissenschaft und-Technologie* 34, 2001: 273–278.
6. H.W. McLean, *HALT, HASS & HASA Explained*, Milwaukee, WI: Quality Press, 2000.
7. P.S. Taoukis and T.P. Labuza, Time–temperature indicators (TTIs), in *Novel Food Packaging Techniques*, ed., R. Ahvenainenpp, Cambridge, U.K.: Woodhead Publishing Limited, 2003, pp. 103–126.
8. R.T. Magari, Uncertainty of measurement and error in stability studies, *Journal of Pharmaceutical and Biomedical Analysis* 45, 2007: 171–175.
9. M.A.J.S. van Boekel, Kinetic modeling of food quality: A critical review, *Comprehensive Reviews in Food Science and Food Safety* 7, 2008: 144–158.
10. S. Mizrahi, Accelerated shelf life tests, in *Understanding and Measuring the Shelf life of Food*, ed., R. Steele, Cambridge, U.K.: Woodhead Publishing, 2004.
11. T.P. Labuza, *Shelf Life Dating of Foods*, Westport, CT: Food and Nutrition Press Inc, 1982.
12. T.P. Labuza and K. Schmidtl, Accelerated shelf life testin, *Food Technology* 39, 1985: 57–62, 134.
13. Institute of Food Science and Technology (IFST), *Shelf Life of Foods-Guidelines for Its Determination and Prediction*, London, UK: IFST, 1993.
14. H.R. Moskowitz, J.B. German, and I.S. Saguy, Unveiling health attitudes and creating good-for-you foods via informatics & innovative web-based technologies, *Critical Reviews in Food Science and Nutrition* 45, 2005: 165–191.
15. M. Peleg, M.G. Corradini, and M.D. Normand, Kinetic models of complex biochemical reactions and biological processes, *Chemie Ingenieur Technik* 76, 2004: 413–423.
16. M.G. Corradini, M.D. Normand, and M. Peleg, Non linear growth and decay kinetics— Principles and potential food applications, in *Food Engineering: Integrated Approaches*, eds., G.F. Gutiérrez Lopez, G.V. Barbosa-Cánovas, J. Welti-Chanes, and E. Parada Arias, New York: Springer, 2008, pp. 47–71.
17. M.G. Corradini and M. Peleg, Estimating non-isothermal bacterial growth in foods from isothermal experimental data, *Journal of Applied Microbiology* 99, 2005: 187–200.
18. M.G. Corradini and M. Peleg, Shelf life estimation from accelerated storage data, *Trends in Food Science & Technology* 18, 2007: 37–47.
19. L.L. Leake, The search for shelf life solutions, *Food Technology* 62, 2007: 66–70.
20. M.G. Corradini and M. Peleg, Prediction of vitamin loss during non-isothermal heat processes and storage with non-linear kinetic models, *Trends in Food Science & Technology* 17, 2006: 24–34.
21. R. McKellar and X. Lu, eds., *Modeling Microbial Responses on Foods*, Boca Raton, FL: CRC Press, 2004.
22. M. Peleg, M.D. Normandand, and M.G. Corradini, Generating microbial survival curves during thermal processing in real time, *Journal of Applied Microbiology* 98, 2005: 406–417.
23. M.G. Corradini, A. Amézquita, M.D. Normand, and M. Peleg, Modeling and predicting non-isothermal microbial growth using general purpose software, *International Journal of Food Microbiology* 106, 2006: 223–328.

26 Commercialization and Manufacturing

J. Peter Clark and Leon Levine

CONTENTS

$$\text{Success} = \boxed{\begin{array}{c}\text{Defining and}\\\text{meeting}\\\text{target}\\\text{consumer}\\\text{needs and}\\\text{expectations}\end{array}} \times \boxed{\begin{array}{c}\text{The}\\\text{right food}\end{array}} \times \boxed{\begin{array}{c}\text{Proper}\\\text{packaging}\\\text{and}\\\text{preparation}\end{array}} \times \boxed{\begin{array}{c}\text{Positioned}\\\text{correctly}\\\text{at the shelf}\\\text{and in the}\\\text{media}\end{array}} \times \boxed{\begin{array}{c}\text{Meet}\\\text{corporate}\\\text{logistics and}\\\text{financial}\\\text{imperatives}\end{array}}$$

26.1 INTRODUCTION

As the last chapter in this book about new product development, we deal with how to make the product and bring it successfully to market. We address the options for manufacturing, some of the pitfalls and hazards of commercialization and how to use internal and external resources. The reader should consult Figure 26.1 for an overall schematic of how one goes about manufacturing a product.

The options for manufacturing are to build a new facility (which includes buying or leasing an existing building and converting it to use as a food plant), expanding an existing facility, or using a third-party comanufacturer. Each option has advantages and disadvantages. Common to all the options are the need to develop a process, design and select equipment, scale-up the process from the laboratory level, demonstrate

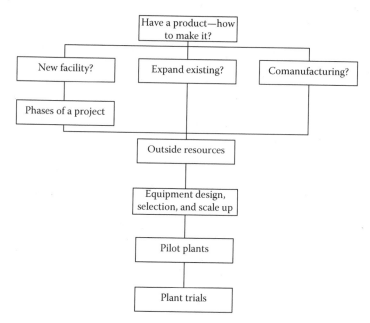

FIGURE 26.1 Flowchart for decision points for commercialization and manufacturing.

the process at increasing scale, and culminating in a plant trial. Along the path, there is often a need to produce various quantities of product for various types of tests, ranging from focus groups to full-scale marketing in several medium-sized cities. The decisions involved in manufacturing test quantities are similar to those decisions involving full-scale production, but are usually less economics-driven and are more concerned with timeliness, consistency, and convenience.

Along the path to commercialization, there are typical hazards and pitfalls, and there are numerous resources for help. Food product development is commonly performed by food scientists with the heavy involvement of marketing people. Subsequent commercialization and manufacturing are in the special province of process engineers. In turn, facility design and modification are often implemented by architects, engineers, and people experienced in construction. Most companies do not have people with all these specializations, and so specialists are usually engaged through consultants when needed.

Among the various disciplines involved in process engineering, facility design and construction and between them and food science are frequent divides in communication and culture. One of our objectives in this chapter is to help bridge those divides using experience from both sides of the divide.

26.2 DO WE NEED A NEW FACILITY?

Before rushing to invest in a new facility, or converting an existing building to food use, it is important to be sure that existing facilities are fully utilized. Food plants have often operated 5 days per week, two shifts per day, with a third shift for sanitation, and weekends for maintenance and occasional "emergency" production. Many companies have demonstrated that they can operate their lines almost 24/7, that is,

24 h per day and 7 days per week. A common strategy is to run 13 days in a row and perform maintenance on the 14th day. This schedule may require careful design for sanitation and skilled operation, but it significantly increases the capacity of an existing line.

Another potential source of unrealized capacity is seasonality. Many agricultural raw materials are only available at certain times in a given area. Plants that use such materials may be largely idle for long periods of time. However, with global sourcing and bulk storage of raw or partially processed raw material, seasonality of manufacturing can be reduced and capacity increased. Bulk aseptic storage of tomatoes, peaches, orange juice, and other fruits and vegetables (for which past IFT President Phil Nelson recently received the World Food Prize) has increased capacity of canneries and other processors of seasonal crops. Potatoes, apples, carrots, and some other crops are routinely stored for long periods, extending the time for processing. Southern hemisphere sources of raw or partially processed materials can extend the season for Northern hemisphere plants. Frozen tomato paste is a good example of such a material. Using it has greatly improved the utilization of existing processing plants.

There are three main reasons a new or converted manufacturing facility might be appropriate:

1. Use of new technology that does not fit in existing facilities
2. Existing facilities are at capacity (or do not exist)
3. Logistics, meaning a location close to raw materials or markets

If a company's existing facilities are operating at capacity, after all efforts to optimize have been made, then a new or converted facility may be justified. If a company has no facilities then, unless it uses a comanufacturer, it must build or convert.

In the United States, early twenty-first century, relatively few new food plants are being constructed. In developing countries, however, new or converted facilities are much more common, driven by the absence of existing facilities and the economic growth that increases demand for processed food products. In fact, one can argue that the increased supply of processed food can drive economic growth by releasing people, especially women, from the tedium of preparing their food and making these women available for wage-earning employment. In any event, new food processing facilities are more likely to be located in developing areas than in the United States. This fact has some consequences in design and project execution.

New technology often requires a new facility simply because it occupies more space than is otherwise available. A good example is the automation of almost any task on a food processing line. Machinery almost always takes up more space than do humans who perform the same task. The fact that the machine may operate faster and more accurately does not change the fact that it takes up more space. New technology may also dictate a different layout than the machine it replaces. For instance, a bakery that converts from batch ovens to a conveyor oven would need a radically different arrangement. A facility to irradiate foods needs a highly specialized design with shielding, and conveyors that are unique compared to any other facility in the food industry. A hydrostatic retort for continuous processing of canned food requires a very different layout than a comparable battery of batch retorts.

High-capacity and highly automated food processes may not be appropriate for every situation. Developing countries need good jobs as much as they need anything, and labor is usually relatively inexpensive. Sophisticated equipment needs skilled maintenance and support, which is not evenly distributed around the world. Older technology is often robust and used equipment is frequently available, lowering the capital cost. In a place where seasonal fruits and vegetables are not preserved at all, old fashioned batch canning would qualify as a great advance.

In developing countries, facilities to use high hydrostatic pressure preservation, irradiation, pulsed electric fields, and other new technologies usually must be newly designed to optimize material flow, adhere to good manufacturing practices (GMP), and take fullest advantage of computer integrated manufacturing (CIM).

Facility location is critical and may be dictated by the sources of raw materials, by distance to new or existing markets, or by distance to new or existing distribution centers. The relative bulk density of raw materials and products, and the shelf life of products often determine whether a product should be manufactured close to markets or can be made farther away. For example, frozen and dehydrated potato products are commonly made near potato growing areas, whereas potato chip snacks are made closer to markets. Products that have frequent, store door deliveries are made close to market. Examples are soft drinks, refrigerated dairy products, most salty snacks, and bread.

26.2.1 CONVERTING AN EXISTING BUILDING

Converting an existing building to use as a food processing facility is fairly common and often tempting because it would seem to be less expensive than building a new facility from scratch. Sadly, it is not always less expensive, and it usually involves some compromises in design and construction. However, done properly, with experienced leadership, creative reuse of an existing building can be a satisfactory solution to a need for additional capacity.

It is rare that a plant previously used for food processing becomes available, but it does happen. Such a building may have floor drains, though usually not exactly where they ought to be. There might be some refrigerated or frozen storage, which most food plants need. Walls and ceilings might be of sanitary design—or might once have been. Walls and doors are unlikely to be where they are wanted, and an older plant, especially if it has been unoccupied for a while, may be in poor shape. Real estate agents, not understanding the uniqueness of food processes, will present an existing food plant as the answer for any food client. This is not often true.

Much more common on most real estate markets are new or used generic warehouses or distribution centers. These can be quite large, 500,000 sq.ft., buildings and even larger are frequently found. If a food plant needs that much space, the company might be well advised to design and build (or have built for it) a green field plant. More likely, the food company needs a portion of such a building. This means finding a smaller candidate building, sharing the large building with one or more other occupants, or taking the entire space and finding another use for the excess area. Controlling the entire building permits the food company more influence over the nature of other tenants, though most leases will allow some right of review so that competitors or potentially noxious neighbors are not introduced.

Most buildings designed as warehouses or for light manufacturing have relatively light roof structures, to save costs. These are typically made from metal trusses or, in California, laminated wooden beams. Such roofs are not designed with any extra load-carrying allowance, whereas food plants often have piping and refrigeration units that are routinely hung from the roof or mounted outside on the roof. One of the more or less inevitable costs of converting an existing building to a food plant is strengthening the roof structure or providing separate structural steel for piping and equipment.

Another predictable cost is cutting the floor and installing process drains, because existing buildings rarely have floor drains, and if they do in some areas, they will not be where they are needed for the new use. A frequent issue is whether to also slope the floors in areas of heavy water use. GMP require that there will be no standing water in food contact areas, so this often means sloping floors toward hub or trench drains. Replacing a large amount of existing floor can be expensive and hard to justify, but may be the correct long-term decision. If it is not done, then measures should be taken to reduce water use and to provide squeegees to workers for the prevention of falls and to reduce the potential biological hazard of standing water.

Finally, lighting, fire protection, and utilities, such as electricity, water, sewage, and fuel, must be evaluated and often need to be upgraded. It is common that walls need new paint or panel covering and, whether new or old, the floor will need a food grade coating. Taken all together, the upgrades and improvements required in converting an existing building to food use are not very different from those encountered in building a new plant. One big difference is that most of the work occurs inside the structure, does not depend on the weather, and can start anytime, whereas site work and steel erection for a new building must wait for construction season where that applies.

26.2.2   PHASES OF A PROJECT: NEW, CONVERSION, OR EXPANSION

The five normal phases of a facility project are

1. Feasibility study
2. Preliminary design
3. Detailed design
4. Construction
5. Commissioning

A feasibility study is intended to confirm the economic attractiveness of a proposed project. The feasibility study requires sufficient scope description and design so that costs can be estimated within about 30%, which is reflected in the contingency allowance. Normally, the proposed selling price, anticipated volumes, and unit variable costs are known from the product definition. The key remaining question is what investment is required to manufacture the product in the desired volume and location? To conduct a credible feasibility study usually requires some outside assistance in the form of architects, engineers, consultants, and people with construction experience. The company needs to assign a team led by an executive with authority to commit the firm. Typically, members of the team include people from Research, Engineering, and Operations. People from Marketing, Finance, Human Resources, Information Technology, and Logistics may be involved in the study at times and must be kept informed and listened to.

Preliminary design should be performed only after the feasibility study confirms the desirability of investing in a new, expanded, or converted facility. This phase adds detail to the design started in the previous phase. More of the work is performed by outside resources, such as an architect/engineering (A/E) firm. A preliminary design takes longer and costs more than does the associated feasibility study. A specific site should be identified, because site-related factors are some of the larger causes of uncertainty in cost estimates. The scope description prepared in the feasibility study is expanded in this phase, but the basic assumptions should not be changed.

Detailed design is the phase in which construction documents—drawings and specifications—are prepared. Naturally, it involves more people, typically engineers for each discipline, and costs more than preliminary design. During detailed design, one of the critical functions is the resolution of potential interferences. Interferences occur when two disciplines try to use the same physical space in the plant for some purpose. Examples are structural steel, process piping, utility piping, electrical conduit, and heating, ventilation and air conditioning (HVAC) ducts. Interferences occur because design and detailed layout typically occur in parallel, sometimes by different outside firms that specialize in a given discipline. One of the major responsibilities of the project manager is to identify and resolve potential interferences quickly. It is much less expensive to resolve these on paper than to do so during construction. A useful strategy is to preassign chases or designated paths and areas in the plant for various purposes. Sensible prioritization is required. It is easier to move an electrical conduit than to move HVAC ducts. A good practice is to have most utility piping outside a process area with only vertical drops to use points. It is common to install cleanable ceilings and run most piping above the ceiling. Sometimes the ceiling panels are strong enough so that people can walk on them to maintain the piping and equipment.

Construction is normally the responsibility of a general contractor (GC) or construction manager (CM), but the owner is ultimately responsible for safety, costs, and quality. A GC may perform some tasks himself, but primarily hires subcontractors in the various trades and disciplines. He makes a profit by marking up the costs of labors and materials from the subcontractors. The owner may only be quoted a fixed price for the project and unit costs for changes. The more complete the scope definition and the design documents, the fewer changes there should be. Sometimes owners are the source of changes, but they should be aware of the consequences, which almost always are to increase the cost of a project. If an owner wants more transparency and control, then the owner can engage a CM, who normally charges a lower fee or profit margin than a GC does, because the CM assumes much less cost risk. A CM advises the owner to hire certain subcontractors and then manages them on the owner's behalf. All the costs should be known in somewhat greater detail than under a GC.

Commissioning begins when construction is complete and consists of operating the process with realistic raw materials, but with no expectation of high yields. The objective is to find and correct deficiencies in the process and equipment, fine tune conveyors, and train operators. The objective of identifying a commissioning phase is to budget time and money for these essential tasks.

26.2.3   EXPANSION OF EXISTING PLANT

Installation of new line in an existing plant has many of the same phases and characteristics as does building a new plant, but there are some advantages and some challenges. Expansion is probably the most common type of capital project in the food industry. Some of the advantages for this approach are that labor, sources of raw materials, and distribution infrastructure may already be in place and familiar to the firm. In many plants, extra space was built in the first place with expansion in mind. Sometimes expansion space is used for storage until it is needed. Additional storage space may need to be built, but that is usually less expensive than building new process space. If no new construction is required, expansion is not subject to weather the way new construction is in many parts of the world. When considering an expansion, the adequacy of the space available, the utilities, the logistics, and the labor supply must be confirmed. Construction, even the installation of equipment in existing space, can be disruptive to an operating plant, so synchronization with the plant's schedule is important. In addition to the outside and internal resources involved in any project, key figures in an expansion are the local plant management team. They are responsible for training operators and construction workers in local safety rules and practices.

26.2.4   COMANUFACTURING

Often, new products are manufactured by third parties, comanufacturers, or copackers, until their success is assured. Some firms are reluctant to expose new ideas to outsiders, whereas others routinely use such arrangements. One approach outsources established products, and performs initial production of new products in a semiwork or small, flexible facility. Some comanufacturing arrangements are long standing and strong partnerships.

When selecting a comanufacturer, there are at least six major considerations.

1. Technical capability
2. Missing pieces of a process, often packaging
3. Quality procedures and achievements
4. Business practices
5. Security
6. Confidentiality

In addition to the participants from research, engineering, and operations, people from purchasing, supply chain management, and quality are likely to be involved. Some comanufacturers are using excess capacity whereas others specialize in constructing and operating third-party manufacturing facilities. Those in this category often accept a lower profit margin than a large corporation seeks, have lower labor costs, and offer additional service, such as managing a customer's inventory. Whether for a risky new product or well-established cash cow, comanufacturing is often an attractive strategy.

26.3   EQUIPMENT DESIGN, SELECTION, AND SCALE-UP

Food companies rarely design processing equipment. Rather, they or their consultants usually select equipment from existing lines offered by numerous vendors. It is important to understand how to specify and evaluate equipment so that it is suited to its intended function and is cost effective. Be careful of depending solely on the manufacturers of the equipment for technical expertise. The key documents in selecting equipment are process flow sheets, material and energy balances, and a process description. The material and energy balances determine the required size and capacity of the equipment, the description says what is being processed and why, and the flow sheet shows how the process units are related to each other.

Equipment for food processing is designed and constructed to conform to sanitary design principles and regulations. These say that the equipment is made from noncorrodible materials, can be easily inspected for cleanliness, and can be disassembled easily with few tools. As a consequence, much food equipment is made from stainless steel, has highly polished surfaces and welds, and uses special piping connections.

When evaluating vendors of food processing equipment, it is important to confirm that they observe industry standards such as Baking Industry Sanitation Standards (BISSC) for the baking industry and 3-A for dairy and many other segments. Cost, quality of components and fabrication, service capability, and delivery time are other factors that affect the choice of one vendor over another.

26.3.1   SCALE-UP

Often processes are developed on laboratory benches, in kitchens, or in pilot plants using available small-scale equipment. Determining the correct size for a given desired capacity is the challenge of scale-up.

The objective of scale-up is to produce the identical product as was produced for the consumer tests which indicated the consumer acceptability of the product under consideration. Having products which are equally acceptable to the consumer is not the same as identical. There have been cases where these measures of acceptability have been confused, with disastrous commercial results.

The changes in products that occur as one progresses from the bench to pilot plant and on to commercial production facilities result from many, often unidentified, factors. In general, these changes are the result of changes in the time/temperature and/ or time/force (e.g., shear, work, pressure, etc.) histories that the raw materials are subjected to as they are transformed into a finished product. Sometimes the history required to produce the same product is obvious. For example, in processes such as baking, frying, and drying, keeping the same time and temperature profiles that the product experiences will ensure that the results on different scales will be the same. However, achieving the same time and temperature profiles may not be as easy as it seems. For example, simulating the time–temperature profiles of the oil used to fry a product in a commercial continuous fryer is virtually impossible to accomplish in a batch fryer.

In many processes, the forces exerted on the product affecting the desired transformation are not always obvious. For example, even for a simple process such as blending powders, what must be designed into the process to ensure that the finished mixture is uniform? As processes become more complicated, the answer to this question becomes even more difficult to discover. For instance, in the production of an emulsion, or the production of an extruded product, what kind of time/temperature/shear histories are required to make the desired product quality? As one's physical understanding of the transformations occurring is reduced, answering this kind of question becomes more difficult and requires more detailed experimentation, often on more than one scale (size) of equipment.

Ultimately things like the time/temperature/force histories are the result of the flow rate through the process, the size of the equipment, the speed of the moving parts, and the geometry of the equipment. The geometry of the equipment is extremely important and is often the ultimate source of scale-up problems. In the simplest sense, "the importance of geometry implies that the pilot plant equipment 'look-like' (i.e., functionally match) the full-scale equipment." This implies that it is impossible to scale even a simple mixing process from a common kitchen type mixer to a ribbon blender or a horizontal bar dough mixer. The concept of geometry extends to more subtle differences. For example, extruders from different manufacturers look superficially the same, but the ratios of screw diameter to thread depth differ from one manufacturer to another. The ratio of screw diameter to thread depth is an important determinant of shear exerted on the product, screw capacity, residence time, etc., so making identical products on different manufacturer's extruders may require considerable effort.

Another area where geometry comes into play is the simple fact that the ratio of heat transfer surface area to equipment size/capacity almost always decreases with scale. This makes maintenance of the temperature history between scales almost impossible. This is frequently a problem associated with difficult or impossible scale-ups.

The speed of the moving parts and the flow rate through the equipment are the other areas of concern. The speed directly affects the shear imparted by the machinery. The flow rate directly affects the residence time (history) the product sees. One common mistake is to run pilot plant equipment at disproportionately lower rate than the commercial equipment will be run. So, although this may be economically undesirable or otherwise inconvenient, pilot plants must be run at reasonable capacities.

26.3.2 SOME PITFALLS AND HAZARDS OF COMMERCIALIZATION

Failure to observe some of the recommendations of this chapter can have unpleasant consequences. Here are some experiences with heat transfer on the one hand, and shear history on the other.

The problem of the ratio of heat transfer surface decreasing with scale is probably the commonest example of scale-up failure. The simplest examples are observed when mixing or extruding viscous materials when the final temperature of the product is a critical factor in the determination of product quality. In a batch cooking or a cooling/mixing operation, the reduced surface area requires that the heating media temperature be decreased when cooling or increased when heating. It does not take a very large increase in scale of the equipment for these temperature changes to be significant. The freezing of water out of the product and the "burn on" of product on the walls of the larger mixing vessel have been observed when scaling these types of processes.

The problem of heat transfer is frequently observed when scaling up extrusion processes. A common extrusion practice is to cook and then cool the product below the boiling point of water before leaving the extruder. It does not take much of an increase in extruder scale to cause this scale-up to fail. In one case, the required coolant temperature, for a moderately larger extruder, turned out to be below absolute zero! In another case, this problem was not recognized prior to the construction of a large-scale confectionary operation. The full-scale equipment was only capable of running about half its design capacity.

There are myriad examples where the shear history has not been properly scaled up. A common example is encountered in dough sheeting equipment. Larger rolls and higher speed lines result in large increases in the work and forces that the rolls put into/exert on the product. This generally results in more "damage" to the dough on larger scale equipment. This kind of problem cannot be readily repaired, which results in the commercial product being very different than the test products originally produced on a small scale. Because of the increased forces experienced on the larger scale, actual physical failure of the sheeting equipment has been observed.

26.3.3 Pilot Plants

Pilot plants can range from very flexible, multipurpose spaces to dedicated miniature simulations of full-scale operations. A pilot plant is intended to provide a step between laboratory or kitchen scale and full-scale plant operation. The pilot plant often is the place where the transition occurs between a culinary approach and a manufacturing approach, in the sense that industrial style equipment is used. Batch operations may be converted to continuous, heating over a flame on a stove becomes heating in a jacketed kettle, pouring from beakers becomes dispensing from loss-in-weight scales, and so on.

A significant characteristic of pilot plants is that they are places for research, not for production, despite the temptation to use them for production. Most pilot plants should be flexible, so that various processes can be simulated and studied. This means plentiful utility connections, equipment that is mounted on casters for easy movement, and ample storage space for equipment and supplies. The principles of scale-up previously discussed need to be applied in selecting equipment for a pilot plant so that results can be reliably translated into specifications for a full-scale plant.

Sometimes pilot plants are used to simulate an operating plant as a trouble shooting tool and so, again, the principles of scale-up are applied. Pilot plant equipment should be heavily instrumented because they are research tools. Today, most instruments are connected to computers for automatic data collection and analysis.

If a food company does not have a pilot plant, then it might be able to perform the same tasks at other facilities. Equipment vendors often have places to demonstrate their own equipment. Often the vendor will have some support equipment, such as tanks, feeders, and even packaging equipment. Many universities have pilot plants in food science, agricultural and biological engineering, and chemical engineering departments. These university pilot plants are usually designed primarily for education and academic research, but they often can be adapted for a specific test or development project. One advantage is that students may be available for inexpensive labor and professors may provide technical advice. Some universities have larger facilities specifically intended to support the food industry, which can be used for a fee. These often have more realistic equipment and experienced technicians.

Pilot plants should rarely, if ever, be used to produce food for human consumption beyond the occasional taste panel. To do so for foods involving meat or poultry, the pilot plant would need to become a USDA establishment (in the United States), for which it probably was not designed. More critically, as a research facility, the staff and equipment are not oriented to the consistency and safety required for quality production.

26.3.4 Plant Trials

Plant trials, meaning the running of a new product on a new or existing line in a plant, using plant operators, is an expensive but necessary evil. It is usually the last step before releasing a new or changed product to commercial production. Some companies that do not have pilot plants routinely perform tests on plant lines, but this can rapidly become prohibitively expensive and time consuming. Plant lines consume large amounts of raw materials and energy. True trials, as distinct from a confirming demonstration, will often fail, resulting in large quantities of waste. Running a trial means the line is not running profitable product.

Often trials are performed on weekends or otherwise unconventional shifts, meaning the operators are paid premium wages. Trials usually are supervised by staff from research, who typically must travel from their normal location. There are many good reasons to limit plant trials and to only do them when there is a high probability of success.

Plant trials require the cooperation of local plant management, who must educate visiting staff to work and safety rules, procedures, and chain of commands. Plant operators often have valuable contributions to make if asked and know the plant equipment better than anyone. Successful transition to full scale depends on the enthusiastic involvement of plant management and operators.

26.4   WHAT IS THE FUTURE OF COMMERCIALIZATION?

The issues discussed in this chapter are relatively timeless and unlikely to change significantly because they are based on physics, geometry, and human nature. However, aspects of the world in which commercialization occurs are changing and will have an effect. Some of these aspects include information and communication technology, attitudes and education of individuals, and computational tools and scientific understanding. In addition, governmental regulation is likely to increase, globalization of markets will have an influence on sources and distribution of materials, and cultural diversity will profoundly affect the character of the workforce.

It may be true that no single technical development has had a greater impact than the creation of the Internet and the accompanying astonishing advances in inexpensive computing power. One obvious consequence is the replacement of paper and telephone communication by electronic means, specifically e-mail, special purpose Web sites, electronic "meetings," and computer-aided drafting (CAD). It is increasingly recognized that these remarkable advances have their disadvantages. Clarity and accuracy of written communications can deteriorate when the emphasis is on the immediacy and speed of e-mail. It is said that the present time, and the future, are the ages of instant gratification, with laptop computers, cellular telephones, and personal digital assistants (PDA). At risk of being lost is the precision of expression that can only occur upon re-reading of a document, even a brief reply to a request for information (ROI) from a contractor to a designer. Misspelled and missing words can dramatically alter the meaning of a sentence, with potentially devastating results.

On the other hand, the ability to share a common data base, the maintenance of a single master drawing, rather than the dozens of individual versions that used to proliferate, and the ability to simultaneously gather the wisdom of people scattered about the globe are uniquely valuable when used carefully and appropriately. The savings in travel costs alone are often enough to pay for the costs of sophisticated equipment and high speed data transmission lines. Tools such as simulation, computer integrated control, and logistics models more than pay for themselves in improved designs, reduction of human error, and improved utilization of resources.

The changes in technology as well as changes in culture, politics, and society in general, are changing the nature of the people involved in design, construction, and operation of food plants. Graduate engineers and food scientists are proficient in the new tools, often from high school, while the generation of people leading and mentoring them were probably educated in an age of slide rules, main frame computers, and hand held calculators. This alone can create a significant cultural divide. The young have come to believe that if information cannot be found with an Internet search engine, it does not exist, while their supervisors may still be found in a library. Books like this are one attempt to pass along accumulated wisdom and experience that are not otherwise documented and thus unlikely to be acquired by the new generation.

The disappearance of mutual loyalty on the part of employers and employees, in which there existed an implied understanding of long-term security and long-term employment, means that employees view themselves as self-employed and must be led and persuaded rather than dictated to and disciplined. The ability to transmit information widely coupled with the general reduction in layers of supervision means that more people have the tools with which to make decisions. This can increase efficiency but can have the risk of poor or malicious choices. Trust and responsibility must be combined with careful attention and diligence.

The factory workforce in the United States is beginning to resemble that of a developing country with many not having English as their first language. The same is often true of the graduating engineers and food scientists. Most engineers are not required to study a foreign language and, until recently, it was difficult for a student engineer to experience a year or semester of study abroad, so few young engineers graduate with much understanding of the cultures and languages of those likely to be their colleagues and subordinates. Any correction of this situation must rely on the self-motivation, discipline, and energy of individuals to educate themselves. Cultural and linguistic divides can have serious consequences for safety, efficiency, and quality.

With rapid advances in science and the increasing ability to access knowledge, it is the individual's responsibility to continually educate him or herself, more so than in the past when it was possible for one to rely more confidently on knowledge acquired in college or by personal experience. Now, the experience of almost anyone in the world can be accessed, but evaluating it critically can be a challenge.

26.5 SUMMING UP

Commercialization and manufacturing of a new food product involve people from many functions and disciplines from within and from outside an organization. Teamwork and good communication are key factors in success. Whereas each participant has his or her role, it is important to understand the overall task, to understand how to contribute and to respect the contributions of others. This chapter has described some of these contributions, which usually are made by other individuals, not the food scientists nor the marketers.

BIBLIOGRAPHY

Clark, J. P. Food manufacturing: Opportunities for improvement. *Food Technology* 41(12), 1987, 56–58.

Clark, J. P. Engineering and manufacturing, in *Food Product Development*, eds. E. Graf and I. S. Saguy (New York: Van Nostrand Reinhold, 1991), pp. 91–103.

Valentas, K. J., Levine, L., and Clark, J. P. *Food Processing Operations and Scale Up* (New York: Marcel Dekker, 1991).

Index